Rüdiger U. Seydel

Tools for Computational Finance

Third Edition

 Springer

Rüdiger U. Seydel

University of Köln
Institute of Mathematics
Weyertal 86-90
50931 Köln, Germany
E-mail: seydel@math.uni-koeln.de

The figure in the front cover illustrates the value of an American put option. The slices are taken from the surface shown in the Figure 1.5.

Mathematics Subject Classification (2000): 65-01, 90-01, 90A09

Library of Congress Control Number: 2005938669

ISBN-10   3-540-27923-7 Springer Berlin Heidelberg New York
ISBN-13   978-3-540-27923-5 Springer Berlin Heidelberg New York

Springer is a part of Springer Science+Business Media
springer.com
© Springer-Verlag Berlin Heidelberg 2006
Printed in The Netherlands

The use of general descriptive names, registered names, trademarks, etc. in this publication does not imply, even in the absence of a specific statement, that such names are exempt from the relevant protective laws and regulations and therefore free for general use.

Typesetting: by the author and TechBooks using a Springer TeX macro package

Cover design: *design & production* GmbH, Heidelberg

Printed on acid-free paper   SPIN: 11431596   40/TechBooks   5 4 3 2 1 0

Preface to the First Edition

Basic principles underlying the transactions of financial markets are tied to probability and statistics. Accordingly it is natural that books devoted to *mathematical finance* are dominated by stochastic methods. Only in recent years, spurred by the enormous economical success of financial derivatives, a need for sophisticated computational technology has developed. For example, to price an American put, quantitative analysts have asked for the numerical solution of a free-boundary partial differential equation. Fast and accurate numerical algorithms have become essential tools to price financial derivatives and to manage portfolio risks. The required methods aggregate to the new field of *Computational Finance*. This discipline still has an aura of mysteriousness; the first specialists were sometimes called *rocket scientists*. So far, the emerging field of computational finance has hardly been discussed in the mathematical finance literature.

This book attempts to fill the gap. Basic principles of computational finance are introduced in a monograph with textbook character. The book is divided into four parts, arranged in six chapters and seven appendices. The general organization is

Part I (Chapter 1): Financial and Stochastic Background
Part II (Chapters 2, 3): Tools for Simulation
Part III (Chapters 4, 5, 6): Partial Differential Equations for Options
Part IV (Appendices A1...A7): Further Requisits and Additional Material.

The first chapter introduces fundamental concepts of financial options and of stochastic calculus. This provides the financial and stochastic background needed to follow this book. The chapter explains the terms and the functioning of standard options, and continues with a definition of the Black-Scholes market and of the principle of risk-neutral valuation. As a first computational method the simple but powerful binomial method is derived. The following parts of Chapter 1 are devoted to basic elements of stochastic analysis, including Brownian motion, stochastic integrals and Itô processes. The material is discussed only to an extent such that the remaining parts of the book can be understood. Neither a comprehensive coverage of derivative products nor an explanation of martingale concepts are provided. For such in-depth coverage of financial and stochastic topics ample references to special literature are given as hints for further study. The focus of this book is on numerical methods.

Chapter 2 addresses the computation of random numbers on digital computers. By means of congruential generators and Fibonacci generators, uniform deviates are obtained as first step. Thereupon the calculation of normally distributed numbers is explained. The chapter ends with an introduction into low-discrepancy numbers. The random numbers are the basic input to integrate stochastic differential equations, which is briefly developed in Chapter 3. From the stochastic Taylor expansion, prototypes of numerical methods are derived. The final part of Chapter 3 is concerned with Monte Carlo simulation and with an introduction into variance reduction.

The largest part of the book is devoted to the numerical solution of those partial differential equations that are derived from the Black-Scholes analysis. Chapter 4 starts from a simple partial differential equation that is obtained by applying a suitable transformation, and applies the finite-difference approach. Elementary concepts such as stability and convergence order are derived. The free boundary of American options —the optimal exercise boundary— leads to variational inequalities. Finally it is shown how options are priced with a formulation as linear complimentarity problem. Chapter 5 shows how a finite-element approach can be used instead of finite differences. Based on linear elements and a Galerkin method a formulation equivalent to that of Chapter 4 is found. Chapters 4 and 5 concentrate on standard options.

Whereas the transformation applied in Chapters 4 and 5 helps avoiding spurious phenomena, such artificial oscillations become a major issue when the transformation does not apply. This is frequently the situation with the non-standard *exotic* options. Basic computational aspects of exotic options are the topic of Chapter 6. After a short introduction into exotic options, Asian options are considered in some more detail. The discussion of numerical methods concludes with the treatment of the advanced total variation diminishing methods. Since exotic options and their computations are under rapid development, this chapter can only serve as stimulation to study a field with high future potential.

In the final part of the book, seven appendices provide material that may be known to some readers. For example, basic knowledge on stochastics and numerics is summarized in the appendices A2, A4, and A5. Other appendices include additional material that is slightly tangential to the main focus of the book. This holds for the derivation of the Black-Scholes formula (in A3) and the introduction into function spaces (A6).

Every chapter is supplied with a set of exercises, and hints on further study and relevant literature. Many examples and 52 figures illustrate phenomena and methods. The book ends with an extensive list of references.

This book is written from the perspectives of an applied mathematician. The level of mathematics in this book is tailored to readers of the advanced undergraduate level of science and engineering majors. Apart from this basic knowledge, the book is self-contained. It can be used for a course on the subject. The intended readership is interdisciplinary. The audience of this book includes professionals in financial engineering, mathematicians, and scientists of many fields.

An expository style may attract a readership ranging from graduate students to practitioners. Methods are introduced as tools for immediate application. Formulated and summarized as algorithms, a straightforward implementation in computer programs should be possible. In this way, the reader may learn by computational experiment. *Learning by calculating* will be a possible way to explore several aspects of the financial world. In some parts, this book provides an algorithmic introduction into computational finance. To keep the text readable for a wide range of readers, some of the proofs and derivations are exported to the exercises, for which frequently hints are given.

This book is based on courses I have given on computational finance since 1997, and on my earlier German textbook *Einführung in die numerische Berechnung von Finanz-Derivaten*, which Springer published in 2000. For the present English version the contents have been revised and extended significantly.

The work on this book has profited from cooperations and discussions with Alexander Kempf, Peter Kloeden, Rainer Int-Veen, Karl Riedel und Roland Seydel. I wish to express my gratitude to them and to Anita Rother, who TEXed the text. The figures were either drawn with xfig or plotted and designed with gnuplot, after extensive numerical calculations.

Additional material to this book, such as hints on exercises and colored figures and photographs, is available at the website address

www.mi.uni-koeln.de/numerik/compfin/

It is my hope that this book may motivate readers to perform own computational experiments, thereby exploring into a fascinating field.

Köln   *Rüdiger Seydel*
February 2002

Preface to the Second Edition

This edition contains more material. The largest addition is a new section on jump processes (Section 1.9). The derivation of a related partial integro-differential equation is included in Appendix A3. More material is devoted to Monte Carlo simulation. An algorithm for the standard workhorse of inverting the normal distribution is added to Appendix A7. New figures and more exercises are intended to improve the clarity at some places. Several further references give hints on more advanced material and on important developments.

Many small changes are hoped to improve the readability of this book. Further I have made an effort to correct misprints and errors that I knew about.

A new domain is being prepared to serve the needs of the computational finance community, and to provide complementary material to this book. The address of the domain is

<code>www.compfin.de</code>

The domain is under construction; it replaces the website address `www.mi.uni-koeln.de/numerik/compfin/`.

Suggestions and remarks both on this book and on the domain are most welcome.

Köln
July 2003

Rüdiger Seydel

Preface to the Third Edition

The rapidly developing field of financial engineering has suggested extensions to the previous editions. Encouraged by the success and the friendly reception of this text, the author has thoroughly revised and updated the entire book, and has added significantly more material. The appendices were organized in a different way, and extended. In this way, more background material, more jargon and terminology are provided in an attempt to make this book more self-contained. New figures, more exercises, and better explanations improve the clarity of the book, and help bridging the gap to finance and stochastics.

The largest addition is a new section on analytic methods (Section 4.8). Here we concentrate on the interpolation approach and on the quadratic approximation. In this context, the analytic method of lines is outined. In Chapter 4, more emphasis is placed on extrapolation and the estimation of the accuracy. New sections and subsections are devoted to risk-neutrality. This includes some introducing material on topics such as the theorem of Girsanov, state-price processes, and the idea of complete markets. The analysis and geometry of early-exercise curves is discussed in more detail. In the appendix, the derivations of the Black-Scholes equation, and of a partial integro-differential equation related to jump diffusion are rewritten. An extra section introduces multidimensional Black-Scholes models. Hints on testing the quality of random-number generators are given. And again more material is devoted to Monte Carlo simulation. The integral representation of options is included as a link to quadrature methods. Finally, the references are updated and expanded.

It is my pleasure to acknowledge that the work on this edition has benefited from helpful remarks of Rainer Int-Veen, Alexander Kempf, Sebastian Quecke, Roland Seydel, and Karsten Urban.

The material of this Third Edition has been tested in courses the author gave recently in Cologne and in Singapore. Parallel to this new edition, the website www.compfin.de is supplied by an option calculator.

Köln
October 2005

Rüdiger Seydel

Contents

Notations

elements of options:

t	time
T	maturity date, time to expiration
S	price of underlying asset
	S_j, S_{ji} specific values of the price S
S_t	price of the asset at time t
K	strike price, exercise price
V	value of an option (V_C value of a call, V_P value of a put, $^\mathrm{am}$ American, $^\mathrm{eur}$ European)
σ	volatility
r	interest rate (Appendix A1)

general mathematical symbols:

$\mathbb{R}$	set of real numbers
$\mathbb{N}$	set of integers > 0
$\mathbb{Z}$	set of integers
$\in$	element in
$\subseteq$	subset of, $\subset$ strict subset
$[a, b]$	closed interval $\{x \in \mathbb{R} : a \leq x \leq b\}$
$[a, b)$	half-open interval $a \leq x < b$ (analogously $(a, b]$, (a, b))
P	probability
E	expectation (Appendix B1)
Var	variance
Cov	covariance
log	natural logarithm
$:=$	defined to be
$\doteq$	equal except for rounding errors
$\equiv$	identical
$\Longrightarrow$	implication
$\Longleftrightarrow$	equivalence
$O(h^k)$	Landau-symbol: for $h \to 0$
	$f(h) = O(h^k) \iff \frac{f(h)}{h^k}$ is bounded
$\sim \mathcal{N}(\mu, \sigma^2)$	normal distributed with expectation μ and variance σ^2
$\sim \mathcal{U}[0, 1]$	uniformly distributed on $[0, 1]$

Δt	small increment in t
tr	transposed; A^{tr} is the matrix where the rows and columns of A are exchanged.
$\mathcal{C}^0[a,b]$	set of functions that are continuous on $[a,b]$
$\in \mathcal{C}^k[a,b]$	k-times continuously differentiable
$\mathcal{D}$	set in $\mathbb{R}^n$ or in the complex plane, $\bar{\mathcal{D}}$ closure of $\mathcal{D}$, $\mathcal{D}^\circ$ interior of $\mathcal{D}$
$\partial \mathcal{D}$	boundary of $\mathcal{D}$
$\mathcal{L}^2$	set of square-integrable functions
$\mathcal{H}$	Hilbert space, Sobolev space (Appendix C3)
$[0,1]^2$	unit square
Ω	sample space (in Appendix B1)
$f^+ := \max\{f,0\}$	
$\dot{u}$	time derivative $\frac{du}{dt}$ of a function $u(t)$

integers:

i,j,k,l,m,n,M,N,ν

various variables:

$X_t, X, X(t)$	random variable
W_t	Wiener process, Brownian motion (Definition 1.7)
$y(x,\tau)$	solution of a partial differential equation for (x,τ)
w	approximation of y
h	discretization grid size
φ	basis function (Chapter 5)
ψ	test function (Chapter 5)
$1_{\mathcal{D}}$	indicator function: $= 1$ on $\mathcal{D}$, $= 0$ elsewhere.

abbreviations:

BDF	Backward Difference Formula, see Section 4.2.1
CFL	Courant-Friedrichs-Lewy, see Section 6.5.1
Dow	Dow Jones Industrial Average
FTBS	Forward Time Backward Space, see Section 6.5.1
FTCS	Forward Time Centered Space, see Section 6.4.2
GBM	Geometric Brownian Motion, see (1.33)
MC	Monte Carlo
ODE	Ordinary Differential Equation
OTC	Over The Counter
PDE	Partial Differential Equation
PIDE	Partial Integro-Differential Equation
PSOR	Projected Successive Overrelaxation
QMC	Quasi Monte Carlo
SDE	Stochastic Differential Equation
SOR	Successive Overrelaxation

TVD	Total Variation Diminishing
i.i.d.	independent and identical distributed
inf	infimum, largest lower bound of a set of numbers
sup	supremum, least upper bound of a set of numbers
supp(f)	support of a function f: $\{x \in \mathcal{D} : f(x) \neq 0\}$

hints on the organization:

(2.6)	number of equation (2.6)
	(The first digit in all numberings refers to the chapter.)
(A4.10)	equation in Appendix A; similarly B, C, D
$\longrightarrow$	hint (for instance to an exercise)

Chapter 1 Modeling Tools for Financial Options

1.1 Options

What do we mean by option? An option is the right (but not the obligation) to buy or sell a risky asset at a prespecified fixed price within a specified period. An option is a financial instrument that allows —amongst other things— to make a bet on rising or falling values of an underlying asset. The **underlying** asset typically is a stock, or a parcel of shares of a company. Other examples of underlyings include stock indices (as the Dow Jones Industrial Average), currencies, or commodities. Since the value of an option depends on the value of the underlying asset, options and other related financial instruments are called *derivatives* ($\longrightarrow$ Appendix A2). An option is a contract between two parties about trading the asset at a certain future time. One party is the *writer*, often a bank, who fixes the terms of the option contract and sells the option. The other party ist the *holder*, who purchases the option, paying the market price, which is called *premium*. How to calculate a fair value of the premium is a central theme of this book. The holder of the option must decide what to do with the rights the option contract grants. The decision will depend on the market situation, and on the type of option. There are numerous different types of options, which are not all of interest to this book. In Chapter 1 we concentrate on standard options, also known as *vanilla options*. This Section 1.1 introduces important terms.

Options have a limited life time. The *maturity date T* fixes the time horizon. At this date the rights of the holder expire, and for later times $(t > T)$ the option is worthless. There are two basic types of option: The **call** option gives the holder the right to *buy* the underlying for an agreed price K by the date T. The **put** option gives the holder the right to *sell* the underlying for the price K by the date T. The previously agreed price K of the contract is called **strike** or **exercise price**[1]. It is important to note that the holder is not obligated to *exercise* —that is, to buy or sell the underlying according to the terms of the contract. The holder may wish to close his position by selling the option. In summary, at time t the holder of the option can choose to

---

[1] The price K as well as other prices are meant as the price of one unit of an asset, say, in $.

- sell the option at its current market price on some options exchange (at $t < T$),
- retain the option and do nothing,
- exercise the option $(t \leq T)$, or
- let the option expire worthless $(t \geq T)$.

In contrast, the writer of the option has the obligation to deliver or buy the underlying for the price K, in case the holder chooses to exercise. The risk situation of the writer differs strongly from that of the holder. The writer receives the premium when he issues the option and somebody buys it. This up-front premium payment compensates for the writer's potential liabilities in the future. The asymmetry between writing and owning options is evident. This book mostly takes the standpoint of the holder.

Not every option can be exercised at any time $t \leq T$. For **European options** exercise is only permitted at expiration T. **American options** can be exercised at any time up to and including the expiration date. For options the labels American or European have no geographical meaning. Both types are traded in every continent. Options on stocks are mostly American style.

The value of the option will be denoted by V. The value V depends on the price per share of the underlying, which is denoted S. This letter S symbolizes stocks, which are the most prominent examples of underlying assets. The variation of the asset price S with time t is expressed by writing S_t or $S(t)$. The value of the option also depends on the remaining time to expiry $T - t$. That is, V depends on time t. The dependence of V on S and t is written $V(S, t)$. As we shall see later, it is not easy to calculate the fair value V of an option for $t < T$. But it is an easy task to determine the terminal value of V at expiration time $t = T$. In what follows, we shall discuss this topic, and start with European options as seen with the eyes of the holder.

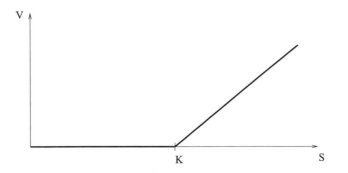

Fig. 1.1. Intrinsic value of a call with exercise price K (payoff function)

The Payoff Function

At time $t = T$, the holder of a European call option will check the current price $S = S_T$ of the underlying asset. The holder will exercise the call (buy the stock for the strike price K), when $S > K$. For then the holder can immediately sell the asset for the spot price S and makes a gain of $S - K$ per share. In this situation the value of the option is $V = S - K$. (This reasoning ignores transaction costs.) In case $S < K$ the holder will not exercise, since then the asset can be purchased on the market for the cheaper price S. In this case the option is worthless, $V = 0$. In summary, the value $V(S, T)$ of a call option at expiration date T is given by

$$V(S_T, T) = \begin{cases} 0 & \text{in case } S_T \leq K \text{ (option expires worthless)} \\ S_T - K & \text{in case } S_T > K \text{ (option is exercised)} \end{cases}$$

Hence

$$V(S_T, T) = \max\{S_T - K, 0\}.$$

Considered for all possible prices $S_t > 0$, $\max\{S_t - K, 0\}$ is a function of S_t. This **payoff function** is shown in Figure 1.1. Using the notation $f^+ := \max\{f, 0\}$, this payoff can be written in the compact form $(S_t - K)^+$. Accordingly, the value $V(S_T, T)$ of a call at maturity date T is

$$V(S_T, T) = (S_T - K)^+. \tag{1.1C}$$

For a European put exercising only makes sense in case $S < K$. The payoff $V(S, T)$ of a put at expiration time T is

$$V(S_T, T) = \begin{cases} K - S_T & \text{in case } S_T < K \text{ (option is exercised)} \\ 0 & \text{in case } S_T \geq K \text{ (option is worthless)} \end{cases}$$

Hence

$$V(S_T, T) = \max\{K - S_T, 0\},$$

or

$$V(S_T, T) = (K - S_T)^+, \tag{1.1P}$$

compare Figure 1.2.

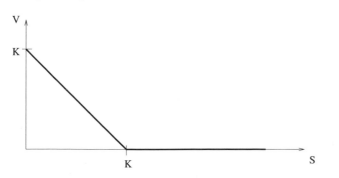

Fig. 1.2. Intrinsic value of a put with exercise price K (payoff function)

The curves in the payoff diagrams of Figures 1.1, 1.2 show the option values from the perspective of the holder. The profit is not shown. For an illustration of the profit, the initial costs paid when buying the option at $t = t_0$ must be subtracted. The initial costs basically consist of the premium and the transaction costs. Since both are paid upfront, they are multiplied by $e^{r(T-t_0)}$ to take account of the time value; r is the continuously compounded interest rate. Substracting this amount leads to shifting the curves in Figures 1.1, 1.2 down. The resulting *profit diagram* shows a negative profit for some range of S-values, which of course means a loss, see Figure 1.3.

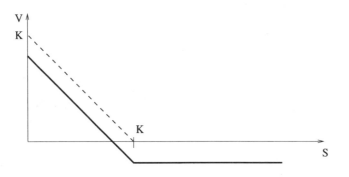

Fig. 1.3. Profit diagram of a put

The payoff function for an American call is $(S_t - K)^+$ and for an American put $(K - S_t)^+$ for any $t \leq T$. The Figures 1.1, 1.2 as well as the equations (1.1C), (1.1P) remain valid for American type options.

The payoff diagrams of Figures 1.1, 1.2 and the corresponding profit diagrams show that a potential loss for the purchaser of an option (long position) is limited by the initial costs, no matter how bad things get. The situation for the writer (short position) is reverse. For him the payoff curves of Figures 1.1, 1.2 as well as the profit curves must be reflected on the S-axis. The writer's profit or loss is the reverse of that of the holder. Multiplying the payoff of a call in Figure 1.1 by (-1) illustrates the potentially unlimited risk of a short call. Hence the writer of a call must carefully design a strategy to compensate for his risks. We will came back to this issue in Section 1.5.

A Priori Bounds

No matter what the terms of a specific option are and no matter how the market behaves, the values V of the options satisfy certain bounds. These bounds are known a priori. For example, the value $V(S, t)$ of an American option can never fall below the payoff, for all S and all t. These bounds follow from the *no-arbitrage principle* ($\longrightarrow$ Appendices A2, A3). To illustrate the strength of these arguments, we assume for an American put that its value is below the payoff. $V < 0$ contradicts the definition of the option. Hence $V \geq 0$, and S and V would be in the triangle seen in Figure 1.2. That is, $S < K$ and $0 \leq V < K - S$. This scenario would allow arbitrage. The strategy would be as follows: Borrow the cash amount of $S + V$, and buy both the underlying and the put. Then immediately exercise the put, selling the underlying for the strike price K. The profit of this arbitrage strategy is $K - S - V > 0$. This is in conflict with the no-arbitrage principle. Hence the assumption that the value of an American put is below the payoff must be wrong. We conclude for the put

$$V_P^{am}(S,t) \geq (K - S)^+ \quad \text{for all } S, t .$$

Similarly, for the call

$$V_C^{am}(S,t) \geq (S - K)^+ \quad \text{for all } S, t .$$

(The meaning of the notations V_C^{am}, V_P^{am}, V_C^{eur}, V_P^{eur} is evident.)

Other bounds are listed in Appendix D1. For example, a European put on an asset that pays no dividends until T may also take values below the payoff, but is always above the lower bound $Ke^{-r(T-t)} - S$. The value of an American option should never be smaller than that of a European option because the American type includes the European type exercise at $t = T$ and in addition *early exercise* for $t < T$. That is

$$V^{am} \geq V^{eur}$$

as long as all other terms of the contract are identical. For European options the values of put and call are related by the *put-call parity*

$$S + V_P^{eur} - V_C^{eur} = Ke^{-r(T-t)} ,$$

which can be shown by applying arguments of arbitrage ($\longrightarrow$ Exercise 1.1).

Options in the Market

The features of the options imply that an investor purchases puts when the price of the underlying is expected to fall, and buys calls when the prices are about to rise. This mechanism inspires speculators. An important application of options is hedging ($\longrightarrow$ Appendix A2).

The value of $V(S,t)$ also depends on other factors. Dependence on the strike K and the maturity T is evident. Market parameters affecting the price are the interest rate r, the **volatility** σ of the price S_t, and dividends in case of a dividend-paying asset. The interest rate r is the risk-free rate, which applies to zero bonds or to other investments that are considered free of risks ($\longrightarrow$ Appendices A1, A2). The important volatility parameter σ can be defined as standard deviation of the fluctuations in S_t, for scaling divided by the square root of the observed time period. The larger the fluctuations, respresented by large values of σ, the harder is to predict a future value of the asset. Hence the volatility is a standard measure of risk. The dependence of V on σ is highly sensitive. On occasion we write $V(S,t;T,K,r,\sigma)$ when the focus is on the dependence of V on the market parameters.

The units of r and σ^2 are per year. Time is measured in years. Writing $\sigma = 0.2$ means a volatility of 20%, and $r = 0.05$ represents an interest rate of 5%. The Table 1.1 summarizes the key notations of option pricing. The notation is standard except for the strike price K, which is sometimes denoted X, or E.

The time period of interest is $t_0 \leq t \leq T$. One might think of t_0 denoting the date when the option is issued and t as a symbol for "today." But this book mostly sets $t_0 = 0$ in the role of "today," without loss of generality. Then the interval $0 \leq t \leq T$ represents the remaining life time of the option. The price S_t is a stochastic process, compare Section 1.6. In real markets, the interest rate r and the volatility σ vary with time. To keep the models and the analysis simple, we mostly assume r and σ to be constant on $0 \leq t \leq T$. Further we suppose that all variables are arbitrarily divisible and consequently can vary continuously —that is, all variables vary in the set $\mathbb{R}$ of real numbers.

Table 1.1. List of important variables

t	current time, $0 \leq t \leq T$
T	expiration time, maturity
$r > 0$	risk-free interest rate
S, S_t	spot price, current price per share of stock/asset/underlying
σ	annual volatility
K	strike, exercise price per share
$V(S,t)$	value of an option at time t and underlying price S

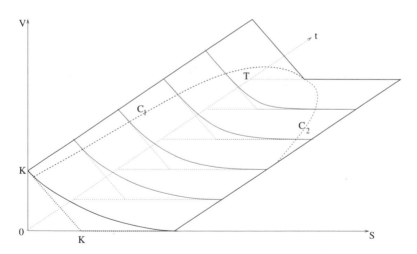

Fig. 1.4. Value $V(S,t)$ of an American put, schematically

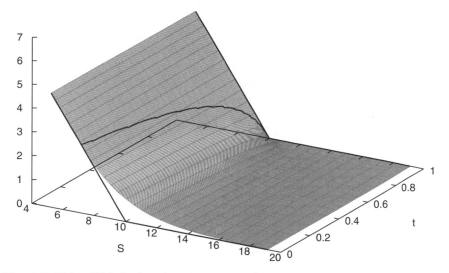

Fig. 1.5. Value $V(S,t)$ of an American put with $r = 0.06$, $\sigma = 0.30$, $K = 10$, $T = 1$

The Geometry of Options

As mentioned, our aim is to calculate $V(S,t)$ for fixed values of K, T, r, σ. The values $V(S,t)$ can be interpreted as a piece of surface over the subset

$$S > 0 \, , \, 0 \leq t \leq T$$

of the (S,t)-plane. The Figure 1.4 illustrates the character of such a surface for the case of an American put. For the illustration assume $T = 1$. The figure depicts six curves obtained by cutting the *option surface* with the planes $t = 0, 0.2, \ldots, 1.0$. For $t = T$ the payoff function $(K - S)^+$ of Figure 1.2 is clearly visible.

Shifting this payoff parallel for all $0 \leq t < T$ creates another surface, which consists of the two planar pieces $V = 0$ (for $S \geq K$) and $V = K - S$ (for $S < K$). This *payoff surface* created by $(K - S)^+$ is a lower bound to the option surface, $V(S,t) \geq (K - S)^+$. The Figure 1.4 shows two curves C_1 and C_2 on the option surface. The curve C_1 is the *early-exercise curve*, because on the planar part with $V(S,t) = K - S$ holding the option is not optimal. (This will be explained in Section 4.5.) The curve C_2 has a technical meaning explained below. Within the area limited by these two curves the option surface is clearly above the payoff surface, $V(S,t) > (K - S)^+$. Outside that area, both surfaces coincide. This is strict above C_1, where $V(S,t) = K - S$, and holds approximately for S beyond C_2, where $V(S,t) \approx 0$ or $V(S,t) < \varepsilon$ for a small value of $\varepsilon > 0$. The location of C_1 and C_2 is not known, these curves are calculated along with the calculation of $V(S,t)$. Of special interest is $V(S,0)$, the value of the option "today." This curve is seen in Figure 1.4

for $t = 0$ as the front edge of the option surface. This front curve may be seen as smoothing the corner in the payoff function. The schematic illustration of Figure 1.4 is completed by a concrete example of a calculated put surface in Figure 1.5. An approximation of the curve C_1 is shown.

The above was explained for an American put. For other options the bounds are different ($\longrightarrow$ Appendix D1). As mentioned before, a European put takes values above the lower bound $Ke^{-r(T-t)} - S$, compare Figure 1.6.

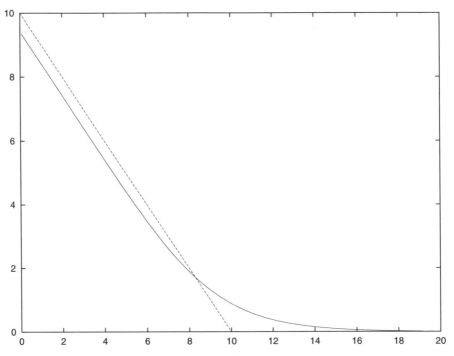

Fig. 1.6. Value of a European put $V(S,0)$ for $T = 1$, $K = 10$, $r = 0.06$, $\sigma = 0.3$. The payoff $V(S,T)$ is drawn with a dashed line. For small values of S the value V approaches its lower bound, here $9.4 - S$.

1.2 Model of the Financial Market

Mathematical models can serve as approximations and idealizations of the complex reality of the financial world. For modeling financial options the models named after the pioneers Black, Merton and Scholes are both successful and widely accepted. This Section 1.2 introduces some key elements of the models.

The ultimate aim is to be able to calculate $V(S,t)$. It is attractive to define the option surfaces $V(S,t)$ on the *half strip* $S > 0$, $0 \leq t \leq T$ as solutions of suitable equations. Then calculating V amounts to solving the equations. In fact, a series of assumptions allows to characterize the *value functions* $V(S,t)$ as solutions of certain partial differential equations or partial differential inequalities. The model is represented by the famous Black-Scholes equation, which was suggested 1973.

Definition 1.1   (Black-Scholes equation)

$$\frac{\partial V}{\partial t} + \frac{1}{2}\sigma^2 S^2 \frac{\partial^2 V}{\partial S^2} + rS\frac{\partial V}{\partial S} - rV = 0 \qquad (1.2)$$

The equation (1.2) is a partial differential equation for the value function $V(S,t)$ of options. This equation may serve as symbol of the market model. But what are the assumptions leading to the Black-Scholes equation?

Assumptions 1.2   (model of the market)
(a) *The market is frictionless.*
   This means that there are no transaction costs (fees or taxes), the interest rates for borrowing and lending money are equal, all parties have immediate access to any information, and all securities and credits are available at any time and in any size. Consequently, all variables are perfectly divisible —that is, may take any real number. Further, individual trading will not influence the price.
(b) *There are no arbitrage opportunities.*
(c) *The asset price follows a geometric Brownian motion.*
   (This stochastic motion will be discussed in Sections 1.6–1.8.)
(d) Technical assumptions (some are preliminary):
   r and σ are constant for $0 \leq t \leq T$. No dividends are paid in that time period. The option is European.

These are the assumptions that lead to the Black-Scholes equation (1.2). A derivation of this partial differential equation is given in Appendix A4. Admitting all real numbers t within the interval $0 \leq t \leq T$ leads to characterize the model as *continuous-time model*. In view of allowing also arbitrary $S > 0$, $V > 0$, we speak of a continuous model.

   Solutions $V(S,t)$ are functions that satisfy this equation for all S and t out of the half strip. In addition to solving the partial differential equation, the function $V(S,t)$ must satisfiy a terminal condition and boundary conditions. The **terminal condition** for $t = T$ is

$$V(S,T) = \text{payoff},$$

with payoff function (1.1C) or (1.1P), depending on the type of option. The boundaries of the half strip $0 < S$, $0 \leq t \leq T$ are defined by $S = 0$ and $S \to \infty$. At these boundaries the function $V(S,t)$ must satisfy **boundary conditions**. For example, a European call must obey

$$V(0,t) = 0; \quad V(S,t) \to S - Ke^{-r(T-t)} \text{ for } S \to \infty. \tag{1.3C}$$

In Chapter 4 we will come back to the Black-Scholes equation and to boundary conditions. For (1.2) an analytic solution is known (equation (A4.10) in Appendix A4). This does not hold for more general models. For example, considering transaction costs as k per unit would add the term

$$-\sqrt{\frac{2}{\pi}} \frac{k\sigma S^2}{\sqrt{\sigma t}} \left| \frac{\partial^2 V}{\partial S^2} \right|$$

to (1.2), see [WDH96], [Kwok98]. In the general case, closed-form solutions do not exist, and a solution is calculated numerically, especially for American options. For numerically solving (1.2) a variant with dimensionless variables can be used ($\longrightarrow$ Exercise 1.2).

At this point, a word on the notation is appropriate. The symbol S for the asset price is used in different roles: First it comes without subscript in the role of an independent real variable $S > 0$ on which $V(S,t)$ depends, say as solution of the partial differential equation (1.2). Second it is used as S_t with subscript t to emphasize its random character as stochastic process. When the subscript t is omitted, the current role of S becomes clear from the context.

1.3 Numerical Methods

Applying numerical methods is inevitable in all fields of technology including financial engineering. Often the important role of numerical algorithms is not noticed. For example, an analytical formula at hand (such as the Black-Scholes formula (A4.10)) might suggest that no numerical procedure is needed. But closed-form solutions may include evaluating the logarithm or the computation of the distribution function of the normal distribution. Such elementary tasks are performed using sophisticated numerical algorithms. In pocket calculators one merely presses a button without being aware of the numerics. The robustness of those elementary numerical methods is so dependable and the efficiency so large that they almost appear not to exist. Even for apparently simple tasks the methods are quite demanding ($\longrightarrow$ Exercise 1.3). The methods must be carefully designed because inadequate strategies might produce inaccurate results ($\longrightarrow$ Exercise 1.4).

Spoilt by generally available black-box software and graphics packages we take the support and the success of numerical workhorses for granted. We make use of the numerical tools with great respect but without further comments. We just assume an elementary education in numerical methods. An introduction into important methods and hints on the literature are given in Appendix C1.

Since financial markets undergo apparently stochastic fluctuations, stochastic approaches will be natural tools to simulate prices. These methods are based on formulating and simulating stochastic differential equations. This leads to Monte Carlo methods ($\longrightarrow$ Chapter 3). In computers, related simulations of options are performed in a deterministic manner. It will be decisive how to simulate randomness ($\longrightarrow$ Chapter 2). Chapters 2 and 3 are devoted to tools for simulation. These methods can be applied even in case the Assumptions 1.2 are not satisfied.

More efficient methods will be preferred provided their use can be justified by the validity of the underlying models. For example it may be advisable to solve the partial differential equations of the Black-Scholes type. Then one has to choose among several methods. The most elementary ones are finite-difference methods ($\longrightarrow$ Chapter 4). A somewhat higher flexibility concerning error control is possible with finite-element methods ($\longrightarrow$ Chapter 5). The numerical treatment of exotic options requires a more careful consideration of stability issues ($\longrightarrow$ Chapter 6). The methods based on differential equations will be described in the larger part of this book.

The various methods are discussed in terms of accuracy and speed. Ultimately the methods must give quick and accurate answers to real-time problems posed in financial markets. Efficiency and reliability are key demands. Internally the numerical methods must deal with diverse problems such as convergence order or stability. So the numerical analyst is concerned in error estimates and error bounds. Technical criteria such as complexity or storage requirements are relevant for the implementation.

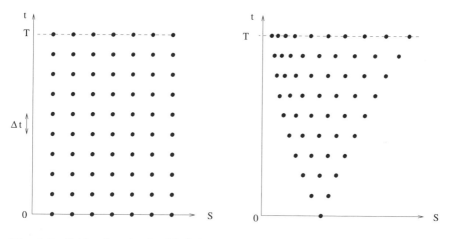

Fig. 1.7. Grid points in the (S, t)-domain

The mathematical formulation benefits from the assumption that all variables take values in the continuum $\mathbb{R}$. This idealization is practical since it avoids initial restrictions of technical nature. This gives us freedom to impose *artificial* discretizations convenient for the numerical methods. The hypothesis of a continuum applies to the (S, t)-domain of the half strip $0 \leq t \leq T$, $S > 0$, and to the differential equations. In contrast to the hypothesis of a continuum, the financial reality is rather discrete: Neither the price S nor the trading times t can take any real value. The artificial discretization introduced by numerical methods is at least twofold:

1.) The (S, t)-domain is replaced by a **grid** of a finite number of (S, t)-points, compare Figure 1.7.
2.) The differential equations are adapted to the grid and replaced by a finite number of algebraic equations.

Another kind of discretization is that computers replace the real numbers by a finite number of rational numbers, namely the floating-point numbers. The resulting rounding error will not be relevant for much of our analysis, except for investigations of stability.

The restriction of the differential equations to the grid causes **discretization errors**. The errors depend on the coarsity of the grid. In Figure 1.7, the distance between two consecutive t-values of the grid is denoted Δt.[2] So the errors will depend on Δt and on ΔS. It is one of the aims of numerical algorithms to control the errors. The left-hand figure in Figure 1.7 shows a simple rectangle grid, whereas the right-hand figure shows a tree-type grid as used in Section 1.4. The type of the grid matches the kind of underlying equations. The values of $V(S, t)$ are primarily approximated at the grid points. Intermediate values can be obtained by interpolation.

The continuous model is an idealization of the discrete reality. But the numerical discretization does not reproduce the original discretization. For example, it would be a rare coincidence when Δt represents a day. The derivations that go along with the twofold transition

$$\text{discrete} \longrightarrow \text{continuous} \longrightarrow \text{discrete}$$

do not compensate.

1.4 The Binomial Method

The major part of the book is devoted to continuous models and their discretizations. With much less effort a discrete approach provides us with a short way to establish a first algorithm for calculating options. The resulting *binomial method* due to Cox, Ross and Rubinstein is robust and widely applicable.

---

[2] The symbol Δt denotes a small increment in t (analogously $\Delta S, \Delta W$). In case Δ would be a number, the product with u would be denoted $\Delta \cdot u$ or $u\Delta$.

In practice one is often interested in the one value $V(S_0, 0)$ of an option at the current spot price S_0. Then it can be unnecessarily costly to calculate the surface $V(S, t)$ for the entire domain to extract the required information $V(S_0, 0)$. The relatively small task of calculating $V(S_0, 0)$ can be comfortably solved using the binomial method. This method is based on a tree-type grid applying appropriate binary rules at each grid point. The grid is not predefined but is constructed by the method. For illustration see the right-hand grid in Figure 1.7, and Figure 1.10.

A Discrete Model

We begin with discretizing the continuous time t, replacing t by equidistant time instances t_i. Let us use the notations

M: number of time steps
$\Delta t := \frac{T}{M}$
$t_i := i \cdot \Delta t, \quad i = 0, ..., M$
$S_i := S(t_i)$

So far the domain of the (S, t) half strip is *semi-discretized* in that it is replaced by parallel straight lines with distance Δt apart, leading to a discrete-time model. The next step of discretization replaces the continuous values S_i along the parallel $t = t_i$ by discrete values S_{ji}, for all i and appropriate j. (Here the indices j, i in S_{ji} mean a matrix-like notation.) For a better understanding of the S-discretization compare Figure 1.8. This figure shows a mesh of the grid, namely the transition from t to $t + \Delta t$, or from t_i to t_{i+1}.

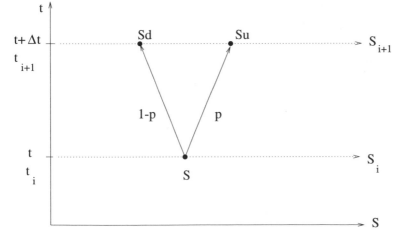

Fig. 1.8. The principle of the binomial method

Assumptions 1.3   (binomial method)

(Bi1) The price S over each period of time Δt can only have two possible outcomes: An initial value S either evolves up to Su or down to Sd with $0 < d < u$. Here u is the factor of an upward movement and d is the factor of a downward movement.

(Bi2) The probability of an up movement is p, $\mathsf{P}(\text{up}) = p$.

The rules (Bi1), (Bi2) represent the framework of a binomial process. Such a process behaves like tossing a biased coin where the outcome "head" (up) occurs with probability p. At this stage of the modeling, the values of the three parameters u, d und p are undetermined. They are fixed in a way such that the model is consistent with the continuous model in case $\Delta t \to 0$. This aim leads to further assumptions. The basic idea of the approach is to equate the expectation and the variance of the discrete model with the corresponding values of the continuous model. This amounts to require

(Bi3) Expectation and variance of S refer to the continuous counterparts, evaluated for the risk-free interest rate r.

This assumption leads to equations for the parameters u, d, p. The resulting probability P of (Bi2) does not reflect the expectations of an individual in the market. Rather P is an artificial risk-neutral probability that matches (Bi3). The expectation E below in (1.4) refers to this probability; this is sometimes written E_P. (We shall return to the assumptions (Bi1)-(Bi3) in the subsequent Section 1.5.) Let us further assume that no dividend is paid within the time period of interest. This assumption simplifies the derivation of the method and can be removed later.

Derivation of Equations

Recall the definition of the expectation for the discrete case, Appendix B1, equation (B1.13), and conclude

$$\mathsf{E}(S_{i+1}) = pS_iu + (1 - p)S_id .$$

Here S_i is an arbitrary value for t_i, which develops randomly to S_{i+1}, following the assumptions (Bi1), (Bi2). In this sense, E is a conditional expectation. As will be seen in Section 1.7.2, the expectation of the continuous model is

$$\mathsf{E}(S_{i+1}) = S_ie^{r\Delta t} \tag{1.4}$$

Equating gives

$$e^{r\Delta t} = pu + (1 - p)d. \tag{1.5}$$

This is the first of three required equations to fix u, d, p. Solved for the risk-neutral probability p we obtain

$$p = \frac{e^{r\Delta t} - d}{u - d}. \tag{1.6}$$

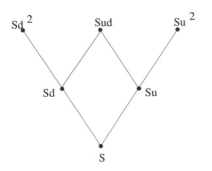

Fig. 1.9. Sequence of several meshes (schematically)

To be a valid model of probability, $0 \le p \le 1$ must hold. This is equivalent to

$$d \le e^{r \Delta t} \le u \ . \tag{1.7}$$

These inequalities relate the upward and downward movements of the asset price to the riskless interest rate r. The inequalities (1.7) are no new assumption but follow from the no-arbitrage principle. The assumption $0 < d < u$ remains valid.

Next we equate variances. Via the variance the volatility σ enters the model. From the continuous model we apply the relation

$$\mathsf{E}(S_{i+1}^2) = S_i^2 e^{(2r+\sigma^2)\Delta t}. \tag{1.8}$$

For the relations (1.4) and (1.8) we refer to Section 1.8 ($\longrightarrow$ Exercise 1.12). Recall that the variance satisfies $\mathsf{Var}(S) = \mathsf{E}(S^2) - (\mathsf{E}(S))^2$ ($\longrightarrow$ Appendix B1). Equations (1.4) and (1.8) combine to

$$\mathsf{Var}(S_{i+1}) = S_i^2 e^{2r\Delta t}(e^{\sigma^2 \Delta t} - 1).$$

On the other hand the discrete model satisfies

$$\mathsf{Var}(S_{i+1}) = \mathsf{E}(S_{i+1}^2) - (\mathsf{E}(S_{i+1}))^2$$
$$= p(S_i u)^2 + (1 - p)(S_i d)^2 - S_i^2(pu + (1 - p)d)^2.$$

Equating variances of the continuous and the discrete model, and applying (1.5) leads to

$$e^{2r\Delta t}(e^{\sigma^2 \Delta t} - 1) = pu^2 + (1 - p)d^2 - (e^{r\Delta t})^2$$
$$e^{2r\Delta t + \sigma^2 \Delta t} = pu^2 + (1 - p)d^2 \tag{1.9}$$

The equations (1.5), (1.9) constitute two relations for the three unknowns u, d, p. We are free to impose an arbitrary third equation. One example is the plausible assumption

$$u \cdot d = 1 \ , \tag{1.10}$$

which reflects a symmetry between upward and downward movement of the asset price. Now the parameters u, d and p are fixed. They depend on r, σ and Δt. So does the grid, which is analyzed next (Figure 1.9).

The above rules are applied to each grid line $i = 0, \ldots, M$, starting at $t_0 = 0$ with the specific value $S = S_0$. Attaching meshes of the kind depicted in Figure 1.8 for subsequent values of t_i builds a tree with values $Su^j d^k$ and $j + k = i$. In this way, specific discrete values S_{ji} of S_i are defined. Since the same constant factors u and d underlie all meshes and since $Sud = Sdu$ holds, after the time period $2\Delta t$ the asset price can only take three values rather than four: The tree is recombining. It does not matter which of the two possible paths we take to reach Sud. This property extends to more than two time periods. Consequently the binomial process defined by Assumption 1.3 is *path independent*. Accordingly at expiration time $T = M\Delta t$ the price S can take only the $(M + 1)$ discrete values $Su^j d^{M-j}$, $j = 0, 1, \ldots, M$. By (1.10) these are the values $Su^{2j-M} =: S_{jM}$. The number of nodes in the tree grows quadratically in M. (Why?)

The symmetry of the choice (1.10) becomes apparent in that after two time steps the asset value S repeats. (Compare also Figure 1.10.) In the (t, S)-plane the tree can be interpreted as a grid of exponential-like curves. The binomial approach defined by (Bi1) with the proportionality between S_i and S_{i+1} reflects exponential growth or decay of S. So all grid points have the desirable property $S > 0$.

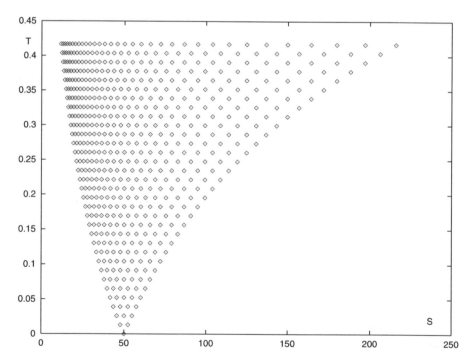

Fig. 1.10. Tree in the (S, t)-plane for $M = 32$ (data of Example 1.6)

Solution of the Equations

Using the abbreviation $\alpha := e^{r\Delta t}$ we obtain by elimination (which the reader may check in more generality in Exercise 1.14) the quadratic equation

$$0 = u^2 - u(\underbrace{\alpha^{-1} + \alpha e^{\sigma^2 \Delta t}}_{=:2\beta}) + 1,$$

with solutions $u = \beta \pm \sqrt{\beta^2 - 1}$. By virtue of $ud = 1$ and Vieta's Theorem, d is the solution with the minus sign. In summary the three parameters u, d, p are given by

$$
\boxed{
\begin{aligned}
\beta :&= \frac{1}{2}(e^{-r\Delta t} + e^{(r+\sigma^2)\Delta t}) \\
u &= \beta + \sqrt{\beta^2 - 1} \\
d &= 1/u = \beta - \sqrt{\beta^2 - 1} \\
p &= \frac{e^{r\Delta t} - d}{u - d}
\end{aligned}
}
\tag{1.11}
$$

A consequence of this approach is that up to terms of higher order the relation $u = e^{\sigma\sqrt{\Delta t}}$ holds ($\longrightarrow$ Exercise 1.6). Therefore the extension of the tree in S-direction matches the volatility of the asset. So the tree will cover the relevant range of S-values.

Forward Phase: Initializing the Tree

Now the factors u and d can be considered as known and the discrete values of S for each t_i until $t_M = T$ can be calculated. The current spot price $S = S_0$ for $t_0 = 0$ is the root of the tree. (To adapt the matrix-like notation to the two-dimensional grid of the tree, this initial price will be also denoted S_{00}.) Each initial price S_0 leads to another tree of values S_{ji}.

$$
\boxed{
\begin{aligned}
&For\ i = 1, 2, ..., M\ calculate: \\
&S_{ji} := S_0 u^j d^{i-j}, \quad j = 0, 1, ..., i
\end{aligned}
}
$$

Now the grid points (t_i, S_{ji}) are fixed, on which the option values $V_{ji} := V(t_i, S_{ji})$ are to be calculated.

Calculating the Option Value, Valuation of the Tree

For t_M the payoff $V(S, t_M)$ is known from (1.1C), (1.1P). This payoff is valid for each S, including $S_{jM} = Su^j d^{M-j}$, $j = 0, ..., M$. This defines the values V_{jM}:

Call: $V(S(t_M), t_M) = \max\{S(t_M) - K, 0\}$, hence:

$$V_{jM} := (S_{jM} - K)^+ \tag{1.12C}$$

Put: $V(S(t_M), t_M) = \max\{K - S(t_M), 0\}$, hence:

$$V_{jM} := (K - S_{jM})^+ \tag{1.12P}$$

The **backward phase** calculates recursively for t_{M-1}, t_{M-2}, ... the option values V for all t_i, starting from V_{jM}. The recursion is based on Assumption 1.3, (Bi3). Repeating the equation that corresponds to (1.5) with double index leads to

$$S_{ji}e^{r\Delta t} = pS_{ji}u + (1-p)S_{ji}d,$$

and

$$S_{ji}e^{r\Delta t} = pS_{j+1,i+1} + (1-p)S_{j,i+1}.$$

Relating the Assumption 1.3, (Bi3) of risk neutrality to V, $V_i = e^{-r\Delta t}\mathsf{E}(V_{i+1})$, we obtain using the double-index notation the recursion

$$V_{ji} = e^{-r\Delta t} \cdot (pV_{j+1,i+1} + (1-p)V_{j,i+1}) . \tag{1.13}$$

So far, this recursion for V_{ji} is merely an analogy, which might be seen as a further assumption. But the following Section 1.5 will give a justification for (1.13), which turns out to be a consequence of the no-arbitrage principle and the risk-neutral valuation.

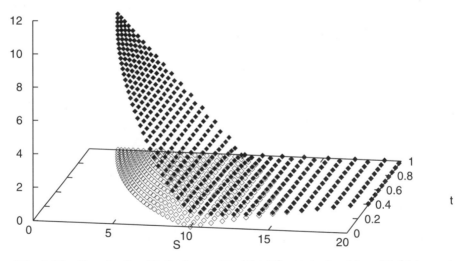

Fig. 1.11. Tree in the (S,t)-plane with (S,t,V)-points for $M = 32$ (data as in Figure 1.5)

For **European options** this is a recursion for $i = M - 1, \ldots, 0$, starting from (1.12), and terminating with V_{00}. (For an illustration see Figure 1.11.) The obtained value V_{00} is an approximation to the value $V(S_0, 0)$ of the continuous model, which results in the limit $M \to \infty$ ($\Delta t \to 0$). The accuracy of the approximation V_{00} depends on M. This is reflected by writing $V_0^{(M)}$ ($\longrightarrow$ Exercise 1.7). The basic idea of the approach implies that the limit of $V_0^{(M)}$ for $M \to \infty$ is the Black-Scholes value $V(S_0, 0)$ ($\longrightarrow$ Exercise 1.8).

For **American options** the above recursion must be modified by adding a test whether early exercise is to be preferred. To this end the value of (1.13) is compared with the value of the payoff. Then the equations (1.12) for i rather than M, combined with (1.13), read as follows:

Call:

$$V_{ji} = \max \left\{ (S_{ji} - K)^+, \ e^{-r\Delta t} \cdot (p V_{j+1,i+1} + (1-p) V_{j,i+1}) \right\} \qquad (1.14\text{C})$$

Put:

$$V_{ji} = \max \left\{ (K - S_{ji})^+, \ e^{-r\Delta t} \cdot (p V_{j+1,i+1} + (1-p) V_{j,i+1}) \right\} \qquad (1.14\text{P})$$

This amounts to a *dynamic programming* procedure. Let us summarize:

Algorithm 1.4   (binomial method)

> *Input:* r, σ, $S = S_0$, T, K, choice of put or call,
>
>   European or American, M
>
> *calculate:* $\Delta t := T/M$, u, d, p from (1.11)
>
> $S_{00} := S_0$
>
> $S_{jM} = S_{00} u^j d^{M-j}$, $j = 0, 1, \ldots, M$
>
>   (for American options, also $S_{ji} = S_{00} u^j d^{i-j}$
>
>   for $0 < i < M$, $j = 0, 1, \ldots, i$)
>
> V_{jM} from (1.12)
>
> V_{ji} for $i < M$ $\begin{cases} \text{from (1.13) for European options} \\ \text{from (1.14) for American options} \end{cases}$
>
> *Output:* V_{00} is the approximation $V_0^{(M)}$ to $V(S_0, 0)$

Example 1.5   European put

     $K = 10$, $S = 5$, $r = 0.06$, $\sigma = 0.3$, $T = 1$.

The Table 1.2 lists approximations $V^{(M)}$ to $V(5, 0)$. The convergence towards the Black-Scholes value $V(S, 0)$ is visible; the latter was calculated by evaluating (A4.10). (In this book the number of printed

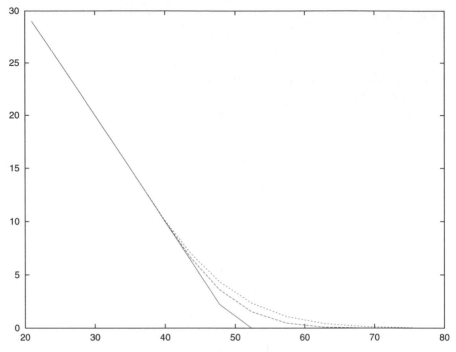

Fig. 1.12. to Example 1.6: Three cuts through the rough approximation of the surface $V(S,t)$ for $t = 0.404$ (solid curve), $t = 0.3$ (dashed), $t = 0.195$ (dotted), approximated with $M = 32$

decimals illustrates at best the attainable accuracy and does not reflect economic practice.) Applying other methods the function $V(S,0)$ can be approximated for an interval of S-values. The Figure 1.6 shows related results obtained by using the methods of Chapter 4. The convergence rate is reflected by the results in Table 1.2. The rate is linear, $O(\Delta t) = O(M^{-1})$, which is seen by plotting $V^{(M)}$ over M^{-1}. In such a plot, the values of $V^{(M)}$ roughly lie close to a straight line, which reflects the linear error decay. The reader may wish to investigate more closely how the error decays with M ($\longrightarrow$ Exercise 1.7). It turns out that for the described version of the binomial method the convergence in M is not monotonic. It will not be recommendable to extrapolate the $V^{(M)}$-data to the limit $M \to \infty$, at least not the data of Table 1.2.

Example 1.6 American put

$K = 50$, $S = 50$, $r = 0.1$, $\sigma = 0.4$, $T = 0.41666...$ ($\frac{5}{12}$ for 5 months), $M = 32$.

The Figure 1.10 shows the tree for $M = 32$. The approximation to V_0 is 4.2719. Although the binomial method is not designed to accurately approximate the surface $V(S,t)$, it provides rough information also for

Table 1.2. Results of Example 1.5

M	$V^{(M)}(5,0)$
8	4.42507
16	4.42925
32	4.429855
64	4.429923
128	4.430047
256	4.430390
2048	4.430451
Black-Scholes	4.43046477621

$t > 0$. Figure 1.12 depicts for three time instances $t = 0.404$, $t = 0.3$, $t = 0.195$ the obtained approximation of $V(S,t)$; the calculated discrete values are interpolated by straight line segments. The function $V(S,0)$ can be approximated with the methods of Chapter 4, compare Figure 4.11.

Extensions

The paying of dividends can be incorporated into the binomial algorithm. If dividends are paid at t_k the price of the asset drops by the same amount. To take into account this jump, the tree is cut at t_k and the S-values are reduced appropriately, see [Hull00, § 16.3], [WDH96].

Correcting the terminal probabilities, which come out of the binomial distribution ($\longrightarrow$ Exercise 1.8), it is possible to adjust the tree to actual market data [Ru94]. Another extension of the binomial model is the *trinomial model*. Here each mesh offers three outcomes, with probabilities p_1, p_2, p_3 and $p_1 + p_2 + p_3 = 1$. The trinomial model allows for higher accuracy. The reader may wish to derive the trinomial method.

1.5 Risk-Neutral Valuation

In the previous section we have used the Assumptions 1.3 to derive an algorithm for valuation of options. This Section 1.5 discusses the assumptions again leading to a different interpretation.

The situation of a path-independent binomial process with the two factors u and d continues to be the basis of the argumentation. The scenario is illustrated in Figure 1.13. Here the time period is the time to expiration T, which replaces Δt in the local mesh of Figure 1.8. Accordingly, this global model is called *one-period model*. The one-period model with only two possible values of S_T has two clearly defined values of the payoff, namely $V^{(d)}$ (corresponds to $S_T = S_0 d$) and $V^{(u)}$ (corresponds to $S_T = S_0 u$). In contrast to the Assumptions 1.3 we neither assume the risk-neutral world (Bi3) nor the corresponding probability $P(\mathrm{up}) = p$ from (Bi2). Instead we derive the probability using another argument. In this section the factors u and d are assumed to be given.

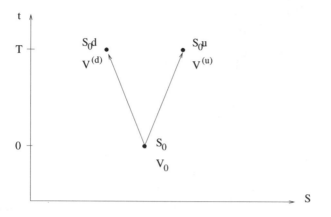

Fig. 1.13. One-period binomial model

Let us construct a portfolio of an investor with a short position in one option and a long position consisting of Δ shares of an asset, where the asset is the underlying of the option. The portfolio manager must **choose the number Δ of shares such that the portfolio is riskless.** That is, a hedging strategy is needed. To discuss the hedging properly we assume that no funds are added or withdrawn.

By Π_t we denote the wealth of this portfolio at time t. Initially the value is

$$\Pi_0 = S_0 \cdot \Delta - V_0 , \qquad (1.15)$$

where the value V_0 of the written option is not yet determined. At the end of the period the value V_T either takes the value $V^{(u)}$ or the value $V^{(d)}$. So the value of the portfolio Π_T at the end of the life of the option is either

$$\Pi^{(u)} = S_0 u \cdot \Delta - V^{(u)}$$

or

$$\Pi^{(d)} = S_0 d \cdot \Delta - V^{(d)} .$$

In case Δ is chosen such that the value Π_T is riskless, all uncertainty is removed and $\Pi^{(u)} = \Pi^{(d)}$ must hold. This is equivalent to

$$(S_0 u - S_0 d) \cdot \Delta = V^{(u)} - V^{(d)} ,$$

which defines the strategy

$$\Delta = \frac{V^{(u)} - V^{(d)}}{S_0(u - d)} . \qquad (1.16)$$

With this value of Δ the portfolio with initial value Π_0 evolves to the final value $\Pi_T = \Pi^{(u)} = \Pi^{(d)}$, regardless of whether the stock price moves up or down. Consequently the portfolio is riskless.

If we rule out early exercise, the final value Π_T is reached with certainty. The value Π_T must be compared to the alternative risk-free investment of an amount of money that equals the initial wealth Π_0, which after the time period T reaches the value $e^{rT}\Pi_0$. Both the assumptions $\Pi_0 e^{rT} < \Pi_T$ and $\Pi_0 e^{rT} > \Pi_T$ would allow a strategy of earning a risk-free profit. This is in contrast to the assumed arbitrage-free world. Hence both $\Pi_0 e^{rT} \geq \Pi_T$ and $\Pi_0 e^{rT} \leq \Pi_T$ and hence equality must hold.[3] Accordingly the initial value Π_0 of the portfolio equals the discounted final value Π_T, discounted at the interest rate r,

$$\Pi_0 = e^{-rT}\Pi_T .$$

This means

$$S_0 \cdot \Delta - V_0 = e^{-rT}(S_0 u \cdot \Delta - V^{(u)}) ,$$

which upon substituting (1.16) leads to the value V_0 of the option:

$$
\begin{aligned}
V_0 &= S_0 \cdot \Delta - e^{-rT}(S_0 u \Delta - V^{(u)}) \\
&= e^{-rT}\{\Delta \cdot [S_0 e^{rT} - S_0 u] + V^{(u)}\} \\
&= \tfrac{e^{-rT}}{u-d}\{(V^{(u)} - V^{(d)})(e^{rT} - u) + V^{(u)}(u - d)\} \\
&= \tfrac{e^{-rT}}{u-d}\{V^{(u)}(e^{rT} - d) + V^{(d)}(u - e^{rT})\} \\
&= e^{-rT}\{V^{(u)}\tfrac{e^{rT}-d}{u-d} + V^{(d)}\tfrac{u-e^{rT}}{u-d}\} \\
&= e^{-rT}\{V^{(u)}q + V^{(d)} \cdot (1 - q)\}
\end{aligned}
$$

with

$$q := \frac{e^{rT} - d}{u - d} . \tag{1.17}$$

We have shown that with q from (1.17) the value of the option is given by

$$V_0 = e^{-rT}\{V^{(u)}q + V^{(d)} \cdot (1 - q)\} . \tag{1.18}$$

The expression for q in (1.17) is identical to the formula for p in (1.6), which was derived in the previous section. Again we have

$$0 < q < 1 \quad \Longleftrightarrow \quad d < e^{rT} < u .$$

Presuming these bounds for u and d, q can be interpreted as a probability Q. Then $qV^{(u)} + (1 - q)V^{(d)}$ is the expected value of the payoff with respect to this probability (1.17),

$$\mathsf{E}_{\mathsf{Q}}(V_T) = qV^{(u)} + (1 - q)V^{(d)} .$$

---

[3] For an American option it is not certain that Π_T can be reached because the holder may choose early exercise. Hence we have only the inequality $\Pi_0 e^{rT} \leq \Pi_T$.

Now (1.18) can be written

$$V_0 = e^{-rT} \mathsf{E}_\mathsf{Q}(V_T) \ . \tag{1.19}$$

That is, the value of the option is obtained by discounting the expected payoff (with respect to q from (1.17)) at the risk-free interest rate r. An analogous calculation shows

$$\mathsf{E}_\mathsf{Q}(S_T) = q S_0 u + (1-q) S_0 d = S_0 e^{rT} \ .$$

The probabilities p of Section 1.4 and q from (1.17) are defined by identical formulas (with T corresponding to Δt). Hence $p = q$, and $\mathsf{E}_\mathsf{P} = \mathsf{E}_\mathsf{Q}$. But the underlying arguments are different. Recall that in Section 1.4 we showed the implication

$$\mathsf{E}(S_T) = S_0 e^{rT} \quad \Longrightarrow \quad p = \mathsf{P}(\text{up}) = \frac{e^{rT} - d}{u - d} \ ,$$

whereas in this section we arrive at the implication

$$p = \mathsf{P}(\text{up}) = \frac{e^{rT} - d}{u - d} \quad \Longrightarrow \quad \mathsf{E}(S_T) = S_0 e^{rT} \ .$$

So both statements must be equivalent. Setting the probability of the up movement equal to p is equivalent to assuming that the expected return on the asset equals the risk-free rate. This can be rewritten as

$$e^{-rT} \mathsf{E}_\mathsf{P}(S_T) = S_0 \ . \tag{1.20}$$

The important property expressed by equation (1.20) is that of a *martingale*: The random variable $e^{-rT} S_T$ of the left-hand side has the tendency to remain at the same level. That is why a martingale is also called "fair game." A martingale displays no trend, where the trend is measured with respect to E_P. In the martingale property of (1.20) the discounting at the risk-free interest rate r exactly matches the risk-neutral probability $\mathsf{P}(= \mathsf{Q})$ of (1.6)/(1.17). The specific probability for which (1.20) holds is also called *martingale measure*.

Summary of results for the one-period model: Under the Assumptions 1.2 of the market model, the choice Δ of (1.16) eliminates the random-dependence of the payoff and makes the portfolio riskless. There is a specific probability Q (= P) with $\mathsf{Q}(\text{up}) = q$, q from (1.17), such that the value V_0 satisfies (1.19) and S_0 the analogous property (1.20). These properties involve the risk-neutral interest rate r. That is, the option is valued in a risk-neutral world, and the corresponding Assumption 1.3 (Bi3) is meaningful.

In the real-world economy, growth rates in general are different from r, and individual subjective probabilities differ from our Q. But the assumption of a risk-neutral world leads to a fair valuation of options. The obtained value V_0 can be seen as a *rational* price. In this sense the resulting value V_0 applies to the real world. The risk-neutral valuation can be seen as a technical tool.

The assumption of risk neutrality is just required to define and calculate a rational price or fair value of V_0. For this specific purpose we do not need actual growth rates of prices, and individual probabilities are not relevant. But note that we do not really assume that financial markets are actually free of risk.

The general principle outlined for the one-period model is also valid for the multi-period binomial model and for the continuous model of Black and Scholes ($\longrightarrow$ Exercise 1.8).

The Δ of (1.16) is the hedge parameter *delta*, which eliminates the risk exposure of our portfolio caused by the written option. In multi-period models and continuous models Δ must be adapted dynamically. The general definition is

$$\Delta = \Delta(S, t) = \frac{\partial V(S, t)}{\partial S} \;;$$

the expression (1.16) is a discretized version.

1.6 Stochastic Processes

Brownian motion originally meant the erratic motion of a particle (pollen) on the surface of a fluid, caused by tiny impulses of molecules. Wiener suggested a mathematical model for this motion, the *Wiener process*. But earlier Bachelier had applied Brownian motion to model the motion of stock prices, which instantly respond to the numerous upcoming informations similar as pollen react to the impacts of molecules. The illustration of the *Dow* in Figure 1.14 may serve as motivation.

A *stochastic process* is a family of random variables X_t, which are defined for a set of parameters t ($\longrightarrow$ Appendix B1). Here we consider the continuous-time situation. That is, $t \in \mathbb{R}$ varies continuously in a time interval I, which typically represents $0 \le t \le T$. A more complete notation for a stochastic process is $\{X_t, t \in I\}$, or $(X_t)_{0 \le t \le T}$. Let the chance play for all t in the interval $0 \le t \le T$, then the resulting function X_t is called *realization* or *path* of the stochastic process.

Special properties of stochastic processes have lead to the following names:

Gaussian process: All finite-dimensional distributions $(X_{t_1}, \ldots, X_{t_k})$ are Gaussian. Hence specifically X_t is distributed normally for all t.

Markov process: Only the present value of X_t is relevant for its future motion. That is, the past history is fully reflected in the present value.[4]

An example of a process that is both Gaussian and Markov, is the Wiener process.

---

[4] This assumption together with the assumption of an immediate reaction of the market to arriving informations are called *hypothesis of the efficient market* [Bo98].

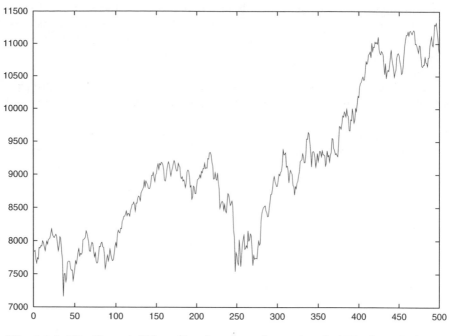

Fig. 1.14. The Dow at 500 trading days from September 8, 1997 through August 31, 1999

1.6.1 Wiener Process

Definition 1.7   (Wiener process, Brownian motion)

A Wiener process (or Brownian motion; notation W_t or W) is a time-continuous process with the properties

(a) $W_0 = 0$ (with probability one)

(b) $W_t \sim \mathcal{N}(0, t)$ for all $t \geq 0$. That is, for each t the random variable W_t is normally distributed with mean $\mathsf{E}(W_t) = 0$ and variance $\mathsf{Var}(W_t) = \mathsf{E}(W_t^2) = t$.

(c) All increments $\Delta W_t := W_{t+\Delta t} - W_t$ on nonoverlapping time intervals are independent: That is, the displacements $W_{t_2} - W_{t_1}$ and $W_{t_4} - W_{t_3}$ are independent for all $0 \leq t_1 < t_2 \leq t_3 < t_4$.

(d) W_t depends continuously on t.

Generally for $0 \leq s < t$ the property $W_t - W_s \sim \mathcal{N}(0, t-s)$ holds, in particular

$$\mathsf{E}(W_t - W_s) = 0 \,, \tag{1.21a}$$

$$\mathsf{Var}(W_t - W_s) = \mathsf{E}((W_t - W_s)^2) = t - s. \tag{1.21b}$$

The relations (1.21a,b) can be derived from Definition 1.7 ($\longrightarrow$ Exercise 1.9). The relation (1.21b) is also known as

$$E((\Delta W_t)^2) = \Delta t \ . \tag{1.21c}$$

The independence of the increments according to Definition 1.7(c) implies for $t_{j+1} > t_j$ the independence of W_{t_j} and $(W_{t_{j+1}} - W_{t_j})$, but not of $W_{t_{j+1}}$ and $(W_{t_{j+1}} - W_{t_j})$. Wiener processes are examples of martingales — there is no drift.

Discrete-Time Model

Let $\Delta t > 0$ be a constant time increment. For the discrete instances $t_j := j\Delta t$ the value W_t can be written as a sum of increments ΔW_k,

$$W_{j\Delta t} = \sum_{k=1}^{j} \underbrace{\left(W_{k\Delta t} - W_{(k-1)\Delta t} \right)}_{=:\Delta W_k} \ .$$

The ΔW_k are independent and because of (1.21) normally distributed with $\mathsf{Var}(\Delta W_k) = \Delta t$. Increments ΔW with such a distribution can be calculated from standard normally distributed random numbers Z. The implication

$$Z \sim \mathcal{N}(0,1) \implies Z \cdot \sqrt{\Delta t} \sim \mathcal{N}(0, \Delta t)$$

leads to the discrete model of a Wiener process

$$\Delta W_k = Z\sqrt{\Delta t} \ \text{ for } \ Z \sim \mathcal{N}(0,1) \text{ for each } k \ . \tag{1.22}$$

We summarize the numerical simulation of a Wiener process as follows:

Algorithm 1.8   (simulation of a Wiener process)

$$
\begin{array}{l}
Start: \quad t_0 = 0, \ W_0 = 0; \ \Delta t \\
loop \ \ j = 1, 2, \dots : \\
\qquad t_j = t_{j-1} + \Delta t \\
\qquad \text{draw } \ Z \sim \mathcal{N}(0,1) \\
\qquad W_j = W_{j-1} + Z\sqrt{\Delta t}
\end{array}
$$

The drawing of Z —that is, the calculation of $Z \sim \mathcal{N}(0,1)$— will be explained in Chapter 2. The values W_j are realizations of W_t at the discrete points t_j. The Figure 1.15 shows a realization of a Wiener process; 5000 calculated points (t_j, W_j) are joined by linear interpolation.

Almost all realizations of Wiener processes are nowhere differentiable. This becomes intuitively clear when the difference quotient

$$\frac{\Delta W_t}{\Delta t} = \frac{W_{t+\Delta t} - W_t}{\Delta t}$$

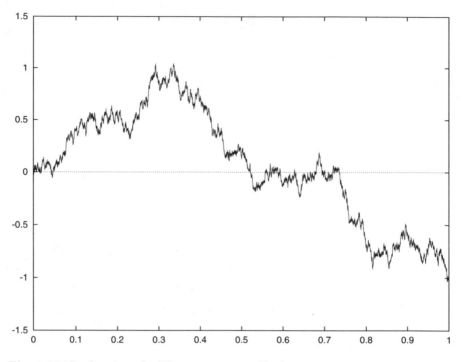

Fig. 1.15. Realization of a Wiener process, with $\Delta t = 0.0002$

is considered. Because of relation (1.21b) the standard deviation of the numerator is $\sqrt{\Delta t}$. Hence for $\Delta t \to 0$ the normal distribution of the difference quotient disperses and no convergence can be expected.

1.6.2 Stochastic Integral

Let us suppose that the price development of an asset is described by a Wiener process W_t. Let $b(t)$ be the number of units of the asset held in a portfolio at time t. We start with the simplifying assumption that trading is only possible at discrete time instances t_j, which define a partition of the interval $0 \le t \le T$. Then the trading strategy b is piecewise constant,

$$
\begin{aligned}
&b(t) = b(t_{j-1}) \quad \text{for} \quad t_{j-1} \le t < t_j \\
&\text{and} \quad 0 = t_0 < t_1 < \ldots < t_N = T \; .
\end{aligned}
\tag{1.23}
$$

Such a function $b(t)$ is called *step function*. The trading gain for the subinterval $t_{j-1} \le t < t_j$ is given by $b(t_{j-1})(W_{t_j} - W_{t_{j-1}})$, and

$$
\sum_{j=1}^{N} b(t_{j-1})(W_{t_j} - W_{t_{j-1}})
\tag{1.24}
$$

represents the trading gain over the time period $0 \le t \le T$. The trading gain (possibly < 0) is determined by the strategy $b(t)$ and the price process W_t.

We now drop the assumption of fixed trading times t_j and allow b to be arbitrary continuous functions. This leads to the question whether (1.24) has a limit when with $N \to \infty$ the size of all subintervals tends to 0. If W_t would be of bounded variation than the limit exists and is called *Riemann-Stieltjes integral*

$$\int_0^T b(t) dW_t \ .$$

In our situation this integral generally does not exist because almost all Wiener processes are not of bounded variation. That is, the *first variation* of W_t, which is the limit of

$$\sum_{j=1}^N |W_{t_j} - W_{t_{j-1}}| \ ,$$

is unbounded even in case the lengths of the subintervals vanish for $N \to \infty$.

Although this statement is not of primary concern for the theme of this book[5], we digress for a discussion because it introduces the important rule $(dW_t)^2 = dt$. For an arbitrary partition of the interval $[0, T]$ into N subintervals the inequality

$$\sum_{j=1}^N |W_{t_j} - W_{t_{j-1}}|^2 \le \max_j(|W_{t_j} - W_{t_{j-1}}|) \sum_{j=1}^N |W_{t_j} - W_{t_{j-1}}| \qquad (1.25)$$

holds. The left-hand sum in (1.25) is the *second variation* and the right-hand sum the first variation of W for a given partition into subintervals. The expectation of the left-hand sum can be calculated using (1.21),

$$\sum_{j=1}^N \mathsf{E}(W_{t_j} - W_{t_{j-1}})^2 = \sum_{j=1}^N (t_j - t_{j-1}) = t_N - t_0 = T \ .$$

But even convergence in the mean holds:

Lemma 1.9   (second variation: convergence in the mean)
Let $t_0 = t_0^{(N)} < t_1^{(N)} < \ldots < t_N^{(N)} = T$ be a sequence of partitions of the interval $t_0 \le t \le T$ with

$$\delta_N := \max_j(t_j^{(N)} - t_{j-1}^{(N)}) \ . \qquad (1.26)$$

Then (dropping the $^{(N)}$)

$$\underset{\delta_N \to 0}{\text{l.i.m.}} \sum_{j=1}^N (W_{t_j} - W_{t_{j-1}})^2 = T - t_0 \qquad (1.27)$$

---

[5] The less mathematically oriented reader may like to skip the rest of this subsection.

Proof: The statement (1.27) means convergence in the mean ($\longrightarrow$ Appendix B1). Because of $\sum \Delta t_j = T - t_0$ we must show

$$\mathsf{E}\left(\sum_j ((\Delta W_j)^2 - \Delta t_j)\right)^2 \to 0 \quad \text{for} \quad \delta_N \to 0 .$$

Carrying out the multiplications and taking the mean gives

$$2 \sum_j (\Delta t_j)^2$$

($\longrightarrow$ Exercise 1.10). This can be bounded by $2(T - t_0)\delta_N$, which completes the proof.

Part of the derivation can be summarized to

$$\mathsf{E}((\Delta W_t)^2 - \Delta t) = 0 \quad , \quad \mathsf{Var}((\Delta W_t)^2 - \Delta t) = 2(\Delta t)^2 ,$$

hence $(\Delta W_t)^2 \approx \Delta t$. This property of a Wiener process is symbolically written

$$\boxed{(dW_t)^2 = dt} \tag{1.28}$$

It will be needed in subsequent sections.

Now we know enough about the convergence of the left-hand sum of (1.25) and turn to the right-hand side of this inequality. The continuity of W_t implies

$$\max_j |W_{t_j} - W_{t_{j-1}}| \to 0 \quad \text{for} \quad \delta_N \to 0 .$$

Convergence in the mean applied to (1.25) shows that the vanishing of this factor must be compensated by an unbounded growth of the other factor, so

$$\sum_{j=1}^{N} |W_{t_j} - W_{t_{j-1}}| \to \infty \quad \text{für} \quad \delta_N \to 0 .$$

In summary, Wiener processes are not of bounded variation, and the integration with respect to W_t can not be defined as an elementary limit of (1.24).

The aim is to construct a stochastic integral

$$\int_{t_0}^{t} f(s) dW_s$$

for general stochastic integrands $f(t)$. For our purposes it suffices to briefly sketch the Itô integral, which is the prototype of a stochastic integral.

For a step function b from (1.23) an integral can be defined via the sum (1.24),

$$\int_{t_0}^{t} b(s)dW_s := \sum_{j=1}^{N} b(t_{j-1})(W_{t_j} - W_{t_{j-1}}) \ . \tag{1.29}$$

This is the Itô integral over a step function b. In case the $b(t_{j-1})$ are random variables, b is called a *simple process*. Then the Itô integral is again defined by (1.29). Stochastically integrable functions f can be obtained as limits of simple processes b_n in the sense

$$\mathsf{E}\left[\int_{t_0}^{t}(f(s) - b_n(s))^2 ds\right] \to 0 \quad \text{for} \quad n \to \infty \ . \tag{1.30}$$

Convergence in terms of integrals $\int ds$ carries over to integrals $\int dW_t$. This is achieved by applying Cauchy convergence $\mathsf{E}\int(b_n - b_m)^2 ds \to 0$ and the *isometry*

$$\mathsf{E}\left[\left(\int_{t_0}^{t} b(s)dW_s\right)^2\right] = \mathsf{E}\left[\int_{t_0}^{t} b(s)^2 ds\right] \ .$$

Hence the integrals $\int b_n(s)dW_s$ form a Cauchy sequence with respect to convergence in the mean. Accordingly the Itô integral of f is defined as

$$\int_{t_0}^{t} f(s)dW_s := \text{l.i.m.}_{n\to\infty} \int_{t_0}^{t} b_n(s)dW_s \ ,$$

for simple processes b_n defined by (1.30). The value of the integral is independent of the choice of the b_n in (1.30). The Itô integral as function in t is a stochastic process with the martingale property.

If an integrand $a(x,t)$ depends on a stochastic process X_t, the function f is given by $f(t) = a(X_t, t)$. For the simplest case of a constant integrand $a(X_t, t) = a_0$ the Itô integral can be reduced via (1.29) to

$$\int_{t_0}^{t} dW_s = W_t - W_{t_0}.$$

For the "first" nontrivial Itô integral consider $X_t = W_t$ and $a(W_t, t) = W_t$. Its solution will be presented in Section 3.2.

1.7 Stochastic Differential Equations

1.7.1 Itô Process

Many phenomena in nature, technology and economy are modeled by means of deterministic differential equations $\dot{x} = \frac{d}{dt}x = a(x,t)$. This kind of modeling neglects stochastic fluctuations and is not appropriate for stock prices. The easiest way to consider stochastic movements is via an additive term,

$$\frac{dx}{dt} = a(x,t) + b(x,t)\xi_t.$$

Here we use the notations

    a: deterministic part,

    $b\xi_t$: stochastic part, ξ_t denotes a generalized stochastic process.

An example of a generalized stochastic process is *white noise*. For a brief definition of white noise we note that to each stochastic process a generalized version can be assigned [Ar74]. For generalized stochastic processes derivatives of any order can be defined. Suppose that W_t is the generalized version of a Wiener process, then W_t can be differentiated. Then white noise ξ_t is defined as $\xi_t = \dot{W}_t = \frac{d}{dt}W_t$, or vice versa,

$$W_t = \int_0^t \xi_s ds.$$

That is, a Wiener process is obtained by smoothing the white noise. The smoother integral version dispenses with using generalized stochastic processes. Hence the integrated form of $\dot{x} = a(x,t) + b(x,t)\xi_t$ is studied,

$$x(t) = x_0 + \int_{t_0}^t a(x(s),s)ds + \int_{t_0}^t b(x(s),s)\xi_s ds,$$

and we replace $\xi_s ds = dW_s$. The first integral in this integral equation is an ordinary (Lebesgue- or Riemann-) integral. The second integral is an Itô integral to be taken with respect to the Wiener process W_t. The resulting stochastic differential equation (SDE) is named after Itô.

Definition 1.10   (Itô stochastic differential equation)

    An Itô stochastic differential equation is

$$dX_t = a(X_t,t)dt + b(X_t,t)dW_t; \tag{1.31a}$$

this together with $X_{t_0} = X_0$ is a symbolic short form of the integral equation

$$X_t = X_{t_0} + \int_{t_0}^t a(X_s,s)ds + \int_{t_0}^t b(X_s,s)dW_s. \tag{1.31b}$$

The terms in (1.31) are named as follows:

    $a(X_t,t)$: drift term or drift coefficient
    $b(X_t,t)$: diffusion term
    solution X_t: Itô process

A Wiener process is a special case of an Itô process, because from $X_t = W_t$ the trivial SDE $dX_t = dW_t$ follows, hence the drift vanishes, $a = 0$, and $b = 1$ in (1.31). If $b \equiv 0$ and X_0 is constant, then the SDE becomes deterministic.

An experimental approach may help developing an intuitive understanding of Itô processes. The simplest numerical method combines the discretized version of the Itô SDE

$$\Delta X_t = a(X_t, t)\Delta t + b(X_t, t)\Delta W_t \tag{1.32}$$

with the Algorithm 1.8 for approximating a Wiener process, using the same Δt for both discretizations. The result is

Algorithm 1.11   (Euler discretization of an SDE)

Approximations y_j to X_{t_j} are calculated by

$Start:$ t_0, $y_0 = X_0$, Δt, $W_0 = 0$.

$loop$ $j = 0, 1, 2, ...$

$\qquad t_{j+1} = t_j + \Delta t$

$\qquad \Delta W = Z\sqrt{\Delta t}$ with $Z \sim \mathcal{N}(0,1)$

$\qquad y_{j+1} = y_j + a(y_j, t_j)\Delta t + b(y_j, t_j)\Delta W$

In the easiest case the *step length* Δt is chosen equidistant, $\Delta t = T/m$ for a suitable integer m. Of course the accuracy of the approximation depends on the choice of Δt ($\longrightarrow$ Chapter 3). The Algorithm 1.11 is sometimes named after Euler and Maruyama. The evaluation is straightforward. When for some example the functions a and b are easily calculated, the greatest effort may be to calculate random numbers $Z \sim \mathcal{N}(0,1)$ ($\longrightarrow$ Section 2.3). Solutions to the SDE or to its discretized version for a given realization of the Wiener process are called *trajectories* or *paths*. By *simulation* of the SDE we understand the calculation of one or more trajectories. For the purpose of visualization, the discrete data are mostly joined by straight lines.

Example 1.12 $dX_t = 0.05X_t\, dt + 0.3X_t\, dW_t$

Without the diffusion term the exact solution would be $X_t = X_0 e^{0.05t}$. For $X_0 = 50$, $t_0 = 0$ and a time increment $\Delta t = 1/300$ the Figure 1.16 depicts a trajectory X_t of the SDE for $0 \le t \le 1$. For another realization of a Wiener process W_t the solution looks different. This is demonstrated for a similar SDE in Figure 1.17.

1.7.2 Application to the Stock Market

Now we discuss one of the most important continuous models for motions of the prices S_t of stocks. This standard model assumes that the relative change (return) dS/S of a security in the time interval dt is composed of a deterministic drift term μ plus stochastic fluctuations in the form σdW_t:

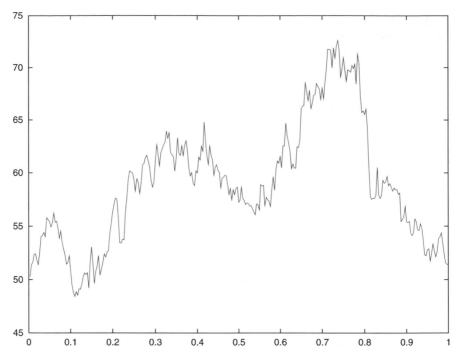

Fig. 1.16. Numerically approximated trajectory of Example 1.12 with $a = 0.05X_t$, $b = 0.3X_t$, $\Delta t = 1/300$, $X_0 = 50$

Model 1.13   (geometric Brownian motion, GBM)

$$dS_t = \mu S_t\, dt + \sigma S_t\, dW_t. \qquad\qquad (1.33, \text{GBM})$$

This SDE is linear in $X_t = S_t$, $a(S_t, t) = \mu S_t$ is the drift rate with the expected *rate of return* μ, $b(S_t, t) = \sigma S_t$, σ is the volatility. (Compare Example 1.12 and Figure 1.16.) The geometric Brownian motion of (1.33) is the reference model on which the Black-Scholes-Merton approach is based. According to Assumption 1.2 we assume that μ and σ are constant.

A theoretical solution of (1.33) will be given in (1.49). The deterministic part of (1.33) is the ordinary differential equation

$$\dot{S} = \mu S$$

with solution $S_t = S_0 e^{\mu(t - t_0)}$. For the linear SDE of (1.33) the expectation $\mathsf{E}(S_t)$ solves $\dot{S} = \mu S$. Hence $S_0 e^{\mu(t-t_0)}$ is the expectation of the stochastic process and μ is the expected continuously compounded return earned by an investor per year. The rate of return μ is also called *growth rate*. The function $S_0 e^{\mu(t-t_0)}$ may be seen as a core about which the process fluctuates.

Accordingly the simulated values S_1 of the ten trajectories in Figure 1.17 group around the value $50 \cdot e^{0.1} \approx 55.26$.

Let us test empirically how the values S_1 distribute about their expected value. To this end we calculate, for example, 10000 trajectories and count how many of the terminal values S_1 fall into the subintervals $k5 \le t < (k+1)5$, for $k = 0, 1, 2 \ldots$. The Figure 1.18 shows the resulting histogram. Apparently the distribution is skewed. We revisit this distribution in the next section.

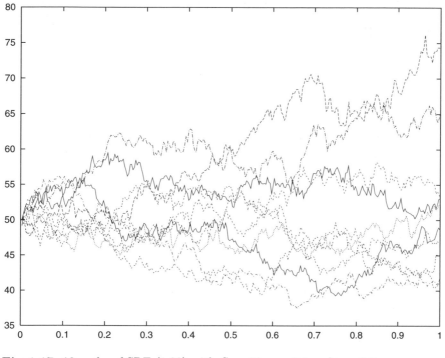

Fig. 1.17. 10 paths of SDE (1.33) with $S_0 = 50$, $\mu = 0.1$ and $\sigma = 0.2$

The discrete version of (1.33) is

$$\frac{\Delta S}{S} = \mu \Delta t + \sigma Z \sqrt{\Delta t}, \tag{1.34a}$$

known from Algorithm 1.11. The ratio $\frac{\Delta S}{S}$ is called one-period *simple return*, where we interpret Δt as one period. According to (1.34a) this return satisfies

$$\frac{\Delta S}{S} \sim \mathcal{N}(\mu \Delta t, \sigma^2 \Delta t). \tag{1.34b}$$

The distribution of the simple return matches actual market data in a crude approximation, see for instance Figure 1.21. This allows to calculate estimates of historical values of the volatility σ.[6] The approximation is valid as long as Δt is small. We will return to this in Section 1.8.

1.7.3 Risk-Neutral Valuation

As Appendix A4 shows for the continuous case, an option can be modeled independent of individual subjective expectations on the growth rate μ. For modeling of $V(S_t, t)$, a risk-neutral world is assumed which allows to replace μ by the risk-free rate r. This was discussed for the discrete one-period model in Section 1.5.

For the continuous-time setting, we start from GBM in (1.33), and get

$$dS_t = rS_tdt + (\mu - r)S_tdt + \sigma S_tdW_t$$
$$= rS_tdt + \sigma S_t\left[\frac{\mu - r}{\sigma}dt + dW_t\right] \tag{1.35}$$

In the reality of the market, an investor expects $\mu > r$ as compensation for the risk that is higher for stocks than for bonds. In this sense, the quotient γ of the *excess return* $\mu - r$ to the risk σ,

$$\gamma := \frac{\mu - r}{\sigma} , \tag{1.36}$$

is called *market price of risk*. With this variable γ, (1.35) is written

$$dS_t = rS_tdt + \sigma S_t[\gamma dt + dW_t] . \tag{1.37}$$

It turns out ($\longrightarrow$ Appendix B2) that there is another probability measure Q and —matching it— another Wiener process W_t^γ depending on γ, such that

$$dW_t^\gamma = \gamma dt + dW_t . \tag{1.38}$$

Then (1.37) becomes
$$dS_t = rS_tdt + \sigma S_tdW_t^\gamma . \tag{1.39}$$

Comparing this SDE to (1.33), notice that the growth rate μ is replaced by the risk-free rate r. Together the transition consists of

$$\begin{array}{ccc} \mu & \to & r \\ P & \to & Q \\ W & \to & W^\gamma \end{array}$$

which is named **risk-neutral valuation principle.** The advantage of the "risk-free measure" Q that corresponds to (1.38) is that the *discounted* process $e^{-rt}S_t$ is drift-free,

---

[6] For the *implied volatility* see Exercise 1.5.

$$d(e^{-rt}S_t) = e^{-rt}\sigma S_t dW_t^\gamma \ .$$

This property of having no drift is an essential ingredient of a no-arbitrage market and a prerequisite to modeling options. For a thorough discussion of the continuous model, martingale theory is used. (More background and explanation is provided by Appendix B3.) Let us summarize the situation in a remark:

Remark 1.14   (risk-neutral valuation principle)
    For modeling options the return rate μ is replaced by the risk-free interest rate r, $\mu = r$.

Of course, in the reality of the market $\mu \neq r$; otherwise nobody would invest in the stock market.

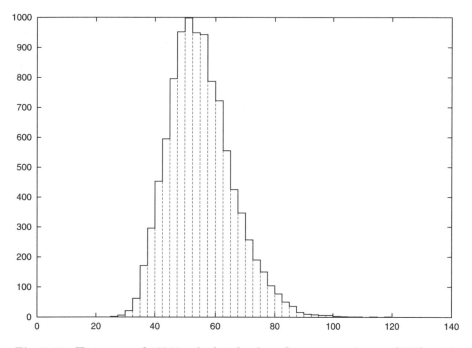

Fig. 1.18. Histogram of 10000 calculated values S_1 corresponding to (1.33), with $S_0 = 50$, $\mu = 0.1$, $\sigma = 0.2$

1.7.4 Mean Reversion

The assumptions of a constant interest rate r and a constant volatility σ are quite restrictive. To overcome this simplification, SDEs for r_t and σ_t have been constructed that control r_t or σ_t stochastically. A class of models is given by the SDE

$$dr_t = \alpha(R - r_t)dt + \sigma_r r_t^\beta dW_t\,,\ \alpha > 0. \tag{1.40}$$

W_t is again a Brownian motion. The drift term in (1.40) is positive for $r_t < R$ and negative for $r_t > R$, which causes a pull to R. This effect is called *mean reversion*. The strength of the reversion can be influenced by the choice of the *frequency* parameter α. The parameter R, which may depend on t, corresponds to a long-run mean of the interest rate over time. For $\beta = 0$ (constant volatility) equation (1.40) specializes to the Vasicek model. The Cox-Ingersoll-Ross model is obtained for $\beta = \frac{1}{2}$. Then the volatility $\sigma_r\sqrt{r_t}$ and with it the stochastic part vanish when r_t tends to zero. Provided $r_0 > 0$, $R > 0$, this guarantees $r_t \geq 0$ for all t. An illustration of the mean reversion is provided by Figure 1.19. In a transient phase (until $t \approx 1$ in the run documented in the figure) the relatively large deterministic term dominates, and the range $r \approx R$ is reached quickly. Thereafter the stochastic term dominates, and r dances about the mean value R. Figure 1.19 shows this for a Cox-Ingersoll-Ross model. For a discussion of related models we refer to [LL96], [Hull00], [Kwok98]. The *calibration* of the models (that is, the adaption of the parameters to the data) is a formidable task.

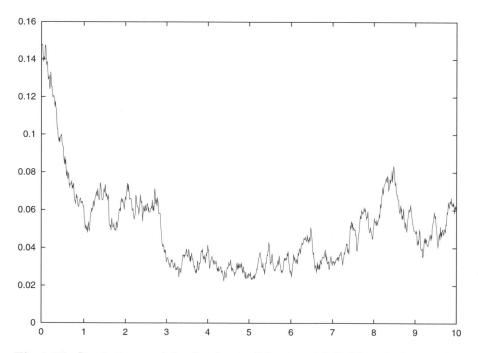

Fig. 1.19. Simulation r_t of the Cox-Ingersoll-Ross model (1.40) with $\beta = 0.5$ for $R = 0.05$, $\alpha = 1$, $\sigma_r = 0.1$, $y_0 = 0.15$, $\Delta t = 0.01$

The SDE (1.40) is of a different kind as (1.33). Coupling the SDE for r_t to that for S_t leads to a system of two SDEs. Even larger systems are obtained when further SDEs are coupled to define a stochasic process R_t or to calculate stochastic volatilities. A related example is given by Example 1.15 below.

1.7.5 Vector-Valued SDEs

The Itô equation (1.31) is formulated as scalar equation; accordingly the SDE (1.33) represents a *one-factor model*. The general *multifactor* version can be written in the same notation. Then $X_t = (X_t^{(1)}, \ldots, X_t^{(n)})$ and $a(X_t, t)$ are n-dimensional vectors. The Wiener processes of each component SDE need not be correlated. In the general situation, the Wiener process can be m-dimensional, with components $W_t^{(1)}, \ldots, W_t^{(m)}$. Then $b(X_t, t)$ is an $(n \times m)$-matrix. The interpretation of the SDE systems is componentwise. The scalar stochastic integrals are sums of m stochastic integrals,

$$X_t^{(i)} = X_0^{(i)} + \int_0^t a_i(X_s, s)ds + \sum_{k=1}^m \int_0^t b_{ik}(X_s, s)dW_s^{(k)} , \qquad (1.41a)$$

for $i = 1, \ldots, n$, and $t_0 = 0$ for convenience. Or in the symbolic SDE notation, this system reads

$$dX_t = a(X_t, t)dt + b(X_t, t)dW_t \quad , \qquad (1.41b)$$

where bdW is a matrix multiplication. When we take the components of the vector dW as uncorrelated,

$$\mathsf{E}\,(dW^{(k)}dW^{(j)}) = \begin{cases} 0 & \text{for } k \neq j \\ dt & \text{for } k = j \end{cases} \qquad (1.42)$$

then possible correlations between the components of dX must be carried by b.

Example 1.15 (mean-reverting volatility tandem)

We consider a three-factor model with stock price S_t, instantaneous spot volatility σ_t and an averaged volatility ζ_t serving as mean-reverting parameter:

$$\begin{cases} dS = \sigma S dW^{(1)} \\ d\sigma = -(\sigma - \zeta)dt + \alpha\sigma dW^{(2)} \\ d\zeta = \beta(\sigma - \zeta)dt \end{cases}$$

Here and sometimes later on, we suppress the subscript t, which may be done when the role of the variables as stochastic processes is clear from the context. The rate of return μ of S is zero; $dW^{(1)}$ and $dW^{(2)}$ may be correlated. The stochastic volatility σ follows the mean volatility ζ and is simultaneously perturbed by a Wiener process. Both σ und ζ provide mutual mean reversion, and stick together. The two SDEs for σ and ζ may be seen as a tandem controlling the dynamics of the volatility. We recommend numerical tests. As motivation see Figure 3.1.

Computational Matters

Stochastic differential equations are simulated in the context of Monte Carlo methods. Here the SDE is integrated N times, with N large, for example, $N = 10000$, or much larger. Then the weight of any single trajectory is almost neglectable. Expectation and variance are calculated over the N trajectories. Generally this costs an enormous amount of computing time. The required instruments are:

1.)   Generating $\mathcal{N}(0,1)$-distributed random numbers ($\longrightarrow$ Chapter 2)
2.)   Integration methods for SDEs ($\longrightarrow$ Chapter 3)

1.8 Itô Lemma and Implications

Itô's lemma is most fundamental for stochastic processes. It may help, for example, to derive solutions of SDEs ($\longrightarrow$ Exercise 1.11). Itô's lemma is the stochastic counterpart of the chain rule for deterministic functions $X(t)$ and $Y(t) := g(X(t), t)$, which is

$$\frac{d}{dt}g(X(t), t) = \frac{\partial g}{\partial x} \cdot \frac{dX}{dt} + \frac{\partial g}{\partial t} ,$$

and can be written

$$dX = a(X(t), t)dt \;\Rightarrow\; dg = \left(\frac{\partial g}{\partial x}a + \frac{\partial g}{\partial t} \right) dt .$$

Here we state the one-dimensional version of the Itô lemma; for the multidimensional version see the Appendix B2.

Lemma 1.16   (Itô)

Suppose X_t follows an Itô process (1.31), $dX_t = a(X_t, t)dt + b(X_t, t)dW_t$, and let $g(x, t)$ be a $\mathcal{C}^{2,1}$-smooth function (continuous $\frac{\partial g}{\partial x}$, $\frac{\partial^2 g}{\partial x^2}$, $\frac{\partial g}{\partial t}$). Then $Y_t := g(X_t, t)$ follows an Itô process with the *same* Wiener process W_t:

$$dY_t = \left(\frac{\partial g}{\partial x}a + \frac{\partial g}{\partial t} + \frac{1}{2}\frac{\partial^2 g}{\partial x^2}b^2 \right) dt + \frac{\partial g}{\partial x}b \; dW_t \qquad (1.43)$$

where the derivatives of g as well as the coefficient functions a and b in general depend on the arguments (X_t, t).

For a proof we refer to [Ar74], [Øk98], [Ste01], [Pro04]. Here we confine ourselves to the basic idea. When t varies by Δt, then X by $\Delta X = a \cdot \Delta t + b \cdot \Delta W$ and Y by $\Delta Y = g(X + \Delta X, t + \Delta t) - g(X, t)$. The Taylor expansion of ΔY begins with the linear part $\frac{\partial g}{\partial x} \Delta X + \frac{\partial g}{\partial t} \Delta t$, in which $\Delta X = a\Delta t + b\Delta W$ is substituted. The additional term with the derivative $\frac{\partial^2 g}{\partial x^2}$ is new and is introduced via the $O(\Delta x^2)$-term of the Taylor expansion, which gives $b^2 \Delta W^2$. Because of (1.28), $(\Delta W)^2 \approx \Delta t$, this term is also of the order $O(\Delta t)$ and belongs to the linear terms. Taking correct limits (similar as in Lemma 1.9) one obtains (1.43).

Consequences for Stocks and Options

Suppose the stock price follows a geometric Brownian motion, hence $X_t = S_t$, $a = \mu S_t$, $b = \sigma S_t$. The value V_t of an option depends on S_t, $V_t = V(S_t, t)$. Assuming $\mathcal{C}^2$-smoothness of V_t depending on S and t, we apply Itô's lemma. For $V(S, t)$ in the place of $g(x, t)$ the result is

$$dV_t = \left(\frac{\partial V}{\partial S} \mu S_t + \frac{\partial V}{\partial t} + \frac{1}{2} \frac{\partial^2 V}{\partial S^2} \sigma^2 S_t^2 \right) dt + \frac{\partial V}{\partial S} \sigma S_t dW_t. \tag{1.44}$$

This SDE is used to derive the Black-Scholes equation, see Appendix A4.

As second application of Itô's lemma consider $Y_t = \log(S_t)$, viz $g(x, t) = \log(x)$, for S_t solving GBM with constant μ, σ. Itô's lemma leads to the linear SDE

$$d \log S_t = (\mu - \frac{1}{2}\sigma^2)dt + \sigma dW_t . \tag{1.45}$$

In view of (1.31) the solution is straightforward:

$$\begin{aligned} Y_t &= Y_{t_0} + (\mu - \frac{1}{2}\sigma^2) \int_{t_0}^t ds + \sigma \int_{t_0}^t dW_s \\ &= Y_{t_0} + (\mu - \frac{1}{2}\sigma^2)(t - t_0) + \sigma(W_t - W_{t_0}) \end{aligned} \tag{1.46}$$

From the properties of the Wiener process W_t we conclude that Y_t is distributed normally. To write down the density function $\hat{f}(Y_t)$ the mean $\hat{\mu} := \mathsf{E}(Y_t)$ and the variance $\hat{\sigma}$ are needed. For this linear SDE (1.45) the expectation $\mathsf{E}(Y_t)$ satisfies the deterministic part

$$\frac{d}{dt} \mathsf{E}(Y_t) = \mu - \frac{\sigma^2}{2} .$$

The solution of $\dot{y} = \mu - \frac{\sigma^2}{2}$ with initial condition $y(t_0) = y_0$ is

$$y(t) = y_0 + (\mu - \frac{\sigma^2}{2})(t - t_0).$$

In other words, the expectation of the Itô process Y_t is

$$\hat{\mu} = \mathsf{E}(\log S_t) = \log S_0 + (\mu - \frac{\sigma^2}{2})(t - t_0) \ .$$

Analogously, we see from the differential equation for $\mathsf{E}(Y_t^2)$ (or from the analytical solution of the SDE for Y_t) that the variance of Y_t is $\sigma^2(t - t_0)$. In view of (1.45) the simple SDE for Y_t implies that the stochastic fluctuation of Y_t is that of σW_t, namely $\hat{\sigma}^2 := \sigma^2(t - t_0)$. So, from (B1.9) with $\hat{\mu}$ and $\hat{\sigma}$ the density of Y_t is

$$\widehat{f}(Y_t) := \frac{1}{\sigma\sqrt{2\pi(t-t_0)}} \exp\left\{ -\frac{\left(Y_t - y_0 - \left(\mu - \frac{\sigma^2}{2}\right)(t-t_0)\right)^2}{2\sigma^2(t-t_0)} \right\}.$$

Back transformation using $Y = \log(S)$ and considering $dY = \frac{1}{S}dS$ and $\widehat{f}(Y)dY = \frac{1}{S}\widehat{f}(\log S)dS = f(S)dS$ yields the density of S_t:

$$f(S; t - t_0, S_0, \mu, \sigma) :=$$

$$\frac{1}{S\sigma\sqrt{2\pi(t-t_0)}} \exp\left\{ -\frac{\left(\log(S/S_0) - \left(\mu - \frac{\sigma^2}{2}\right)(t-t_0)\right)^2}{2\sigma^2(t-t_0)} \right\} \qquad (1.47)$$

This is the density of the *lognormal* distribution. The stock price S_t is lognormally distributed under the basic assumption of a geometric Brownian motion (1.33). The distribution is skewed, see Figure 1.20. Now the skewed behavior coming out of the experiment reported in Figure 1.18 is clear. Note that the parameters of Figures 1.18 and 1.20 match. Figure 1.18 is an approximation of the solid curve in Figure 1.20. — Having derived the density (1.47), we now can prove equation (1.8), with $\mu = r$ according to Remark 1.14 ($\longrightarrow$ Exercise 1.12).

Another important application of the lognormal density is that it allows for an integral representation of European options. This will be revisited in Subsection 3.5.1, where we show for a European put

$$V(S_0, 0) = e^{-rT} \int_0^\infty (K - S_T)^+ f(S_T; T, S_0, r, \sigma)\, dS_T \ .$$

It is inspiring to test the idealized Model 1.13 of a geometric Brownian motion against actual empirical data. Suppose the time series $S_1, ..., S_M$ represents consecutive quotations of a stock price. To test the data, histograms of the returns are helpful ($\longrightarrow$ Figure 1.21). The transformation $y = \log(S)$ is most practical. It leads to the notion of the *log return*, defined by[7]

$$R_{i,i-1} := \log \frac{S_i}{S_{i-1}} \ . \qquad (1.48)$$

---

[7] Since $S_i = S_{i-1}\exp(R_{i,i-1})$, the log return is also called *continuously compounded return* in the ith time interval [Tsay02].

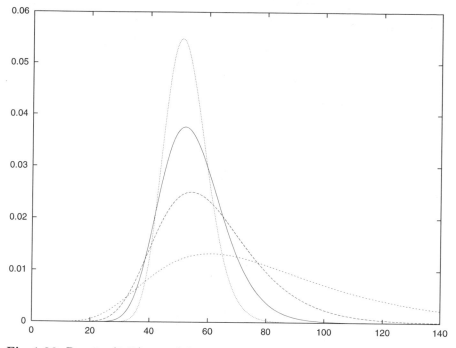

Fig. 1.20. Density (1.47) over S for $\mu = 0.1$, $\sigma = 0.2$, $S_0 = 50$, $t_0 = 0$ and $t = 0.5$ (dotted curve with steep gradient), $t = 1$ (solid curve), $t = 2$ (dashed) and $t = 5$ (dotted with flat gradient)

Let Δt be the equally spaced sampling time interval between the quotations S_{i-1} and S_i, measured in years. Then (1.47) leads to

$$R_{i,i-1} \sim \mathcal{N}((\mu - \frac{\sigma^2}{2})\Delta t \, , \; \sigma^2 \Delta t) \; .$$

Comparing with (1.34) we realize that the variances of the simple return and of the log return are identical. The sample variance $\sigma^2 \Delta t$ of the data allows to calculate estimates of the historical volatility σ ($\longrightarrow$ Exercise 1.13). But the shape of actual market histograms is usually not in good agreement with the well-known bell shape of the Gaussian density. The symmetry may be perturbed, and in particular the tails of the data are not well modeled by the hypothesis of a geometric Brownian motion: The exponential decay expressed by (1.47) amounts to *thin tails*. This underestimates extreme events and hence does not match reality. It is questionable whether plain geometric Brownian motion is suitable to model risks.

We conclude this section with deriving the analytical solution of the basic linear constant-coeffficient SDE (1.33)

$$dS_t = \mu S_t \, dt + \sigma S_t \, dW_t$$

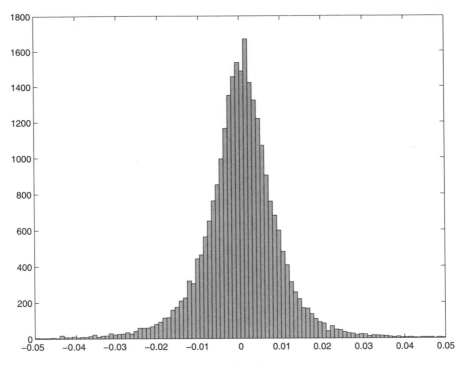

Fig. 1.21. Histogram (compare Exercise 1.13): frequency of daily log returns $R_{i,i-1}$ of the Dow in the time period 1901-1999.

of Section 1.7.2. Here we take the exponential of (1.46), or restart the derivation by again applying Itô's lemma. For an arbitrary Wiener process W_t set $X_t := W_t$ and

$$Y_t = g(X_t, t) := S_0 \exp\left(\left(\mu - \frac{\sigma^2}{2}\right) t + \sigma X_t\right).$$

From $X_t = W_t$ follows the trivial SDE with coefficients $a = 0$ and $b = 1$. By Itô's lemma

$$dY_t = \left(\mu - \frac{\sigma^2}{2}\right) Y_t dt + \frac{\sigma^2}{2} Y_t dt + \sigma Y_t dW_t$$

$$= \mu Y_t dt + \sigma Y_t dW.$$

Consequently the process

$$S_t := S_0 \exp\left(\left(\mu - \frac{\sigma^2}{2}\right) t + \sigma W_t\right) \tag{1.49}$$

solves the linear constant-coefficient SDE (1.33). We will return to this in Chapter 3. Equation (1.49) generalizes to the case of nonconstant coefficients ($\longrightarrow$ Exercise 1.18). As a consequence we note that $S_t > 0$ for all t, provided $S_0 > 0$.

1.9 Jump Processes

The geometric Brownian motion Model 1.13 has continuous paths S_t. As noted before, the continuity is at variance with those rapid asset price movements that can be considered almost instantaneous. Such rapid changes can be modeled as **jumps**. This section introduces the basic building block of a jump process, namely the Poisson process. Related simulations (like that of Figure 1.22) may look more authentic than continuous paths. But one has to pay a price: With a jump process the risk of an option in general can not be hedged away to zero.

To define a jump process, denote the time instances for which a jump occurs τ_j, with

$$\tau_1 < \tau_2 < \tau_3 < \dots$$

Let the number of jumps be counted by the counting variable J_t, where

$$\tau_j = \inf\{t \geq 0 \ , \ J_t = j\}.$$

We introduce the probability that a jump occurs via a Bernoulli experiment. For this local discussion, consider a subinterval of length $\Delta t := \frac{t}{n}$ and allow for only two outcomes, jump *yes* or *no*, with the probabilities

$$
\begin{aligned}
\mathsf{P}(J_t - J_{t-\Delta t} = 1) &= \lambda \Delta t \\
\mathsf{P}(J_t - J_{t-\Delta t} = 0) &= 1 - \lambda \Delta t
\end{aligned}
\tag{1.50}
$$

for some λ such that $0 < \lambda \Delta t < 1$. The parameter λ is referred to as the *intensity* of the jump process. Consequently k jumps in $0 \leq \tau \leq t$ have the probability

$$\mathsf{P}(J_t - J_0 = k) = \binom{n}{k} (\lambda \Delta t)^k (1 - \lambda \Delta t)^{n-k} \ ,$$

where we consider the trials in each subinterval as independent. A little reasoning reveals that for $n \to \infty$ this probability converges to

$$\frac{(\lambda t)^k}{k!} e^{-\lambda t} \ ,$$

which is known as the Poisson distribution with parameter $\lambda > 0$ ($\longrightarrow$ Appendix B1). This leads to the Poisson process.

Definition 1.17   (Poisson process)

The stochastic process $\{J_t \ , \ t \geq 0\}$ is called Poisson process if the following conditions hold:

(a) $J_0 = 0$

(b) $J_t - J_s$ are integer-valued for $0 \leq s < t < \infty$ and

$$\mathsf{P}(J_t - J_s = k) = \frac{\lambda^k (t-s)^k}{k!} e^{-\lambda(t-s)} \text{ for } k = 0, 1, 2 \dots$$

(c)   The increments $J_{t_2} - J_{t_1}$ and $J_{t_4} - J_{t_3}$ are independent for all $0 \le t_1 < t_2 < t_3 < t_4$.

As consequence of this definition, several properties hold.

Properties 1.18   (Poisson process)

(d) J_t is right-continuous and nondecreasing.

(e)   The times between successive jumps are independent and exponentially distributed with parameter λ. Thus,

$$P(\tau_{j+1} - \tau_j > \Delta\tau) = e^{-\lambda\Delta\tau} \quad \text{for each } \Delta\tau .$$

(f) J_t is a Markov process.

(g) $E(J_t) = \lambda t$, $Var(J_t) = \lambda t$

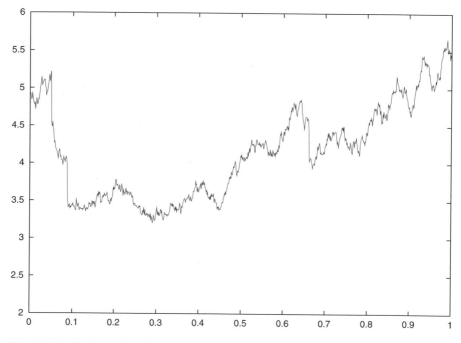

Fig. 1.22. Path of (1.52) with $\mu = 0.06$, $\sigma = 0.3$, $\lambda = 5$, $q = Z \cdot 0.1 + 1$ for $Z \sim \mathcal{N}(0, 1)$, $\Delta t = 0.001$.

Simulating Jumps

Following the above introduction into Poisson processes, there are two possibilities to calculate jump instances τ_j such that the above probabilities are met. First, the equation (1.50) may be used together with uniform deviates ($\longrightarrow$ Chapter 2). In this way a Δt-discretization of a t-grid can be easily exploited to decide whether a jump occurs in a subinterval. The other alternative is to calculate exponentially distributed random numbers $h_1, h_2, \ldots$ ($\longrightarrow$ Section 2.2.2) to simulate the intervals $\Delta\tau$ between consecutive jump instances, and set

$$\tau_{j+1} := \tau_j + h_j.$$

In addition to the jump instances τ_j another random variable is required to simulate the jump *sizes*. The unit amplitudes of the jumps of the Poisson counting process J_t are not relevant for our purpose of establishing a market model. The jump sizes of the price of a financial asset will be considered random.

Let the random variable S_t jump at τ_j, and denote τ^+ the moment after one particular jump and τ^- the moment before. Then the absolute size of the jump is

$$\Delta S = S_{\tau^+} - S_{\tau^-} \;,$$

which we model as a *proportional jump*,

$$S_{\tau^+} = q S_{\tau^-} \quad \text{with} \quad q > 0 \;. \tag{1.51}$$

So, $\Delta S = q S_{\tau^-} - S_{\tau^-} = (q-1)S_{\tau^-}$. The jump sizes equal $q-1$ times the current asset price. Accordingly, a jump process depends on a random variable q_t and is written

$$dS_t = (q_t - 1)S_t dJ_t \;, \quad \text{where } J_t \text{ is a Poisson process.}$$

We assume that $q_{\tau_1}, q_{\tau_2}, \ldots$ are i.i.d. This process is called **compound Poisson**. Next we superimpose the jump process to the continuous Wiener process. The combined geometric Brownian und jump process is given by

$$dS_t = \mu S_t dt + \sigma S_t dW_t + (q_t - 1)S_t dJ_t. \tag{1.52}$$

Here σ is the same as for the GBM, hence conditional on no Poisson jump. We further assume that q is independent of W. Such a combined model represented by (1.52) is called **jump-diffusion process**.

The Figure 1.22 shows a simulation of the SDE (1.52). The choice $\lambda\Delta t = 0.005$ has taken care of the jump times, see (1.50). For this simulation, the jump sizes q are simply chosen $\sim \mathcal{N}(1, 0.01)$. In Figure 1.22, large jumps are seen, for instance, for $\tau_1 = 0.05$ and $\tau_2 = 0.088$. Since the jump size is random, the process (1.52) involves three different stochastic ingredients, namely W_t, J_t and q_t.

An analytical solution of (1.52) can be calculated on each of the jump-free subintervals $\tau_j < t < \tau_{j+1}$ where the SDE is just the GBM diffusion $dS = S(\mu dt + \sigma dW)$. For example, in the first subinterval until τ_1 the solution is given by (1.49). At τ_1 a jump of size

$$(\Delta S)_1 := (q_{\tau_1} - 1)S_{\tau_1^-}$$

occurs, and thereafter the solution continues with

$$S_t = S_0 \cdot \exp\left(\left(\mu - \frac{\sigma^2}{2}\right)t + \sigma W_t\right) + (q_{\tau_1} - 1)S_{\tau_1^-} \ ,$$

until τ_2. The interchange of continuous parts and jumps proceeds in this way, all jumps are added. So the SDE can be written as

$$S_t = S_0 + \int_0^t S_s(\mu ds + \sigma dW_s) + \sum_{j=1}^{J_t} S_{\tau_j^-}(q_{\tau_j} - 1) \ . \tag{1.53}$$

This is the model based on Merton's paper [Mer76]. The equation (1.53) can be rewritten in the log-framework, with $Y_t := \log S_t$. The jump sizes according to model (1.51) match the log-scenario,

$$\begin{aligned} Z_\tau &:= Y_{\tau+} - Y_{\tau-} = \log(qS_{\tau-}) - \log S_{\tau-} \\ &= \log q_\tau \ . \end{aligned}$$

Following (1.49), the model can be written

$$Y_t = Y_0 + \left(\mu - \frac{\sigma^2}{2}\right)t + \sigma W_t + \sum_{j=1}^{J_t} Z_{\tau_j} \tag{1.54}$$

Merton assumes normally distributed Z, which amounts to lognormal q. In summary we emphasize again that the jump-diffusion process has three driving processes, namely W, J, and q. As shown in Appendix A4, the task of minimimizing risks leads to a partial integro-differential equation (A4.14). This equation

$$\frac{\partial V}{\partial t} + \frac{1}{2}\sigma^2 S^2 \frac{\partial^2 V}{\partial S^2} + (r - \lambda c)S\frac{\partial V}{\partial S} - (\lambda + r)V + \lambda \mathsf{E}(V(qS, t)) = 0$$

reduces to the Black-Scholes equation in the no-jump special case for $\lambda = 0$.

Notes and Comments

on Section 1.1:

This section presents a brief introduction into standard options. For more comprehensive studies of financial derivatives we refer, for example, to [CR85], [WDH96], [Hull00]. Mathematical detail can be found in [MR97], [KS98], [LL96], [Shi99], [Epps00], [Ste01]. (All hints on the literature are examples; an extensive overview on the many good books in this rapidly developing field is hardly possible.)

on Section 1.2:

Black, Merton and Scholes developed their approaches concurrently, with basic papers in 1973 ([BS73], [Mer73]; compare also [Me90]). Merton and Scholes were awarded the Nobel Prize in economics in 1997. (Black had died in 1995.) One of the results of these authors is the so-called Black-Scholes equation (1.2) with its analytic solution formula (A4.10). For the Assumption 1.2(c) of a geometric Brownian motion see also the notes and comments on Sections 1.7/1.8. For reference on discrete-time models, see [Pli97], [FöS02].

on Section 1.3:

References on specific numerical methods are given where appropriate. As computational finance is concerned, most quotations refer to research papers. A general text book discussing computational issues is [WDH96]; further hints can be found in [RT97].

on Section 1.4:

The binomial method can sometimes be found under the heading *tree method* or *lattice method*. The binomial method was introduced by Cox, Ross and Rubinstein in 1979 [CRR79], later than the approach of Black, Merton and Scholes. Table 1.2 might suggest that it is easy to obtain high accuracy with binomial methods. This is not the case; flaws were observed in particular close to the early-exercise curve [CoLV02]. As illustrated by Figure 1.10, the described standard version wastes many nodes S_{ji} close to zero and far away from the strike region. Alternatively to the choice $ud = 1$ in equation (1.10) the choice $p = \frac{1}{2}$ is possible, see [Hull00], §16.5. When the strike K is not well grasped by the tree and its grid points, the error depending on M may oscillate. To facilitate extrapolation, it is advisable to have the strike value K on the medium grid point, $S_T = K$, no matter what (even) value of M is chosen. The error can be smoothed by special choices of u and d ($\longrightarrow$ Exercise 1.15). For advanced binomial methods and speeding up convergence see [Br91], [Kl01]. For a detailed account of the binomial method see also [CR85]. [HoP02] explains how to implement the binomial method in spreadsheets. Many applications of binomial trees are found in [Lyuu02].

on Section 1.5:

As shown in Section 1.5, a valuation of options based on a hedging strategy is equivalent to the risk-neutral valuation described in Section 1.4. Another equivalent valuation is obtained by a *replication* portfolio. This basically amounts to including the risk-free investment, to which the hedged portfolio of Section 1.5 was compared, into the portfolio. To this end, the replication portfolio includes a bond with the initial value $B_0 := -(\Delta \cdot S_0 - V_0) = -\Pi_0$ and interest rate r. The portfolio consists of the bond and Δ shares of the asset. At the end of the period T the final value of the portfolio is $\Delta \cdot S_T + e^{rT}(V_0 - \Delta \cdot S_0)$. The hedge parameter Δ and V_0 are determined such that the value of the portfolio is V_T, independent of the price evolution. By adjusting B_0 and Δ in the right proportion we are able to replicate the option position. This strategy is *self-financing*: No initial net investment is required. The result of the self-financing strategy with the replicating portfolio is the same as what was derived in Section 1.5. The reader may like to check this. For the continuous-time case, see Appendix A4.

Frequently discounting is done with the factor $(1 + r \cdot \Delta t)^{-1}$. Our $e^{-r\Delta t}$ or e^{-rT} is consistent with the approach of Black, Merton and Scholes. For references on risk-neutral valuation we mention [Hull00], [MR97], [Kwok98] and [Shr04].

on Section 1.6:

Introductions into stochastic processes and further hints on advanced litera-ture may be found in [Doob53], [Fr71], [Ar74], [Bi79], [RY91], [KP92], [Shi99], [Sato99]. The requirement (a) of Definition 1.7 ($W_0 = 0$) is merely a conven-tion of technical relevance; it serves as normalization. This Brownian motion ist called *standard* Brownian motion. For a proof of the nondifferentiability of Wiener processes, see [HuK00]. For more hints on martingales, see Appendix B2.

In contrast to the results for Wiener processes, differentiable functions W_t satisfy for $\delta_N \to 0$

$$\sum |W_{t_j} - W_{t_{j-1}}| \longrightarrow \int |W'_s| ds \quad , \quad \sum (W_{t_j} - W_{t_{j-1}})^2 \longrightarrow 0 \ .$$

The Itô integral and the alternative Stratonovich integral are explained in [Doob53], [Ar74], [CW83], [RY91], [KS91], [KP92], [Mik98], [Øk98], [Sc80], [Shr04]. The difference between the two integrals is easy to illustrate for diffusion terms $b(t)dW_t$, for which $b(t)$ is a step function, see (1.23). Then a definition of the stochastic integral $\int_{t_0}^{t} b(s)dW_s$ can be constructed by means of Riemann-Stieltjes sums

$$\sum_{i=1}^{n} b(\tau_i) \left(W_{t_i} - W_{t_{i-1}} \right)$$

for intermediate values $\tau_i \in [t_{i-1}, t_i]$. Itô chooses *nonanticipating* $\tau_i = t_{i-1}$, whereas the Stratonovich integral is defined by choosing $\tau_i = \frac{1}{2}(t_{i-1} + t_i)$. Both integrals have different properties. The Stratonovich integral does not satisfy the martingale property and is not suitable for applications in finance. Itô's nonanticipating evaluation of b reflects the fact that looking into the future is impossible; the future value can not affect the upcoming move of the underlying's price. The class of (Itô-)stochastically integrable functions is characterized by the properties $f(t)$ is $\mathcal{F}_t$ adapted and $\mathsf{E} \int f(s)^2 ds < \infty$. We assume that all integrals occuring in the text exist. The integrator W_t needs not be a Wiener process. The stochastic integral can be extended to semimartingales [HuK00].

on Sections 1.7, 1.8:

The connection between white noise and Wiener processes is discussed in [Ar74]. White noise is a Gaussian process ξ_t with $\mathsf{E}(\xi_t) = 0$ and a spectral density that is constant on the entire real axis. White noise is an analogy to white light where all frequencies have the same energy.

The general linear SDE is of the form

$$dX_t = (a_1(t)X_t + a_2(t))dt + (b_1(t)X_t + b_2(t))dW_t.$$

The expectation $\mathsf{E}(X_t)$ of a solution process X_t of a linear SDE satisfies the differential equation

$$\frac{d}{dt}\mathsf{E}(X_t) = a_1\mathsf{E}(X_t) + a_2,$$

and for $\mathsf{E}(X_t^2)$ we have

$$\frac{d}{dt}\mathsf{E}(X_t^2) = (2a_1 + b_1^2)\mathsf{E}(X_t^2) + 2(a_2 + b_1b_2)\mathsf{E}(X_t) + b_2^2 .$$

This is obtained by taking the expection of the SDEs for X_t and X_t^2, the latter one derived by Itô's lemma [KP92], [Mik98]. Combining both differential equations allows to calculate the variance. — The Example 1.15 with a system of three SDEs is taken from [HPS92]. [KP92] in Section 4.4 gives a list of SDEs that are analytically solvable or reducible.

The model of a geometric Brownian motion of equation (1.33) is the classical model describing the dynamics of stock prices. It goes back to Samuelson (1965; Nobel Prize in economics in 1970). Already in 1900 Bachelier had suggested to model stock prices with Brownian motion. Bachelier used the arithmetic version, which can be characterized by replacing the left-hand side of (1.33) by the absolute change dS. This amounts to the process $S_t = S_0 + \mu t + \sigma W_t$. Here the stock price can become negative. Main advantages of the geometric Brownian motion are the success of the approaches of Black, Merton and Scholes, which is based on that motion, and the existence of moments (as the expectation). For positive S, the form (1.33) of GBM is not as restrictive as it might seem, see Exercise 1.18.— Recently, many books on financial markets have been published, see for instance [DaJ03], [ElK99], [Gem00], [MeVN02].

In view of their continuity, Wiener processes are not appropriate to model jumps, which are characteristic for the evolution of stock prices. The jumps lead to relatively *heavy tails* in the distribution of empirial returns (see Figure 1.21)[8]. As already mentioned, the tails of the lognormal distribution are too thin. Other distributions match empirical data better. One example is the Pareto distribution, which has tails behaving like $x^{-\alpha}$ for large

---

[8] The thickness is measured by the *kurtosis* $\mathsf{E}((X - \mu)^4)/\sigma^4$. The normal distribution has kurtosis 3. So the *excess kurtoris* is the difference to 3. Frequently, data of returns are characterized by large values of excess kurtosis.

x and a constant $\alpha > 0$. A correct modeling of the tails is an integral basis for *value at risk* (VaR) calculations. For the risk aspect compare [BaN97], [Dowd98], [EKM97], [ArDEH99]. For distributions that match empirical data see [EK95], [Shi99], [BP00], [MRGS00], [BTT00]. Estimates of future values of the volatility are obtained by (G)ARCH methods, which work with different weights of the returns [Shi99], [Hull00], [Tsay02], [FHH04], [Rup04]. For calibration, the method of [CaM99] is recommendable. Of great promise are models that consider the market as *dynamical system* [Lux98], [BH98], [CDG00], [BV00], MCFR00], [Sta01], [DBG01]. These systems experience the nonlinear phenomena *bifurcation* and *chaos*, which require again numerical methods. Such methods exist, but are explained elsewhere [Se94].

on Section 1.9:

Section 1.9 concentrates on jump-diffusion processes. This class of processes extends to the large class of Lévy processes. Brownian motion and Poisson jumps are just simple special cases, combined to the jump-diffusion model. A second category of Lévy processes consists of models with an infinite number of jumps. For building such models we refer to [ConT04], [Sato99].

The calender time may not be the appropriate measure of time in financial markets. A *business time* can be introduced based on the number of trades, which allows to recover normality for returns [AnéG00]. To set up a business time, increasing Lévy processes are introduced. Call such a *subordinator* process $\tau(t)$, then the model of a financial process is $W_{\tau(t)}$, with a Brownian motion W_t. Changing the clock in such a way has been successfully applied to match empirical data. Again we refer to [ConT04].

Exercises

Exercise 1.1   Put-Call Parity

Consider a portfolio consisting of three positions related to the same asset, namely one share (price S), one European put (value V_P), plus a short position of one European call (value V_C). Put and call have the same expiration date T, and no dividends are paid.

a)   Assume a no-arbitrage market without transaction costs. Show

$$S + V_P - V_C = Ke^{-r(T-t)}$$

for all t, where K is the strike and r the risk-free interest rate.

b)   Use the put-call parity to show

$$V_C(S,t) \geq S - Ke^{-r(T-t)}$$
$$V_P(S,t) \geq Ke^{-r(T-t)} - S \ .$$

Exercise 1.2   Transforming the Black-Scholes Equation

Show that the Black-Scholes equation (1.2)

$$\frac{\partial V}{\partial t} + \frac{\sigma^2}{2}S^2\frac{\partial^2 V}{\partial S^2} + rS\frac{\partial V}{\partial S} - rV = 0$$

for $V(S,t)$ with constant σ and r is equivalent to the equation

$$\frac{\partial y}{\partial \tau} = \frac{\partial^2 y}{\partial x^2}$$

for $y(x,\tau)$. For proving this, you may proceed as follows:

a)   Use the transformation $S = Ke^x$ and a suitable transformation $t \leftrightarrow \tau$ to show that (1.2) is equivalent to

$$-\dot{V} + V'' + \alpha V' + \beta V = 0$$

with $\dot{V} = \frac{\partial V}{\partial \tau}$, $V' = \frac{\partial V}{\partial x}$, α, β depending on r and σ.

b)   The next step is to apply a transformation of the type

$$V = K\exp(\gamma x + \delta\tau)y(x,\tau)$$

for suitable γ, δ.

c)   Transform the boundary conditions and the terminal condition of the Black-Scholes equation accordingly.

Exercise 1.3   Standard Normal Distribution Function

Establish an algorithm to calculate

$$F(x) = \frac{1}{\sqrt{2\pi}}\int_{-\infty}^{x}\exp(-\frac{t^2}{2})dt.$$

Hint: Construct an algorithm to calculate the *error function*

$$\text{erf}(x) := \frac{2}{\sqrt{\pi}}\int_{0}^{x}\exp(-t^2)dt$$

and use $\text{erf}(x)$ to calculate $F(x)$. Use quadrature methods ($\longrightarrow$ Appendix C1).

Exercise 1.4   Calculating an Estimate of the Variance

An estimate of the variance of M numbers $x_1,...,x_M$ is

$$s_M^2 := \frac{1}{M-1}\sum_{i=1}^{M}(x_i - \bar{x})^2, \quad \text{with } \bar{x} := \frac{1}{M}\sum_{i=1}^{M}x_i$$

The alternative formula

$$s_M^2 = \frac{1}{M-1}\left(\sum_{i=1}^{M} x_i^2 - \frac{1}{M}\left(\sum_{i=1}^{M} x_i\right)^2\right)$$

can be evaluated with only one loop $i = 1, ..., M$, but should be avoided because of the danger of cancellation. The following single-loop algorithm is recommended:

$$\alpha_1 := x_1, \ \beta_1 := 0$$
$$for \ \ i = 2, ..., M :$$
$$\alpha_i := \alpha_{i-1} + \frac{x_i - \alpha_{i-1}}{i}$$
$$\beta_i := \beta_{i-1} + \frac{(i-1)(x_i - \alpha_{i-1})^2}{i}$$

a)   Show $\bar{x} = \alpha_M$, $s_M^2 = \frac{\beta_M}{M-1}$.

b)   For the ith $update$ in the algorithm carry out a rounding error analysis. What is your judgement on the algorithm?

Exercise 1.5   Implied Volatility

For European options we take the valuation formula of Black and Scholes of the type $V = v(S, \tau, K, r, \sigma)$, where τ denotes the time to maturity, $\tau := T - t$. For the definition of the function v see Appendix A4, equation (A4.10). If actual market data of the price V are known, then one of the parameters considered known so far can be viewed as unknown and fixed via the implicit equation

$$V - v(S, \tau, K, r, \sigma) = 0. \tag{$*$}$$

In this calibration approach the unknown parameter is calculated iteratively as solution of equation $(*)$. Consider σ to be in the role of the unknown parameter. The volatility σ determined in this way is called $implied\ volatility$ and is zero of $f(\sigma) := V - v(S, \tau, K, r, \sigma)$.

Assignment:

a)   Implement the evaluation of V_C and V_P according to (A4.10).

b)   Design, implement and test an algorithm to calculate the implied volatility of a call. Use Newton's method to construct a sequence $x_k \to \sigma$. The derivative $f'(x_k)$ can be approximated by the difference quotient

$$\frac{f(x_k) - f(x_{k-1})}{x_k - x_{k-1}}.$$

For the resulting $secant\ iteration$ invent a stopping criterion that requires smallness of both $|f(x_k)|$ and $|x_k - x_{k-1}|$.

c)   Calculate the implied volatilities for the data

$$T - t = 0.211 \ , \ S_0 = 5290.36 \ , \ r = 0.0328$$

and the pairs K, V from Table 1.3 (for more data see www.compfin.de).
Enter for each calculated value of σ the point (K, σ) into a figure, joining
the points with straight lines. (You will notice a convex shape of the
curve. This shape has lead to call this phenomenon *volatility smile.*)

Table 1.3. Calls on the DAX on 4. Jan 1999

K	6000	6200	6300	6350	6400	6600	6800
V	80.2	47.1	35.9	31.3	27.7	16.6	11.4

Exercise 1.6   Price Evolution for the Binomial Method
For β from (1.11) and $u = \beta + \sqrt{\beta^2 - 1}$ show

$$u = \exp\left(\sigma\sqrt{\Delta t}\right) + O\left(\sqrt{(\Delta t)^3}\right) .$$

Exercise 1.7   Implementing the Binomial Method
Design and implement an algorithm for calculating the value $V^{(M)}$ of a European or American option. Use the binomial method of Algorithm 1.4.

INPUT: r (interest rate), σ (volatility), T (time to expiration in years),
         K (strike price), S (price of asset), and the choices
         put or *call*, and *European* or *American*.

Control the mesh size $\Delta t = T/M$ adaptively. For example, calculate V for $M = 8$ and $M = 16$ and in case of a significant change in V use $M = 32$ and possibly $M = 64$.

Test examples:
a)   put, European, $r = 0.06$, $\sigma = 0.3$, $T = 1$, $K = 10$, $S = 5$
b)   put, American, $S = 9$, otherwise as in a)
c)   call, otherwise as in a)
d)   The mesh size control must be done carefully and has little relevance to error control. To make this evident, calculate for the test numbers a) a sequence of $V^{(M)}$ values, say for $M = 100, 101, 102, \ldots, 150$, and plot the error $|V^{(M)} - 4.430465|$.

Exercise 1.8   Limiting Case of the Binomial Model
Consider a European Call in the binomial model of Section 1.4. Suppose the calculated value is $V_0^{(M)}$. In the limit $M \to \infty$ the sequence $V_0^{(M)}$ converges to the value $V_C(S_0, 0)$ of the continuous Black-Scholes model given by (A4.10) ($\longrightarrow$ Appendix A4). To prove this, proceed as follows:
a)   Let j_K be the smallest index j with $S_{jM} \geq K$. Find an argument why

$$\sum_{j=j_K}^{M} \binom{M}{j} p^j (1-p)^{M-j} (S_0 u^j d^{M-j} - K)$$

is the expectation $\mathsf{E}(V_T)$ of the payoff. (For an illustration see Figure 1.23.)

b)   The value of the option is obtained by discounting, $V_0^{(M)} = e^{-rT}\mathsf{E}(V_T)$. Show

$$V_0^{(M)} = S_0 B_{M,\tilde{p}}(j_K) - e^{-rT} K B_{M,p}(j_K) \; .$$

Here $B_{M,p}(j)$ is defined by the binomial distribution ($\longrightarrow$ Appendix B1), and $\tilde{p} := pue^{-r\Delta t}$.

c)   For large M the binomial distribution is approximated by the normal distribution with distribution $F(x)$. Show that $V_0^{(M)}$ is approximated by

$$S_0 F\left(\frac{M\tilde{p}-\alpha}{\sqrt{M\tilde{p}(1-\tilde{p})}}\right) - e^{-rT} K F\left(\frac{Mp-\alpha}{\sqrt{Mp(1-p)}}\right) \; ,$$

where

$$\alpha := -\frac{\log\frac{S_0}{K} + M\log d}{\log u - \log d} \; .$$

d)   Substitute the p, u, d by their expressions from (1.11) to show

$$\frac{Mp-\alpha}{\sqrt{Mp(1-p)}} \longrightarrow \frac{\log\frac{S_0}{K} + (r-\frac{\sigma^2}{2})T}{\sigma\sqrt{T}}$$

for $M \to \infty$. Hint: Use Exercise 1.6: Up to terms of high order the approximations $u = e^{\sigma\sqrt{\Delta t}}$, $d = e^{-\sigma\sqrt{\Delta t}}$ hold. (In an analogous way the other argument of F can be analyzed.)

Exercise 1.9

In Definition 1.7 the requirement (a) $W_0 = 0$ is dispensable. Then the requirement (b) reads

$$\mathsf{E}(W_t - W_0) = 0 \quad , \quad \mathsf{E}((W_t - W_0)^2) = t \; .$$

Use these relations to deduce (1.21).
Hint: $(W_t - W_s)^2 = (W_t - W_0)^2 + (W_s - W_0)^2 - 2(W_t - W_0)(W_s - W_0)$

Exercise 1.10

a)   Suppose that a random variable X_t satisfies $X_t \sim \mathcal{N}(0, \sigma^2)$. Use (B1.4) to show

$$\mathsf{E}(X_t^4) = 3\sigma^4 \; .$$

b)   Apply a) to show the assertion in Lemma 1.9,

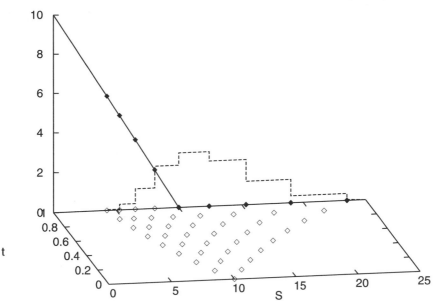

Fig. 1.23. Illustration of a binomial tree and payoff for Exercise 1.8, here for a put, (S, t) points for $M = 8$, $K = S_0 = 10$. The binomial density of the risk-free probability is shown, scaled with factor 10.

$$\mathsf{E}\left(\sum_j ((\Delta W_j)^2 - \Delta t_j)\right)^2 = 2\sum_j (\Delta t_j)^2$$

Exercise 1.11   Analytical Solution of Special SDEs

Apply Itô's lemma to show

a) $X_t = \exp\left(W_t - \frac{1}{2}t\right)$ solves $dX_t = X_t dW_t$
b) $X_t = \exp\left(2W_t - t\right)$ solves $dX_t = X_t dt + 2X_t dW_t$

Hint: Use suitable functions g with $Y_t = g(X_t, t)$. In (a) start with $X_t = W_t$ and $g(x, t) = \exp(x - \frac{1}{2}t)$.

Exercise 1.12   Moments of the Lognormal Distribution

For the density function $f(S; t - t_0, S_0)$ from (1.47) show

a) $\int_0^\infty Sf(S; t - t_0, S_0)dS = S_0 e^{\mu(t - t_0)}$
b) $\int_0^\infty S^2 f(S; t - t_0, S_0)dS = S_0^2 e^{(\sigma^2 + 2\mu)(t - t_0)}$

Hint: Set $y = \log(S/S_0)$ and transform the argument of the exponential function to a squared term.
In case you still have strength afterwards, calculate the value of S for which f is maximal.

Exercise 1.13   Return of the Underlying

Let a time series $S_1, ..., S_M$ of a stock price be given (for example data in the domain www.compfin.de).

The simple return

$$\hat{R}_{i,j} := \frac{S_i - S_j}{S_j} \, ,$$

an index number of the success of the underlying, lacks the desirable property of additivity

$$R_{M,1} = \sum_{i=2}^{M} R_{i,i-1}. \tag{$*$}$$

The log return

$$R_{i,j} := \log S_i - \log S_j \, .$$

has better properties.
a)   Show $R_{i,i-1} \approx \hat{R}_{i,i-1}$, and
b) $R_{i,j}$ satisfies $(*)$.
c)   For empirical data calculate the $R_{i,i-1}$ and set up histograms. Calculate sample mean and sample variance.
d)   Suppose S is lognormally distributed. How can a value of the volatility be obtained from an estimate of the variance?
e)   The mean of the 26866 log returns of the time period of 98.66 years of Figure 1.21 is 0.000199 and the standard deviation is 0.01069. Calculate an estimate of the historical volatility σ.

Exercise 1.14   Solution to the Binomial Model

Derive from equations (1.5), (1.9) and $ud = \gamma$ for some constant γ (not necessarily $\gamma = 1$ as in (1.10)) the relation

$$u = \beta + \sqrt{\beta^2 - \gamma} \quad \text{for} \quad \beta := \frac{1}{2}(\gamma e^{-r\Delta t} + e^{(r+\sigma^2)\Delta t}) \, .$$

Exercise 1.15   Anchoring the Binomial Grid at K

The equation (1.10) has established a kind of symmetry for the grid. As an alternative, one may anchor the grid in another way by choosing (for even M)

$$S_0 u^{M/2} d^{M/2} = K \, .$$

a)   Give a geometrical interpretation.
b)   Derive the relevant formula for u and d.
Hint: Use Exercise 1.14.

Exercise 1.16   Portfolios

Figure 1.24 sketches some payoffs over S. For each of these payoffs, construct portfolios out of vanilla options such that the payoff is met.

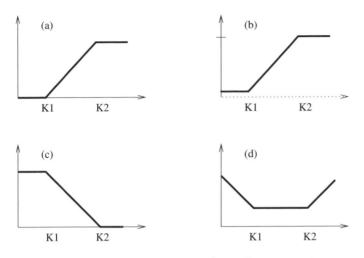

Fig. 1.24. Four payoffs, value over S; see Exercise 1.16

Exercise 1.17   Bounds and Arbitrage

Using arbitrage arguments, show the following bounds for the values V_C of vanilla call options:
a) $0 \leq V_C$
b) $(S - K)^+ \leq V_C^{\text{am}} \leq S$

Exercise 1.18   Positive Itô Process

Let X_t be a positive one-dimensional Itô process for $t \geq 0$.
Show that there exist functions α and β such that

$$dX_t = X_t(\alpha_t dt + \beta_t dW_t)$$

and

$$X_t = X_0 \exp \left\{ \int_0^t (\alpha_s - \frac{1}{2}\beta_s^2)ds + \int_0^t \beta_s dW_s \right\}$$

Exercise 1.19   General Black-Scholes Equation

Assume a portfolio

$$\Pi_t = \alpha_t S_t + \beta_t B_t$$

for a stock S_t and a bond B_t, which obey

$$dS_t = \mu(S_t, t)dt + \sigma(S_t, t)dW_t$$
$$dB_t = r(t)B_t dt$$

The functions μ, σ, and r are assumed to be known, and $\sigma > 0$.   Further assume the portfolio is *self-financing* in the sense

$$d\Pi_t = \alpha_t dS_t + \beta_t dB_t \ .$$

Derive the Black-Scholes equation for this scenario, assuming $\Pi_t = g(S_t, t)$ with g sufficiently often differentiable.

Hint: coefficient matching

Chapter 2 Generating Random Numbers with Specified Distributions

Simulation and valuation of finance instruments require numbers with specified distributions. For example, in Section 1.6 we have used numbers Z drawn from a standard normal distribution, $Z \sim \mathcal{N}(0,1)$. If possible the numbers should be random. But the generation of "random numbers" by digital computers, after all, is done in a deterministic and entirely predictable way. If this point is to be stressed, one uses the term *pseudo-random*[1].

Computer-generated random numbers mimic the properties of true random numbers as much as possible. This is discussed for uniformly distributed numbers in Section 2.1. Suitable transformations generate normally distributed numbers (Sections 2.2, 2.3). Another approach is to dispense with randomness and to generate *quasi-random numbers*, which aim at avoiding one serious disadvantage of random numbers, namely the potential lack of equidistributedness. The resulting *low-discrepancy* numbers will be discussed in Section 2.5. These numbers are used for the deterministic Monte Carlo integration (Section 2.4).

Definition 2.1   (sample from a distribution)
We call a sequence of numbers to be a *sample from* F if the numbers are independent realizations of a random variable with distribution function F.

If F is the uniform distribution over the interval $[0,1)$ or $[0,1]$, then we call the samples from F *uniform deviates (variates)*, notation $\sim \mathcal{U}[0,1]$. If F is the standard normal distribution then we call the samples from F *standard normal deviates (variates)*; as notation we use $\sim \mathcal{N}(0,1)$. The basis of the random-number generation is to draw uniform deviates.

---

[1] Since in our context the predictable origin is clear we omit the modifier "pseudo," and hereafter use the term "random number." Similarly we talk about randomness of these numbers when we mean apparent randomness.

2.1 Uniform Deviates

A standard approach to calculate uniform deviates is provided by linear congruential generators.

2.1.1 Linear Congruential Generators

Choose integers M, a, b. Then a sequence of integers N_i is defined by

Algorithm 2.2   (linear congruential generator)

$$
\begin{array}{l}
\text{Choose } N_0. \\
\text{For } i = 1, 2, \ldots \text{ calculate} \\
N_i = (aN_{i-1} + b) \mod M
\end{array}
\tag{2.1}
$$

The *modulo* congruence $N = Y \mod M$ between two numbers N and Y is an equivalence relation [Ge98]. In Algorithm 2.2 all varibles are integers in the range $a, b, N_0 \in \{0, 1, ..., M - 1\}$, $a \neq 0$. The number N_0 is called the *seed*. Numbers $U_i \in [0, 1)$ are defined by

$$
U_i = N_i/M.
\tag{2.2}
$$

The numbers U_i will be taken as uniform deviates. Whether they are suitable will depend on the choice of M, a, b and will be discussed next.

Properties 2.3   (periodicity)
 (a) $N_i \in \{0, 1, ..., M - 1\}$
 (b) The N_i are periodic with period $\leq M$.
   (Because there are not $M + 1$ different N_i. So two in $\{N_0, ..., N_M\}$ must be equal, $N_i = N_{i+p}$ with $p \leq M$.)

Obviously, some peculiarities must be excluded. For example, $N = 0$ must be ruled out in case $b = 0$, because otherwise $N_i = 0$ would repeat. In case $a = 1$ the generator settles down to $N_n = (N_0 + nb) \mod M$. This sequence is too easily predictable. Various other properties and requirements are discussed in the literature, in particular in [Kn95]. In case the period is M, the numbers U_i are uniformly distributed when exactly M numbers are needed. Then each grid point on a mesh on [0,1] with mesh size $\frac{1}{M}$ is occupied once.

   After these observations we start searching for good choices of M, a, b. But for serious computations we recommend to rely on the many suggestions in the literature. [PTVF92] presents a table of "quick and dirty" generators, for example, $M = 244944$, $a = 1597$, $b = 51749$. But which of the many possible generators are recommendable?

2.1.2 Quality of Generators

What are good random numbers? A practical answer is the requirement that the numbers should meet "all" aims, or rather pass as many tests as possible. The requirements on good number generators can roughly be divided into three groups.

The first requirement is that of a large period. In view of Property 2.3 the number M must be as large as possible, because a small set of numbers makes the outcome easier to predict— a contrast to randomness. This leads to select M close to the largest integer machine number. But a period p close to M is only achieved if a and b are chosen properly. Criteria for relations among M, p, a, b have been derived by number-theoretic arguments. This is outlined in [Kn95], [Ri87]. A common choice for 32-bit computers is $M = 2^{31} - 1$, $a = 16807$, $b = 0$.

A second group of requirements are the *statistical tests* that check whether the numbers are distributed as intended. The simplest of such tests evaluates the sample mean $\hat{\mu}$ and the sample variance $\hat{s}^2$ (B1.11) of the calculated random variates, and compares to the desired values of μ and σ^2. (Recall $\mu = 1/2$ and $\sigma^2 = 1/12$ for the uniform distribution.) Another simple test is to check correlations. For example, it would not be desirable if small numbers are likely to be followed by small numbers.

A slightly more involved test checks how well the probability distribution is approximated. This works for general distributions ($\longrightarrow$ Exercise 2.14). Here we briefly summarize an approach for uniform deviates. Calculate j samples from a random number generator, and investigate how the samples distribute on the unit interval. To this end, divide the unit interval into subintervals of equal length ΔU, and denote by j_k the number of samples that fall into the kth subinterval

$$k\Delta U \leq U < (k+1)\Delta U .$$

Then j_k/j should be close the desired probability, which for this setup is ΔU. Hence a plot of the quotients

$$\frac{j_k}{j\Delta U} \quad \text{for all } k$$

against $k\Delta U$ should be a good approximation of 1, the density of the uniform distribution. This procedure is just the simplest statistical test; for more ambitious tests, see [Kn95].

The third group of tests is to check how well the random numbers distribute in higher-dimensional spaces. This issue of the *lattice structure* is discussed next. We derive a priori analytical results on *where* the random numbers produced by Algorithm 2.2 are distributed.

2.1.3 Random Vectors and Lattice Structure

Random numbers N_i can be arranged in m-tupels $(N_i, N_{i+1}, ..., N_{i+m-1})$ for $i \geq 1$. Then the tupels or the corresponding points $(U_i, ..., U_{i+m-1}) \in [0, 1)^m$ are analyzed with respect to correlation and distribution. The sequences defined by Algorithm 2.2 lie on $(m - 1)$-dimensional hyperplanes. This statement is trivial since it holds for the M parallel planes through $U = i/M$, $i = 0, ..., M - 1$. But the statement becomes exciting in case it is valid for a family of parallel planes with large distances between neighboring planes. Next we attempt to construct such planes.

Analysis for the case m = 2:

$$N_i = (aN_{i-1} + b) \mod M$$
$$= aN_{i-1} + b - kM \quad \text{for} \quad kM \leq aN_{i-1} + b < (k+1)M ,$$

k an integer. A side calculation for arbitrary z_0, z_1 shows

$$z_0 N_{i-1} + z_1 N_i = z_0 N_{i-1} + z_1(aN_{i-1} + b - kM)$$
$$= N_{i-1}(z_0 + az_1) + z_1 b - z_1 kM$$
$$= M \cdot \underbrace{\{N_{i-1}\frac{z_0 + az_1}{M} - z_1 k\}}_{=:c} + z_1 b.$$

We divide by M and obtain the equation of a straight line in the (U_{i-1}, U_i)-plane, namely

$$z_0 U_{i-1} + z_1 U_i = c + z_1 b M^{-1}. \tag{2.3}$$

The points calculated by Algorithm 2.2 lie on these straight lines. To eliminate the seed we take $i > 1$. Fixing one tupel (z_0, z_1), the equation (2.3) defines a family of parallel straight lines, one for each number out of the finite set of c's. The question is whether there exists a tupel (z_0, z_1) such that only few of the straight lines cut the square $[0, 1)^2$? In this case wide areas of the square would be free of random points, which violates the requirement of a uniform distribution of the points. The minimum number of parallel straight lines (hyperplanes) cutting the square, or equivalently the maximum distance between them serve as measures of the equidistributedness. We now analyze the number of straight lines, searching for the worst case.

When we admit only integer $(z_0, z_1) \in \mathbb{Z}^2$, and require

$$z_0 + az_1 = 0 \mod M, \tag{2.4}$$

then c is integer. By solving (2.3) for $c = z_0 U_{i-1} + z_1 U_i - z_1 b M^{-1}$ and applying $0 \leq U_i < 1$ we obtain the maximal interval I_c such that for each integer $c \in I_c$ its straight line cuts or touches the square $[0, 1)^2$. We count how many such c exist, and have the information we need. For some constellations of a, M, z_0 and z_1 it may be possible that the points (U_{i-1}, U_i) lie on very few of these straight lines!

Example 2.4 $N_i = 2N_{i-1} \mod 11$ (that is, $a = 2$, $b = 0$, $M = 11$)
We choose $z_0 = -2$, $z_1 = 1$, which is one tuple satisfying (2.4), and investigate the family (2.3) of straight lines

$$-2U_{i-1} + U_i = c$$

in the (U_{i-1}, U_i)-plane. For $U_i \in [0,1)$ we have $-2 < c < 1$. In view of (2.4) c is integer and so only the two integers $c = -1$ and $c = 0$ remain. The two corresponding straight lines cut the interior of $[0,1)^2$. As Figure 2.1 illustrates, the points generated by the algorithm form a lattice. All points on the lattice lie on these two straight lines. The figure lets us discover also other parallel straight lines such that all points are caught (for other tupels z_0, z_1). The practical question is: What is the largest gap? ($\longrightarrow$ Exercise 2.1)

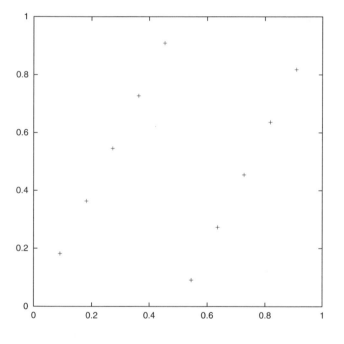

Fig. 2.1. The points (U_{i-1}, U_i) of Example 2.4

Example 2.5 $N_i = (1229N_{i-1} + 1) \mod 2048$
The requirement of equation (2.4)

$$\frac{z_0 + 1229z_1}{2048} \in \mathbb{Z}$$

is satisfied by $z_0 = -1$, $z_1 = 5$, because

$$-1 + 1229 \cdot 5 = 6144 = 3 \cdot 2048$$

The distance between straight lines measured along the vertical U_i–axis is $\frac{1}{z_1} = \frac{1}{5}$. All points (U_{i-1}, U_i) lie on only six straight lines, with $c \in \{-1, 0, 1, 2, 3, 4\}$, see Figure 2.2. On the "lowest" straight line ($c = -1$) there is only one point.

Higher-dimensional vectors ($m > 2$) are analyzed analogously. The generator called RANDU

$$N_i = aN_{i-1} \mod M, \quad \text{with } a = 2^{16} + 3, \ M = 2^{31}$$

may serve as example. Its random points in the cube $[0, 1)^3$ lie on only 15 planes ($\longrightarrow$ Exercise 2.2). For many applications this must be seen as a severe defect.

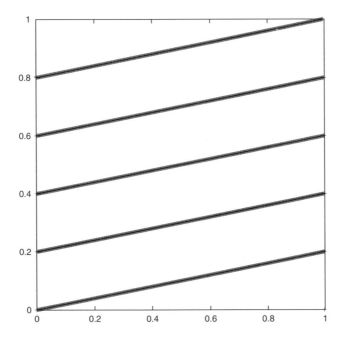

Fig. 2.2. The points (U_{i-1}, U_i) of Example 2.5

In Example 2.4 we asked what the maximum gap between the parallel straight lines is. In other words, we have searched for strips of maximum size in which no point (U_{i-1}, U_i) falls. Alternatively we can directly analyze the lattice formed by consecutive points. For illustration consider again Figure 2.1. We follow the points starting with $(\frac{1}{11}, \frac{2}{11})$. By vectorwise adding an appropriate multiple of $\begin{pmatrix} 1 \\ a \end{pmatrix} = \begin{pmatrix} 1 \\ 2 \end{pmatrix}$ the next two points are obtained. Proceeding in this way one has to take care that upon leaving the unit square each component with value ≥ 1 must be reduced to $[0, 1)$ to observe mod M. The reader may verify this with Example 2.4 and numerate the points of the lattice in Figure 2.1 in the correct sequence. In this way the lattice can be defined. This process of defining the lattice can be generalized to higher dimensions $m > 2$. ($\longrightarrow$ Exercise 2.3)

A disadvantage of the linear congruential generators of Algorithm 2.2 is the boundedness of the period by M and hence by the word length of the computer. The situation can be improved by *shuffling* the random numbers in a random way. For practical purposes, the period gets close enough to infinity. (The reader may test this on Example 2.5.) For practical advice we refer to [PTVF92].

2.1.4 Fibonacci Generators

The original Fibonacci recursion motivates trying the formula

$$N_{i+1} := (N_i + N_{i-1}) \mod M .$$

It turns out that this first attempt of a three-term recursion is not suitable for generating random numbers ($\longrightarrow$ Exercise 2.15). The modified approach

$$N_{i+1} := (N_{i-\nu} - N_{i-\mu}) \mod M \tag{2.5}$$

for suitable ν, $\mu \in \mathbb{N}$ is called *lagged Fibonacci* generator. For many choices of ν, μ the approach (2.5) leads to recommendable generators.

Example 2.6
$$U_i := U_{i-17} - U_{i-5},$$
$$\text{in case } U_i < 0 \text{ set } U_i := U_i + 1.0$$

The recursion of Example 2.6 immediately produces floating-point numbers $U_i \in [0, 1)$. This generator requires a prologue in which 17 initial U's are generated by means of another method. The generator can be run with varying lags ν, μ. [KMN89] recommends

Algorithm 2.7   (Fibonacci generator)

$$
\begin{aligned}
&\textit{Repeat:}\ \ \zeta := U_i - U_j \\
&\qquad\qquad \text{if } \zeta < 0,\ \text{set } \zeta := \zeta + 1 \\
&\qquad\qquad U_i := \zeta \\
&\qquad\qquad i := i - 1 \\
&\qquad\qquad j := j - 1 \\
&\qquad\qquad \text{if } i = 0,\ \text{set } i := 17 \\
&\qquad\qquad \text{if } j = 0,\ \text{set } j := 17
\end{aligned}
$$

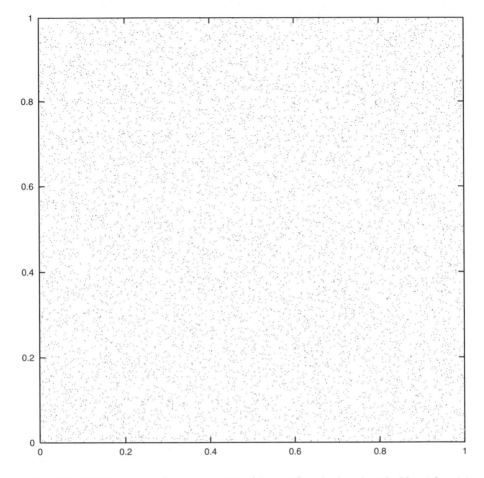

Fig. 2.3. 10000 (pseudo-)random points (U_{i-1}, U_i), calculated with Algorithm 2.7

Initialization: Set $i = 17$, $j = 5$, and calculate $U_1, ..., U_{17}$ with a congruential generator, for instance with $M = 714025$, $a = 1366$, $b = 150889$. Set the seed $N_0 =$ your favorite dream number, possibly inspired by the system clock of your computer.

Figure 2.3 depicts 10000 random points calculated by means of Algorithm 2.7. Visual inspection suggests that the points are not arranged in some apparent structure. The points appear to be sufficiently random. But the generator provided by Example 2.6 is not sophisticated enough for ambitious applications; its pseudo-random numbers are rather correlated. See *Notes and Comments* for hints on other generators.

2.2 Transformed Random Variables

Frequently normal variates are needed. Their generation is based on uniform deviates. The simplest strategy is to calculate

$$X := \sum_{i=1}^{12} U_i - 6, \quad \text{for} \quad U_i \sim \mathcal{U}[0, 1] \,.$$

X has expectation 0 and variance 1. The Central Limit Theorem ($\longrightarrow$ Appendix B1) assures that X is approximately normally distributed ($\longrightarrow$ Exercise 2.4). But this crude attempt is not satisfying. Better methods calculate nonuniformly distributed random variables by a suitable transformation out of a uniformly distributed random variable [Dev86]. But the most obvious approach inverts the distribution function.

2.2.1 Inversion

The following theorem is the basis for inversion methods.

Theorem 2.8   (inversion)
   Suppose $U \sim \mathcal{U}[0, 1]$ and F be a continuous strictly increasing distribution function. Then $F^{-1}(U)$ is a sample from F.

   Proof: Let P denote the underlying probability.
      $U \sim \mathcal{U}[0, 1]$ means $\mathsf{P}(U \leq \xi) = \xi$ for $0 \leq \xi \leq 1$.
      Consequently

$$\mathsf{P}(F^{-1}(U) \leq x) = \mathsf{P}(U \leq F(x)) = F(x).$$

Application
Following Theorem 2.8 the inversion method takes uniform deviates $u \sim \mathcal{U}[0, 1]$ and sets $x = F^{-1}(u)$ ($\longrightarrow$ Exercise 2.5). To judge the inversion method we consider the normal distribution as the most important example. Neither for its distribution function F nor for its inverse F^{-1} there is a closed-form expression ($\longrightarrow$ Exercise 1.3). So numerical methods are used. We discuss two approaches.
   Numerical inversion means to calculate iteratively a solution x of the equation $F(x) = u$ for prescribed u. This iteration requires tricky termination criteria, in particular when x is large. Then we are in the situation $u \approx 1$, where tiny changes in u lead to large changes in x (Figure 2.4). The approximation of the solution x of $F(x) - u = 0$ can be calculated with bisection, or Newton's method, or the secant method ($\longrightarrow$ Appendix C1).
   Alternatively the inversion $x = F^{-1}(u)$ can be approximated by a suitably constructed function $G(u)$,

$$G(u) \approx F^{-1}(u) \,.$$

Then only $x = G(u)$ needs to be evaluated. Constructing such an approximation formula G, it is important to realize that $F^{-1}(u)$ has "vertical" tangents at $u = 1$ and $u = 0$ (horizontal in Figure 2.4). This pole behavior must be reproduced correctly by the approximating function G. This suggests to use rational approximation ($\longrightarrow$ Appendix C1), which allows incorporating the point symmetry with respect to $(u, x) = (\frac{1}{2}, 0)$, and the pole at $u = 1$ (and hence at $u = 0$) in the *ansatz* for G ($\longrightarrow$ Exercise 2.6). Rational approximation of $F^{-1}(u)$ with a sufficiently large number of terms leads to high accuracy [Moro95]. The formulas are given in Appendix D2.

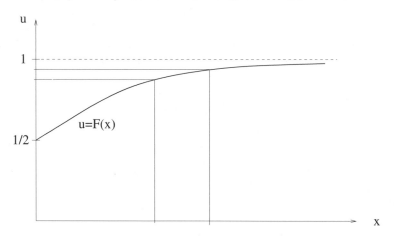

Fig. 2.4. Small changes in u leading to large changes in x

2.2.2 Transformations in $\mathbf{R}^1$

Another class of methods uses transformations between random variables. We start the discussion with the scalar case. If we have a random variable X with *known* density and distribution, what can we say about the density and distribution of a transformed $h(X)$?

Theorem 2.9

Suppose X is a random variable with density $f(x)$ and distribution $F(x)$. Further assume $h : S \longrightarrow B$ with $S, B \subseteq \mathbb{R}$, where S is the support[2] of $f(x)$, and let h be strictly monotonous.

(a) Then $Y := h(X)$ is a random variable. Its distribution F_Y in case $h' > 0$ is $F_Y(y) = F(h^{-1}(y))$.

(b) If h^{-1} is absolutely continuous then for almost all y the density of $h(X)$ is

$$f(h^{-1}(y)) \left| \frac{dh^{-1}(y)}{dy} \right|. \tag{2.6}$$

---

[2] f is zero outside S. (In this section S is no asset price.)

Proof:

(a) For $h' > 0$ we have $\mathsf{P}(h(X) \leq y) = \mathsf{P}(X \leq h^{-1}(y)) = F(h^{-1}(y))$.

(b) h^{-1} absolutely continuous $\Longrightarrow$ The density of $Y = h(X)$ is equal to the derivative of the distribution function almost everywhere. Evaluating the derivative $\frac{dF(h^{-1}(y))}{dy}$ with the chain rule implies the assertion. The absolute value in (2.6) is necessary such that a positive density comes out, because in case $h' < 0$ the distribution is $1 - F(h^{-1}(y))$. (See for instance [Fisz63], § 2.4 C.)

Application

Since we are able to calculate uniform deviates, we start from $X \sim \mathcal{U}[0, 1]$ with f being the density of the uniform distribution,

$$f(x) = 1 \quad \text{for} \quad 0 \leq x \leq 1, \quad \text{otherwise} \quad f = 0.$$

Here the support S is the unit interval. What we need are random numbers Y matching a prespecified density $g(y)$. It remains to find a transformation h such that $g(y)$ is identical to the density in (2.6), $1 \cdot \left| \frac{dh^{-1}(y)}{dy} \right| = g(y)$. Then we only evaluate $h(X)$.

Example 2.10   (exponential distribution)

The exponential distribution with parameter $\lambda > 0$ has the density

$$g(y) = \begin{cases} \lambda e^{-\lambda y} & \text{for } y \geq 0 \\ 0 & \text{for } y < 0. \end{cases}$$

Here the range B consists of the nonnegative numbers. The aim is to generate an exponentially distributed random variable Y out of a $\mathcal{U}[0, 1]$-distributed random variable X. To this end we define the monotonous transformation from the unit interval S into B by

$$y = h(x) := -\frac{1}{\lambda} \log x$$

with the inverse function $h^{-1}(y) = e^{-\lambda y}$ for $y \geq 0$. For this h verify

$$f(h^{-1}(y)) \left| \frac{dh^{-1}(y)}{dy} \right| = 1 \cdot \left| (-\lambda) e^{-\lambda y} \right| = \lambda e^{-\lambda y} = g(y)$$

as density of $h(X)$. Hence $h(X)$ is distributed exponentially.

Application:

In case $U_1, U_2, \ldots$ are nonzero uniform deviates, the numbers $h(U_i)$

$$-\frac{1}{\lambda} \log(U_1), \quad -\frac{1}{\lambda} \log(U_2), \quad \ldots$$

are distributed exponentially.

Attempt to Generate a Normal Distribution

Starting from the uniform distribution ($f = 1$) a transformation $y = h(x)$ is sought such that its density equals that of the standard normal distribution,

$$1 \cdot \left| \frac{dh^{-1}(y)}{dy} \right| = \frac{1}{\sqrt{2\pi}} \exp\left(-\frac{1}{2} y^2 \right).$$

This is a differential equation for h^{-1} without analytical solution. As we will see, a transformation can be applied successfully in $\mathbb{R}^2$. To this end we need a generalization of the scalar transformation of Theorem 2.9 into $\mathbb{R}^n$.

2.2.3 Transformation in $\mathbf{R}^n$

Theorem 2.11

Suppose X is a random variable in $\mathbb{R}^n$ with density $f(x) > 0$ on the support S. The transformation $h : S \to B$, $S, B \subseteq \mathbb{R}^n$ is assumed to be invertible and the inverse be continuously differentiable on B. $Y := h(X)$ is the transformed random variable. Then Y has the density

$$f(h^{-1}(y)) \left| \frac{\partial(x_1, ..., x_n)}{\partial(y_1, ..., y_n)} \right|, \quad y \in B, \tag{2.7}$$

where $x = h^{-1}(y)$ and $\frac{\partial(x_1,...,x_n)}{\partial(y_1,...,y_n)}$ is the determinant of the Jacobian matrix of all first-order derivatives of $h^{-1}(y)$.

(Theorem 4.2 in [Dev86])

2.3 Normally Distributed Random Variables

In this section the focus is on applying the transformation method in $\mathbb{R}^2$ to generate normal variates. Keep in mind that inversion is a valid alternative.

2.3.1 Method of Box and Muller

To apply Theorem 2.11 we start with the unit square $S := [0, 1]^2$ and the density (2.7) of the bivariate uniform distribution. The transformation is

$$\begin{cases} y_1 = \sqrt{-2\log x_1} \cos 2\pi x_2 =: h_1(x_1, x_2) \\ y_2 = \sqrt{-2\log x_1} \sin 2\pi x_2 =: h_2(x_1, x_2). \end{cases} \tag{2.8}$$

The function $h(x)$ is defined on $[0, 1]^2$ with values in $\mathbb{R}^2$. The inverse function h^{-1} is given by

$$\begin{cases} x_1 = \exp\left\{ -\frac{1}{2}(y_1^2 + y_2^2) \right\} \\ x_2 = \frac{1}{2\pi} \arctan \frac{y_2}{y_1} \end{cases}$$

where we take the main branch of arctan. The determinant of the Jacobian matrix is

$$\frac{\partial(x_1, x_2)}{\partial(y_1, y_2)} = \det \begin{pmatrix} \frac{\partial x_1}{\partial y_1} & \frac{\partial x_1}{\partial y_2} \\ \frac{\partial x_2}{\partial y_1} & \frac{\partial x_2}{\partial y_2} \end{pmatrix} =$$

$$= \frac{1}{2\pi} \exp\left\{ -\tfrac{1}{2}(y_1^2 + y_2^2) \right\} \left(-y_1 \frac{1}{1 + \frac{y_2^2}{y_1^2}} \frac{1}{y_1} - y_2 \frac{1}{1 + \frac{y_2^2}{y_1^2}} \frac{y_2}{y_1^2} \right)$$

$$= -\frac{1}{2\pi} \exp\left\{ -\tfrac{1}{2}(y_1^2 + y_2^2) \right\}.$$

This shows that $\left| \frac{\partial(x_1, x_2)}{\partial(y_1, y_2)} \right|$ is the density (2.7) of the bivariate standard normal distribution. Since this density is the product of the two one-dimensional densities,

$$\left| \frac{\partial(x_1, x_2)}{\partial(y_1, y_2)} \right| = \left[\frac{1}{\sqrt{2\pi}} \exp\left(-\tfrac{1}{2} y_1^2 \right) \right] \cdot \left[\frac{1}{\sqrt{2\pi}} \exp\left(-\tfrac{1}{2} y_2^2 \right) \right],$$

the two components of the vector y are independent. So, when the components of the vector X are $\sim \mathcal{U}[0,1]$, the vector $h(X)$ consists of two independent standard normal variates. Let us summarize the application of this transformation:

Algorithm 2.12   (Box-Muller)

> (1) generate $U_1 \sim \mathcal{U}[0,1]$ and $U_2 \sim \mathcal{U}[0,1]$.
> (2) $\theta := 2\pi U_2$, $\quad \rho := \sqrt{-2 \log U_1}$
> (3) $Z_1 := \rho \cos \theta$ is a normal variate
> $\quad$ (as well as $Z_2 := \rho \sin \theta$).

The variables U_1, U_2 stand for the components of X. Each application of the algorithm provides two standard normal variates. Note that a line structure in $[0,1]^2$ as in Example 2.5 is mapped to curves in the (Z_1, Z_2)-plane. This underlines the importance of excluding an evident line structure.

2.3.2 Variant of Marsaglia

The variant of Marsaglia prepares the input in Algorithm 2.12 such that trigonometric functions are avoided. For $U \sim \mathcal{U}[0,1]$ we have $V := 2U - 1 \sim \mathcal{U}[-1,1]$. (Temporarily we misuse also the financial variable V for local purposes.) Two values V_1, V_2 calculated in this way define a point in the (V_1, V_2)-plane. Only points within the unit disk are accepted:

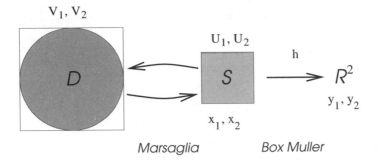

Fig. 2.5. The transformations of the Box-Muller-Marsaglia approach, schematically

$$\mathcal{D} := \{(V_1, V_2) \ : \ V_1^2 + V_2^2 < 1\}; \text{ accept only } (V_1, V_2) \in \mathcal{D}.$$

In case of rejectance both values V_1, V_2 must be rejected. As a result, the surviving (V_1, V_2) are uniformly distributed on $\mathcal{D}$ with density $f(V_1, V_2) = \frac{1}{\pi}$ for $(V_1, V_2) \in \mathcal{D}$. A transformation from the disk $\mathcal{D}$ into the unit square $S := [0, 1]^2$ is defined by

$$\begin{pmatrix} x_1 \\ x_2 \end{pmatrix} = \begin{pmatrix} V_1^2 + V_2^2 \\ \frac{1}{2\pi} \arctan \frac{V_2}{V_1} \end{pmatrix}.$$

That is, the Cartesian coordinates V_1, V_2 on $\mathcal{D}$ are mapped to the squared radius and the angle. For illustration, see Figure 2.5.

Then $\begin{pmatrix} x_1 \\ x_2 \end{pmatrix}$ is uniformly distributed on S ($\longrightarrow$ Exercise 2.7).

Application

For input in (2.8) use $V_1^2 + V_2^2$ as x_1 and $\frac{1}{2\pi} \arctan \frac{V_2}{V_1}$ as x_2. With these variables the relations

$$\cos 2\pi x_2 = \frac{V_1}{\sqrt{V_1^2 + V_2^2}}, \quad \sin 2\pi x_2 = \frac{V_2}{\sqrt{V_1^2 + V_2^2}},$$

hold, which means that it is no longer necessary to evaluate trigonometric functions. The resulting algorithm of Marsaglia has modified the Box-Muller method by constructing input values x_1, x_2 in a clever way.

Algorithm 2.13   (polar method)

> (1) *Repeat:* generate $U_1, U_2 \sim \mathcal{U}[0, 1]$; $V_1 := 2U_1 - 1$,
> $V_2 := 2U_2 - 1$, *until* $W := V_1^2 + V_2^2 < 1$.
> (2) $Z_1 := V_1\sqrt{-2\log(W)/W}$
> $Z_2 := V_2\sqrt{-2\log(W)/W}$
> are both standard normal variates.

The probability that $W < 1$ holds is given by the ratio of the areas, $\pi/4 = 0.785...$ So in about 21% of all $\mathcal{U}[0,1]$ drawings the (V_1, V_2)-tupel is rejected because of $W \geq 1$. Nevertheless the savings of the trigonometric evaluations makes Marsaglia's polar method more efficient than the Box-Muller method. Figure 2.6 illustrates normally distributed random numbers ($\longrightarrow$ Exercise 2.8).

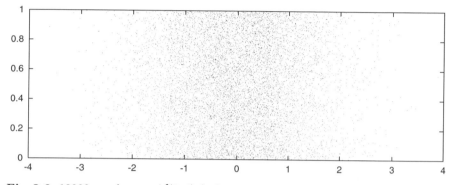

Fig. 2.6. 10000 numbers $\sim \mathcal{N}(0,1)$ (values entered horizontally and separated vertically with distance 10^{-4})

2.3.3 Correlated Random Variables

The above algorithms provide independent normal deviates. In some applications random variables are required that depend on each other in a prescribed way. Let us first recall the general n-dimensional density function.

Multivariate normal distribution (notations):

$$X = (X_1, ..., X_n), \quad \mu = \mathsf{E}X = (\mathsf{E}X_1, ..., \mathsf{E}X_n)$$

The covariance matrix (B1.8) of X is denoted Σ, and has elements

$$\Sigma_{ij} = (\mathsf{Cov}X)_{ij} := \mathsf{E}\left((X_i - \mu_i)(X_j - \mu_j)\right); \quad \sigma_i^2 = \Sigma_{ii}$$

Using this notation, the correlation coefficients are

$$\rho_{ij} := \frac{\Sigma_{ij}}{\sigma_i \sigma_j} \qquad (\Rightarrow \; \rho_{ii} = 1) \;. \tag{2.9}$$

The density function $f(x_1, ..., x_n)$ corresponding to $\mathcal{N}(\mu, \Sigma)$ is

$$f(x) = \frac{1}{(2\pi)^{n/2}} \frac{1}{(\det \Sigma)^{1/2}} \exp\left\{ -\frac{1}{2}(x - \mu)^{tr} \Sigma^{-1}(x - \mu) \right\}. \tag{2.10}$$

The matrix Σ is symmetric positive definite in case $\det \Sigma \neq 0$. From numerical mathematics we know that for such matrices the Cholesky decomposition $\Sigma = LL^{tr}$ exists, with a lower triangular matrix L ($\longrightarrow$ Appendix C1).

Transformation

Suppose $Z \sim \mathcal{N}(0, I)$ and $x = Az$, $A \in \mathbb{R}^{n \times n}$, where z is a realization of Z and I the identity matrix. We see from

$$\exp\left\{-\frac{1}{2}z^{tr}z\right\} = \exp\left\{-\frac{1}{2}(A^{-1}x)^{tr}(A^{-1}x)\right\} = \exp\left\{-\frac{1}{2}x^{tr}A^{-tr}A^{-1}x\right\}$$

and from $dx = |\det A|\,dz$ that

$$\frac{1}{|\det A|}\exp\left\{-\frac{1}{2}x^{tr}(AA^{tr})^{-1}x\right\}dx = \exp\left\{-\frac{1}{2}z^{tr}z\right\}dz$$

holds for arbitrary nonsingular matrices A. In case A is specifically the matrix L of the Cholesky decomposition, $\Sigma = AA^{tr}$ and $|\det A| = (\det \Sigma)^{1/2}$. In this way the densities with the respect to x and z are converted. In view of the general density $f(x)$ recalled above in (2.10), AZ is normally distributed with

$$AZ \sim \mathcal{N}(0, AA^{tr}).$$

Finally, translation implies

$$\mu + AZ \sim \mathcal{N}(\mu, AA^{tr}). \tag{2.11}$$

Application

Suppose we need a normal variate $X \sim \mathcal{N}(\mu, \Sigma)$ for given mean vector μ and covariance matrix Σ. Such a random variable is calculated with the following algorithm:

Algorithm 2.14   (correlated random variable)

> (1) Calculate the Cholesky decomposition $AA^{tr} = \Sigma$
> (2) Calculate $Z \sim \mathcal{N}(0, I)$ componentwise
>   by $Z_i \sim \mathcal{N}(0, 1)$, $i = 1, ..., n$, for instance,
>   with Marsaglia's polar algorithm
> (3) $\mu + AZ$ has the desired distribution $\sim \mathcal{N}(\mu, \Sigma)$

Special case $n = 2$: In this case, in view of (2.9), only one correlation number is involved, namely $\rho := \rho_{12} = \rho_{21}$, and the correlation matrix must be of the form

$$\Sigma = \begin{pmatrix} \sigma_1^2 & \rho\sigma_1\sigma_2 \\ \rho\sigma_1\sigma_2 & \sigma_2^2 \end{pmatrix}. \tag{2.12}$$

In this two-dimensional situation it makes sense to carry out the Cholesky decomposition analytically ($\longrightarrow$ Exercise 2.9). Figure 2.7 illustrates a highly correlated two-dimensional situation, with $\rho = 0.85$.

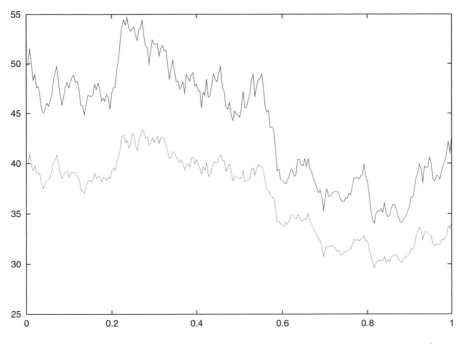

Fig. 2.7. Simulation of a correlated vector process with two components, and $\mu = 0.05$, $\sigma_1 = 0.3$, $\sigma_2 = 0.2$, $\rho = 0.85$, $\Delta t = 1/250$

2.4 Monte Carlo Integration

A classical application of random numbers is the Monte Carlo integration. The discussion in this section will serve as background for Quasi Monte Carlo, a topic of the following Section 2.5.

Let us begin with the one-dimensional situation. Assume a probability distribution with density g. Then the expectation of a function f is

$$\mathsf{E}(f) = \int_{-\infty}^{\infty} f(x)g(x)dx ,$$

compare (B1.4). Specifically for the uniform distribution on an interval $\mathcal{D} = [a, b]$, the density is

$$g = \frac{1}{b-a} \cdot 1_{\mathcal{D}} = \frac{1}{\lambda_1(\mathcal{D})} \cdot 1_{\mathcal{D}} ,$$

where $\lambda_1(\mathcal{D})$ denotes the length of the interval $\mathcal{D}$. This leads to

$$\mathsf{E}(f) = \frac{1}{\lambda_1(\mathcal{D})} \int_{a}^{b} f(x)dx ,$$

or

$$\int_a^b f(x)dx = \lambda_1(\mathcal{D}) \cdot \mathsf{E}(f) \ .$$

This equation is the basis of *Monte Carlo integration*. It remains to approximate $\mathsf{E}(f)$. For independent samples $x_i \sim \mathcal{U}[a, b]$ the law of large numbers ($\longrightarrow$ Appendix B1) establishes the estimator

$$\frac{1}{N}\sum_{i=1}^N f(x_i)$$

as approximation to $\mathsf{E}(f)$. The approximation improves as the number of trials N goes to infinity; the error is characterized by the Central Limit Theorem.

This principle of the Monte Carlo Integration extends to the higher-dimensional case. Let $\mathcal{D} \subset \mathbb{R}^m$ be a domain on which the integral

$$\int_{\mathcal{D}} f(x)dx$$

is to be calculated. For example, $\mathcal{D} = [0, 1]^m$. Such integrals occur in finance, for example, when mortgage-backed securities (CMO, collateralized mortgage obligations) are valuated [CaMO97]. The classical or *stochastic Monte Carlo integration* draws random samples $x_1, ..., x_N \in \mathcal{D}$ which should be independent and uniformly distributed. Then

$$\theta_N := \lambda_m(\mathcal{D})\frac{1}{N}\sum_{i=1}^N f(x_i) \tag{2.13}$$

is an approximation of the integral. Here $\lambda_m(\mathcal{D})$ is the volume of $\mathcal{D}$ (or the m-dimensional Lebesgue measure [Ni92]). We assume $\lambda_m(\mathcal{D})$ to be finite. From the law of large numbers follows convergence of θ_N to $\lambda_m(\mathcal{D})\mathsf{E}(f) = \int_{\mathcal{D}} f(x)\ dx$ for $N \to \infty$. The variance of the error

$$\delta_N := \int_{\mathcal{D}} f(x)dx - \theta_N$$

satisfies

$$\mathsf{Var}(\delta_N) = \mathsf{E}(\delta_N^2) - (\mathsf{E}(\delta_N))^2 = \frac{\sigma^2(f)}{N}(\lambda_m(\mathcal{D}))^2, \tag{2.14a}$$

with the variance of f

$$\sigma^2(f) := \int_{\mathcal{D}} f(x)^2 dx - \left(\int_{\mathcal{D}} f(x)dx\right)^2 \ . \tag{2.14b}$$

Hence the standard deviation of the error δ_N tends to 0 with the order $O(N^{-1/2})$. This result follows from the Central Limit Theorem or from other arguments ($\longrightarrow$ Exercise 2.10). The deficiency of the order $O(N^{-1/2})$ is the slow convergence ($\longrightarrow$ Exercise 2.11 and the second column in Table 2.1). To improve the accuracy by a factor of 10, the costs (that is the number of trials, N) increase by a factor of 100. Another disadvantage is the lack of a genuine error *bound*. The probabilistic error of (2.14) does not rule out the risk that the result may be completely wrong. And the Monte Carlo integration responds sensitively to changes of the initial state of the used random-number generator. This may be explained by the potential clustering of random points. In many applications the above deficiencies are balanced by two good features of Monte Carlo integration: A first advantage is that the order $O(N^{-1/2})$ holds independently of the dimension m. Another good feature is that the integrands f need not be smooth, square integrability suffices ($f \in \mathcal{L}^2$, see Appendix C3).

So far we have described the basic version of Monte Carlo integration, stressing the slow decline of the probabilistic error with growing N. The variance of the error δ can also be diminished by decreasing the numerator in (2.14a). This variance of the problem can be reduced by suitable methods. (We will come back to this issue in Chapter 3.) We conclude the excursion into the stochastic Monte Carlo integration with the variant for those cases in which $\lambda_m(\mathcal{D})$ is hard to calculate. For $\mathcal{D} \subseteq [0,1]^m$ and $x_1, ..., x_N \in [0,1]^m$ use

$$\int_{\mathcal{D}} f(x)dx \approx \frac{1}{N} \sum_{\substack{i=1 \\ x_i \in \mathcal{D}}}^{N} f(x_i). \tag{2.15}$$

2.5 Sequences of Numbers with Low Discrepancy

One difficulty with random numbers is that they may fail to distribute uniformly. Here, "uniform" is not meant in the stochastic sense of a distribution $\sim \mathcal{U}[0,1]$, but has the meaning of equidistributedness. The aim is to generate numbers for which the deviation from uniformity is minimal. This deviation is called "discrepancy." Another objective is to obtain good convergence for some important applications.

2.5.1 Discrepancy

The bad convergence behavior of the stochastic Monte Carlo integration is not inevitable. For example, for $m = 1$ and $\mathcal{D} = [0,1]$ an equidistant x-grid with mesh size $1/N$ leads to a formula (2.13) that resembles the trapezoidal sum ((C1.2) in Appendix C1). For smooth f, the order of the error is at least $O(N^{-1})$. (Why?) But such a grid-based evaluation procedure is somewhat inflexible because the grid must be prescribed in advance and the number N that matches the desired accuracy is unknown beforehand. In contrast, the free placing of sample points with Monte Carlo integration can be performed until some termination criterion is met. It would be desirable to find a compromise picking sample points in some scheme such that the fineness advances but clustering is avoided. The sample points should fill the integration domain $\mathcal{D}$ as uniformly as possible. To this end we require a measure of the equidistributedness.

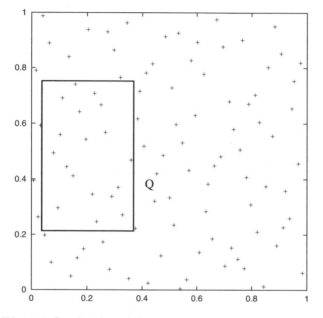

Fig. 2.8 On the idea of discrepancy

Let $Q \subseteq [0,1]^m$ be an arbitrary axially parallel m-dimensional rectangle in the unit cube $[0,1]^m$ of $\mathbb{R}^m$. That is, Q is a product of m intervals. Suppose a set of points $x_1, ..., x_N \in [0,1]^m$. The decisive idea behind discrepancy is that for an evenly distributed point set the fraction of the points lying within the rectangle Q should correspond to the volume of the rectangle (see Figure 2.8). Let # denote the number of points, then the goal is

$$\frac{\# \text{ of } x_i \in Q}{\# \text{ of all points}} \approx \frac{\text{vol}(Q)}{\text{vol}([0,1]^m)}$$

for as many rectangles as possible. This leads to the following definition:

Definition 2.15   (discrepancy)

The discrepancy of the point set $\{x_1, ..., x_N\}$ is

$$D_N := \sup_Q \left| \frac{\# \text{ of } x_i \in Q}{N} - \text{vol}(Q) \right|.$$

Analogously the variant D_N^* *(star discrepancy)* is obtained when the set of rectangles is restricted to those Q^*, for which one corner is the origin:

$$Q^* = \prod_{i=1}^{m} [0, y_i)$$

where $y \in \mathbb{R}^m$ denotes the corner diagonally opposite the origin.

The more evenly the points of a sequence are distributed, the closer the discrepancy D_N is to zero. Here D_N refers to the first N points of a sequence of points $(x_i), i \geq 1$. The discrepancies D_N and D_N^* satisfy ($\longrightarrow$ Exercise 2.12b)

$$D_N^* \leq D_N \leq 2^m D_N^*.$$

Table 2.1 Comparison of different convergence rates to zero

N	$\dfrac{1}{\sqrt{N}}$	$\sqrt{\dfrac{\log\log N}{N}}$	$\dfrac{\log N}{N}$	$\dfrac{(\log N)^2}{N}$	$\dfrac{(\log N)^3}{N}$
10^1	.31622777	.28879620	.23025851	.53018981	1.22080716
10^2	.10000000	.12357911	.04605170	.21207592	.97664572
10^3	.03162278	.04396186	.00690776	.04771708	.32961793
10^4	.01000000	.01490076	.00092103	.00848304	.07813166
10^5	.00316228	.00494315	.00011513	.00132547	.01526009
10^6	.00100000	.00162043	.00001382	.00019087	.00263694
10^7	.00031623	.00052725	.00000161	.00002598	.00041874
10^8	.00010000	.00017069	.00000018	.00000339	.00006251
10^9	.00003162	.00005506	.00000002	.00000043	.00000890

The discrepancy allows to find a deterministic bound on the error δ_N of the Monte Carlo integration,

$$|\delta_N| \leq v(f) D_N^*; \tag{2.16}$$

here $v(f)$ is the variation of the function f with $v(f) < \infty$, and the domain of integration is $\mathcal{D} = [0,1]^m$ [Ni92], [TW92], [MC94]. This result is known as Theorem of Koksma and Hlawka. The bound in (2.16) underlines the importance to find numbers $x_1, ..., x_N$ with small value of the discrepancy D_N. After all, a set of N *random* numbers satisfies

$$\mathsf{E}(D_N) = O\left(\sqrt{\frac{\log\log N}{N}}\right).$$

This is in accordance with the $O(N^{-1/2})$ law. The order of magnitude of these numbers is shown in Table 2.1 (third column).

Definition 2.16   (low-discrepancy point sequence)

A sequence of points or numbers $x_1, x_2, ..., x_N, ... \in \mathbb{R}^m$ is called low-discrepancy sequence if

$$D_N = O\left(\frac{(\log N)^m}{N}\right). \tag{2.17}$$

Deterministic sequences of numbers satisfying (2.17) are called *quasi*-random numbers, although they are fully deterministic. Table 2.1 reports on the orders of magnitude. Since $\log(N)$ grows only modestly, a low discrepancy essentially means $D_N \approx O(N^{-1})$ as long as m is not too large. The equation (2.17) expresses some dependence on the dimension m, contrary to Monte Carlo methods. But the dependence on m in (2.17) is far less stringent than with classical quadrature.

2.5.2 Examples of Low-Discrepancy Sequences

In the one-dimensional case ($m = 1$) the point set

$$x_i = \frac{2i - 1}{2N}, \quad i = 1, ..., N \tag{2.18}$$

has the value $D_N^* = \frac{1}{2N}$; this value can not be improved ($\longrightarrow$ Exercise 2.12c). The monotonous sequence (2.18) can be applied only when a reasonable N is known and fixed; for $N \to \infty$ the x_i would be newly placed and an integrand f evaluated again. Since N is large it is essential that the previously calculated results can be used when N is growing. This means that the points $x_1, x_2, ...$ must be placed "dynamically" so that they are preserved and the fineness improves when N grows. This is achieved by the sequence

$$\frac{1}{2}, \ \frac{1}{4}, \frac{3}{4}, \ \frac{1}{8}, \frac{5}{8}, \frac{3}{8}, \frac{7}{8}, \ \frac{1}{16}, \cdots$$

This sequence is known as van der Corput sequence. To motivate such a dynamical placing of points imagine that you are searching for some item in the interval $[0, 1]$ (or in the cube $[0, 1]^m$). The searching must be fast and successful, and is terminated as soon as the object is found. This defines N dynamically by the process.

The formula that defines the van der Corput sequence can be formulated as algorithm. We first study an example, say, $x_6 = \frac{3}{8}$. The index $i = 6$ is written as binary number

$$6 = (110)_2 =: (d_2 \ d_1 \ d_0)_2 \quad \text{with} \quad d_i \in \{0, 1\}.$$

Then reverse the binary digits and put the radix point in front of the sequence:

$$(. \, d_0 \, d_1 \, d_2)_2 = \frac{d_0}{2} + \frac{d_1}{2^2} + \frac{d_3}{2^3} = \frac{1}{2^2} + \frac{1}{2^3} = \frac{3}{8}$$

If this is done for all indices $i = 1, 2, 3, \dots$ the van der Corput sequence $x_1, x_2, x_3, \dots$ results. These numbers can be defined with the following function:

Definition 2.17   (radical-inverse function)
For $i = 1, 2, \dots$ let

$$i = \sum_{k=0}^{j} d_k b^k$$

be the expansion in base b (integer ≥ 2), with digits $d_k \in \{0, 1, \dots, b-1\}$. Then the radical-inverse function is defined by

$$\phi_b(i) := \sum_{k=0}^{j} d_k b^{-k-1} \, .$$

The function $\phi_b(i)$ is the digit-reversed fraction of i. This mapping may be seen as reflecting with respect to the radix point. To each index i a rational number in the interval $0 < x < 1$ is assigned. Every time the number of digits j increases by one, the mesh becomes finer by a factor $1/b$. This means that the algorithm fills all mesh points on the sequence of meshes with increasing fineness ($\longrightarrow$ Exercise 2.13). The above classical van der Corput sequence is obtained by

$$x_i := \phi_2(i).$$

The radical-inverse function can be applied to construct points x_i in the m-dimensional cube $[0, 1]^m$. The simplest construction is the Halton sequence.

Definition 2.18   (Halton sequence)
Let $p_1, \dots, p_m$ be pairwise prime integers. The Halton sequence is defined as the sequence of vectors

$$x_i := (\phi_{p_1}(i), \dots, \phi_{p_m}(i)), \quad i = 1, 2, \dots$$

Usually one takes $p_1, \dots, p_m$ to be the first m prime numbers. Figure 2.9 shows for $m = 2$ and $p_1 = 2$, $p_2 = 3$ the first 10000 Halton points. Compared to the pseudo-random points of Figure 2.3, the Halton points are distributed more evenly.

Further sequences were developed by Sobol, Faure and Niederreiter, see [Ni92], [MC94], [PTVF92]. All these sequences are of low discrepancy, with

$$N \cdot D_N^* \leq C_m (\log N)^m + O\left((\log N)^{m-1}\right).$$

The Table 2.1 shows how fast the relevant terms $(\log N)^m / N$ tend to zero. If m is large, extremely large values of the denominator N are needed before the terms become small. But it is assumed that the bounds are unrealistically

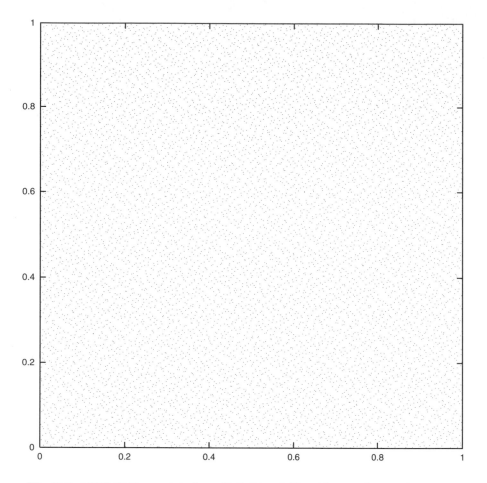

Fig. 2.9. 10000 Halton points from Definition 2.18, with $p_1 = 2$, $p_2 = 3$

large and overestimate the real error. For the Halton sequence in case $m = 2$ the constant is $C_2 = 0.2602$.

Quasi Monte Carlo (QMC) methods approximate the integrals with the arithmetic mean θ_N of (2.13), but use low-discrepancy numbers x_i instead of random numbers. QMC is a *deterministic* method. Practical experience with low-discrepancy sequences are better than might be expected from the bounds known so far. This also holds for the bound (2.16) by Koksma and Hlawka; apparently a large class of functions f satisfy $|\delta_N| \ll v(f)D_N^*$, see [SM94].

Notes and Comments

on Section 2.1:

The linear congruential method is sometimes called Lehmer generator. Easily accessible and popular generators are RAN1 and RAN2 from [PTVF92]. Further references on linear congruential generators are [Ma68], [Ri87], [Ni92], [LEc99]. Nonlinear congruential generators are of the form

$$N_i = f(N_{i-1}) \mod M.$$

Hints on the algorithmic implementation are found in [Ge98]. Generally it is advisable to run the generator in integer arithmetic in order to avoid rounding errors that may spoil the period, see [Lehn02]. For Fibonacci generators we refer to [Br94]. The version of (2.5) is a subtractive generator. Additive versions (with a plus sign instead of the minus sign) are used as well [Kn95], [Ge98]. The codes in [PTVF92] are recommendable. A truely remarkably long period is provided by the Mersenne twister [MaN98]. For simple statistical tests with illustrations see [Hig04].

There are multiplicative Fibonacci generators of the form

$$N_{i+1} := N_{i-\nu} N_{i-\mu} \mod M.$$

Hints on parallelization are given in [Mas99]. For example, parallel Fibonacci generators are obtained by different initializing sequences. Note that computer systems and software packages often provide built-in random number generators.

on Sections 2.2, 2.3:

The inversion result of Theorem 2.8 can be formulated placing less or no restrictions on F, see [Ri87], p. 59, [Dev86], p. 28, or [La99], p. 270. There are numerous other methods to calculate normal and nonnormal variates; for a detailed overview with many references see [Dev86]. The Box-Muller approach was suggested in [BoM58]. Marsaglia's modification was published in a report quoted in [MaB64]. For simulating Lévy processes, see [ConT04].

on Section 2.4:

The bounds on errors of the Monte Carlo integration refer to arbitrary functions f; for smooth functions better bounds can be expected. In the one-dimensional case the variation is defined as the supremum of $\sum_j |f(t_j) - f(t_{j-1})|$ over all partitions, see Section 1.6.2. This definition can be generalized to higher-dimensional cases. A thorough discussion is [Ni78], [Ni92].

An advanced application of Monte Carlo integration uses one or more methods of *reduction of variance*, which allows to improve the accuracy in many cases [HH64], [Ru81], [Ni92], [PTVF92], [Fi96], [Kwok98], [La99]. For example, the integration domain can be split into subsets *(stratified sampling)*

[RiW02]. Another technique is used when for a *control variate* g with $g \approx f$ the exact integral is known. Then f is replaced by $(f-g)+g$ and Monte Carlo integration is applied to $f-g$. Another alternative, the method of *antithetic variates*, will be described in Section 3.5 together with the control-variate technique.

on Section 2.5:

Besides the supremum discrepancy of Definition 2.15 the $\mathcal{L}^2$-analogy of an integral version is used. Hints on speed and preliminary comparison are found in [MC94]. For application on high-dimensional integrals see [PT95]. For large values of the dimension m, the error (2.17) takes large values, which might suggest to discard its use. But the notion of an *effective dimension* and practical results give a favorable picture at least for CMO applications of order $m = 360$ [CaMO97]. The error bound of Koksma and Hlawka (2.16) is not necessarily recommendable for practical use, see the discussion in [SM94]. The analogy of the equidistant lattice in (2.18) in higher-dimensional space has unfavorable values of the discrepancy, $D_N = O\left(\frac{1}{\sqrt[m]{N}}\right)$. For $m > 2$ this is worse than Monte Carlo, compare [Ri87]. — Monte Carlo does not take advantage of smoothness of integrands. In the case of smooth integrands, sparse-grid approaches are highly competitive. These most refined quadrature methods break the *curse of the dimension*, see [GeG98], [GeG03].

Van der Corput sequences can be based also on other bases. Computer programs that generate low-discrepancy numbers are available. For example, Sobol numbers are calculated in [PTVF92] and Sobol- and Faure numbers in the computer program FINDER [PT95] and in [Te95]. At the current state of the art it is open which point set has the smallest discrepancy in the m-dimensional cube. There are generalized Niederreiter sequences, which include Sobol- and Faure sequences as special cases [Te95]. In several applications deterministic Monte Carlo seems to be superior to stochastic Monte Carlo [PT96]. A comparison based on finance applications has shown good performance of Sobol numbers [Gla04]. Chapter 5 in [Gla04] provides more discussion and many references.

Besides volume integration, Monte Carlo is needed to integrate over possibly high-dimensional probability distributions. Drawing samples from the required distribution can be done by running a cleverly constructed Markov chain. This kind of method is called Markov Chain Monte Carlo (MCMC). That is, a chain of random variables $X_0, X_1, X_2, \ldots$ is constructed where for given X_j the next state X_{j+1} does not depend on the history of the chain $X_0, X_1, X_2, \ldots, X_{j-1}$. By suitable construction criteria, convergence to any chosen target distribution is obtained. For MCMC we refer to the literature, for example to [GiRS96], [La99], [Beh00], [Tsay02], [Häg02].

Exercises

Exercise 2.1

Consider the random number generator $N_i = 2N_{i-1} \mod 11$. For $(N_{i-1}, N_i) \in \{0, 1, ..., 10\}^2$ and integer tupels with $z_0 + 2z_1 = 0 \mod 11$ the equation

$$z_0 N_{i-1} + z_1 N_i = 0 \mod 11$$

defines families of parallel straight lines, on which all points (N_{i-1}, N_i) lie. These straight lines are to be analyzed. For which of the families of parallel straight lines are the gaps maximal?

Exercise 2.2   Deficient Random Number Generator

For some time the generator

$$N_i = aN_{i-1} \mod M, \quad \text{with } a = 2^{16} + 3, \ M = 2^{31}$$

was in wide use. Show for the sequence $U_i := N_i/M$

$$U_{i+2} - 6U_{i+1} + 9U_i \quad \text{is integer!}$$

What does this imply for the distribution of the tripels (U_i, U_{i+1}, U_{i+2}) in the unit cube?

Exercise 2.3   Lattice of the Linear Congruential Generator

a) Show by induction over j

$$N_{i+j} - N_j = a^j (N_i - N_0) \mod M$$

b) Show for integer $z_0, z_1, ..., z_{m-1}$

$$\begin{pmatrix} N_i \\ N_{i+1} \\ \vdots \\ N_{i+m-1} \end{pmatrix} - \begin{pmatrix} N_0 \\ N_1 \\ \vdots \\ N_{m-1} \end{pmatrix} = (N_i - N_0) \begin{pmatrix} 1 \\ a \\ \vdots \\ a^{m-1} \end{pmatrix} + M \begin{pmatrix} z_0 \\ z_1 \\ \vdots \\ z_{m-1} \end{pmatrix}$$

$$= \begin{pmatrix} 1 & 0 & \cdots & 0 \\ a & M & \cdots & 0 \\ \vdots & \vdots & \ddots & \vdots \\ a^{m-1} & 0 & \cdots & M \end{pmatrix} \begin{pmatrix} z_0 \\ z_1 \\ \vdots \\ z_{m-1} \end{pmatrix}$$

Exercise 2.4   Coarse Approximation of Normal Deviates

Let $U_1, U_2, ...$ be independent random numbers $\sim \mathcal{U}[0, 1]$, and

$$X_k := \sum_{i=k}^{k+11} U_i - 6.$$

Calculate mean and variance of the X_k.

Exercise 2.5   Cauchy-Distributed Random Numbers

A Cauchy-distributed random variable has the density function

$$f_c(x) := \frac{c}{\pi} \frac{1}{c^2 + x^2} .$$

Show that its distribution function F_c and its inverse F_c^{-1} are

$$F_c(x) = \frac{1}{\pi} \arctan \frac{x}{c} + \frac{1}{2} \quad , \quad F_c^{-1}(y) = c \tan(\pi(y - \frac{1}{2})) .$$

How can this be used to generate Cauchy-distributed random numbers out of uniform deviates?

Exercise 2.6   Inverting the Normal Distribution

Suppose $F(x)$ is the standard normal distribution function. Construct a rough approximation $G(u)$ to $F^{-1}(u)$ for $0.5 \le u < 1$ as follows:

a) Construct a rational function $G(u)$ ($\longrightarrow$ Appendix C1) with correct asymptotic behavior, point symmetry with respect to $(u, x) = (0.5, 0)$, using only one parameter.
b) Fix the parameter by interpolating a given point $(x_1, F(x_1))$.
c) What is a simple criterion for the error of the approximation?

Exercise 2.7   Uniform Distribution

Let U_1 and U_2 be uniformly distributed random variables on $[-1, 1]$. For U_1, U_2 with $U_1^2 + U_2^2 < 1$ consider the transformation

$$\begin{pmatrix} X_1 \\ X_2 \end{pmatrix} = \begin{pmatrix} U_1^2 + U_2^2 \\ \frac{1}{2\pi} \arctan(U_2/U_1) \end{pmatrix}$$

(main branch of arctan). Show that X_1 and X_2 are uniform deviates.

Exercise 2.8   Programming Assignment: Normal Deviates

a) Write a computer program that implements the *Fibonacci generator*

$$U_i := U_{i-17} - U_{i-5}$$
$$U_i := U_i + 1 \text{ in case } U_i < 0$$

in the form of Algorithm 2.7.
Tests: Visual inspection of 10000 points in the unit square.
b) Write a computer program that implements *Marsaglia's Polar Algorithm*. Use the uniform deviates from a).

Tests:

1.) For a sample of 5000 points calculate estimates of mean and variance.
2.) For the discretized SDE

$$\Delta x = 0.1\Delta t + Z\sqrt{\Delta t}, \quad Z \sim \mathcal{N}(0,1)$$

calculate some trajectories for $0 \le t \le 1$, $\Delta t = 0.01$, $x_0 = 0$.

Exercise 2.9   Correlated Distributions

Suppose we need a two-dimensional random variable (X_1, X_2) that must be normally distributed with mean 0, and given variances σ_1^2, σ_2^2 and prespecified correlation ρ. How is X_1, X_2 obtained out of $Z_1, Z_2 \sim \mathcal{N}(0,1)$?

Exercise 2.10   Error of the Monte Carlo Integration

The domain for integration is $Q = [0,1]^m$. For

$$\Theta_N := \frac{1}{N}\sum_{i=1}^{N} f(x_i), \quad \mathsf{E}(f) := \int f \, dx, \quad g := f - \mathsf{E}(f)$$

and $\sigma^2(f)$ from (2.14b) show
a) $\mathsf{E}(g) = 0$
b) $\sigma^2(g) = \sigma^2(f)$
c) $\sigma^2(\delta_N) = \mathsf{E}(\delta_N^2) = \frac{1}{N^2}\int(\sum g(x_i))^2 dx = \frac{1}{N}\sigma^2(f)$
  Hint on (c): When the random points x_i are i.i.d. (independent identical distributed), then also $f(x_i)$ and $g(x_i)$ are i.i.d. A consequence is $\int g(x_i)g(x_j) = 0$ for $i \ne j$.

Exercise 2.11   Experiment on Monte Carlo Integration

To approximate the integral

$$\int_0^1 f(x)dx$$

calculate a Monte Carlo sum

$$\frac{1}{N}\sum_{i=1}^{N} f(x_i)$$

for $f(x) = 5x^4$ and, for example, $N = 100000$ random numbers $x_i \sim \mathcal{U}[0,1]$. The absolute error behaves like $cN^{-1/2}$. Compare the approximation with the exact integral for several N and seeds to obtain an estimate of c.

Exercise 2.12   Bounds on the Discrepancy

(Compare Definition 2.15) Show
a) $0 \le D_N \le 1$,
b) $D_N^* \le D_N \le 2^m D_N^*$ (show this at least for $m \le 2$),
c) $D_N^* \ge \frac{1}{2N}$ for $m = 1$.

Exercise 2.13   Algorithm for the Radical-Inverse Function

Use the idea

$$i = \left(d_k b^{k-1} + \ldots + d_1\right) b + d_0$$

to formulate an algorithm that obtains $d_0, d_1, \ldots, d_k$ by repeated divison by b. Reformulate $\phi_b(i)$ from Definition 2.17 into the form $\phi_b(i) = z/b^{j+1}$ such that the result is represented as rational number. The numerator z should be calculated in the same loop that establishes the digits $d_0, \ldots, d_k$.

Exercise 2.14   Testing the Distribution

Let X be a random variate with density f and let $a_1 < a_2 < \ldots < a_l$ define a partition of the support of f into subintervals, including the unbounded intervals $x < a_1$ and $x > a_l$. Recall from (B1.1), (B1.2) that the probability of a realization of X falling into $a_k \leq x < a_{k+1}$ is given by

$$p_k := \int\limits_{a_k}^{a_{k+1}} f(x)dx \ , \quad k = 1, 2, \ldots, l - 1 \ ,$$

which can be approximated by $(a_{k+1} - a_k)f\left(\frac{a_k + a_{k+1}}{2}\right)$. Perform a sample of j realizations $x_1, \ldots, x_j$ of a random number generator, and denote j_k the number of samples falling into $a_k \leq x < a_{k+1}$. For normal variates with density f from (B1.9) design an algorithm that performs a simple statistical test of the quality of the $x_1, \ldots, x_j$.

Hints: See Section 2.1 for the special case of uniform variates. Argue for what choices of a_1 and a_l the probabilities p_0 and p_l may be neglected. Think about a reasonable relation between l and j.

Exercise 2.15   Quality of Fibonacci-Generated Numbers

Analyze and visualize the planes in the unit cube, on which all points fall that are generated by the Fibonacci recursion

$$U_{i+1} := (U_i + U_{i-1}) \bmod 1 \ .$$

Chapter 3 Simulation
with Stochastic Differential Equations

This chapter provides an introduction into the numerical integration of stochastic differential equations (SDEs). Again X_t denotes a stochastic process and solution of an SDE,

$$dX_t = a(X_t, t)dt + b(X_t, t)dW_t \quad \text{for } 0 \le t \le T,$$

where the driving process W is a Wiener process. The solution of a discrete version of the SDE is denoted y_j. That is, y_j should be an approximation to X_{t_j}, or y_t an approximation to X_t. From Section 1.7 we know the Euler discretization from Algorithm 1.11

$$\begin{cases} y_{j+1} = y_j + a(y_j, t_j)\Delta t + b(y_j, t_j)\Delta W_j, \quad t_j = j\Delta t, \\ \Delta W_j = W_{t_{j+1}} - W_{t_j} = Z\sqrt{\Delta t} \quad \text{with } Z \sim \mathcal{N}(0,1). \end{cases} \qquad (3.1)$$

The step length Δt is assumed equidistant. As is common usage in numerical analysis, we also use the h-notation, $h := \Delta t$. For $\Delta t = h = T/m$ the index j in (3.1) runs from 0 to $m-1$. The initial value for $t = 0$ is assumed a given constant,

$$y_0 = X_0 .$$

From numerical methods for deterministic ODEs ($b \equiv 0$) we know the discretization error of Euler's method. Its order is $O(h)$,

$$X_T - y_T = O(h).$$

The Algorithm 1.11 (repeated in equation (3.1)) is an *explicit* method in that in every step $j \to j+1$ the values of a and b are evaluated at the previous approximation (y_j, t_j). Evaluating b at the left-hand mesh point (y_j, t_j) is consistent with the Itô integral and the Itô process, compare the notes at the end of Chapter 1.

After we have seen in Chapter 2 how $Z \sim \mathcal{N}(0,1)$ can be calculated, all elements of Algorithm 1.11 are known and we are equipped with a first method to numerically integrate an SDE ($\longrightarrow$ Exercise 3.1). In this chapter we learn about other methods and discuss the accuracy of numerical solutions of SDEs. The exposition of Sections 3.1 through 3.3 follows [KP92].

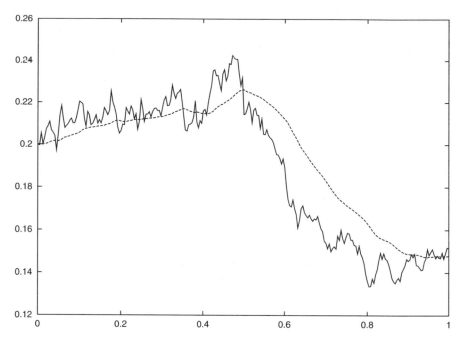

Fig. 3.1. Example 1.15, $\alpha = 0.3$, $\beta = 10$, $\sigma_0 = \zeta_0 = 0.2$, realization of the volatility tandem σ_t, ζ_t (dashed) for $0 \leq t \leq 1$, $\Delta t = 0.004$

3.1 Approximation Error

To study the accuracy of numerical approximations, we choose the example of a linear SDE

$$dX_t = \alpha X_t \, dt + \beta X_t \, dW_t, \quad \text{initial value } X_0 \text{ for } t = 0.$$

For this equation with constant coefficients α, β we derived in Section 1.8 the analytical solution

$$X_t = X_0 \exp \left(\left(\alpha - \tfrac{1}{2} \beta^2 \right) t + \beta W_t \right). \tag{3.2}$$

For a given realization of the Wiener process W_t we obtain as solution a trajectory *(sample path)* X_t. For another realization of the Wiener process the same theoretical solution (3.2) takes other values. If a Wiener process W_t is given, we call a solution X_t of the SDE a *strong solution*. In this sense the solution in (3.2) is a strong solution. If one is free to select a Wiener process, then a solution of the SDE is called *weak solution*. For a weak solution, only the distribution of X is of interest, not its path.

Assuming an identical sample path of a Wiener process for the SDE and for the numerical approximation, a pathwise comparison of the trajectories

X_t of (3.2) and y from (3.1) is possible for t_j. For example, for $t_m = T$ the absolute error for a given Wiener process is $|X_T - y_T|$. Since the approximation y_T also depends on the chosen step length h, we also write y_T^h. For another Wiener process the error is somewhat different. We average the error over "all" sample paths of the Wiener process:

Definition 3.1   (absolute error)
    The absolute error at T is $\epsilon(h) := \mathsf{E}(|X_T - y_T^h|)$.

In practice we represent the set of all sample paths of a Wiener process by N different simulations.

Example 3.2 $X_0 = 50$, $\alpha = 0.06$, $\beta = 0.3$, $T = 1$.
    We want to investigate experimentally how the absolute error depends on h. Starting with a first choice of h we calculate $N = 50$ simulations and for each realization the values of X_T and y_T —that is $X_{T,k}$, $y_{T,k}$ for $k = 1, ..., N$. Again: to obtain pairs of comparable trajectories, also the theoretical solution (3.2) is fed with the same Wiener process from (3.1). Then we calculate the estimate $\widehat{\epsilon}$ of the absolute error ϵ,

$$\widehat{\epsilon}(h) := \frac{1}{N} \sum_{k=1}^{N} |X_{T,k} - y_{T,k}^h|.$$

Such an experiment was performed for five values of h. In this way the first series of results were obtained (first line in Table 3.1). Such a series of experiments was repeated twice, using other seeds. As Table 3.1 shows, $\widehat{\epsilon}(h)$ decreases with decreasing h but slower than one would expect from the behavior of the Euler method applied to deterministic differential equations. We might determine the order by fitting the values of the table. To speed up, let us test the order $O(h^{1/2})$. For this purpose divide each $\widehat{\epsilon}(h)$ of the table by the corresponding $h^{1/2}$. This shows that the order $O(h^{1/2})$ is correct, because each entry of the table leads essentially to the same constant value, here 2.8. Apparently this example satisfies $\widehat{\epsilon}(h) \approx 2.8 \, h^{1/2}$. For another example we expect a different constant.

Table 3.1. Results of Example 3.2

Table of the $\widehat{\epsilon}(h)$	$h = 0.01$	$h = 0.005$	$h = 0.002$	$h = 0.001$	$h = 0.0005$
series 1 (with seed$_1$)	0.2825	0.183	0.143	0.089	0.070
series 2 (with seed$_2$)	0.2618	0.195	0.126	0.069	0.062
series 3 (with seed$_3$)	0.2835	0.176	0.116	0.096	0.065

These results obtained for the estimates $\widehat{\epsilon}$ are assumed to be valid for ϵ. This leads to postulate

$$\epsilon(h) \le c\,h^{1/2} = O(h^{1/2}).$$

The order of convergence is worse than the order $O(h)$, which Euler's method (3.1) achieves for deterministic differential equations ($b \equiv 0$). But in view of (1.28), $(dW)^2 = h$, the order $O(h^{1/2})$ is no surprise. For a proof of the order, see [KP92].

Definition 3.3   (strong convergence)

   y_T^h converges strongly to X_T with order $\gamma > 0$,
   if $\epsilon(h) = \mathsf{E}(|X_T - y_T^h|) = O(h^\gamma)$.
   y_T^h converges strongly, if

$$\lim_{h \to 0} \mathsf{E}(|X_T - y_T^h|) = 0.$$

Hence the Euler method applied to SDEs converges strongly with order $1/2$. Note that convergence refers to fixed finite intervals, here for a fixed value T. For long-time integration ($T \to \infty$), see the notes at the end of this chapter.

   Strongly convergent methods are appropriate when the trajectory itself is of interest. This was the case for Figures 1.16 and 1.17. Often the pointwise approximation of X_t is not our real aim but only an intermediate result in the effort to calculate a **moment**. For example, many applications in finance need to approximate $\mathsf{E}(X_T)$. A first conclusion from this situation is that of all calculated y_i only the last is required, namely y_T. A second conclusion is that for the expectation a single sample value of y_T is of little interest. The same holds true if the ultimate interest is $\mathsf{Var}(X_T)$ rather than X_T. In this situation the primary interest is not strong convergence with the demanding requirement $y_T \approx X_T$ and even less $y_t \approx X_t$ for $t < T$. Instead the concern is the weaker requirement to approximate moments or other functionals of X_T. The aim is to achieve $\mathsf{E}(y_T) \approx \mathsf{E}(X_T)$, or $\mathsf{E}(|y_T|^q) \approx \mathsf{E}(|X_T|^q)$, or more general $\mathsf{E}(g(y_T)) \approx \mathsf{E}(g(X_T))$ for an appropriate function g.

Definition 3.4   (weak convergence)

   y_T^h converges weakly to X_T with respect to g with order $\beta > 0$,
   if $|\mathsf{E}(g(X_T)) - \mathsf{E}(g(y_T^h))| = O(h^\beta)$.

The Euler scheme is weakly $O(h^1)$ convergent with respect to all polynomials g provided the coefficient functions a and b are four times continuously differentiable ([KP92], Chapter 14). For the special polynomial $g(x) = x$, (B1.4) implies convergence of the mean $\mathsf{E}(x)$. For $g(x) = x^2$ the relation $\mathsf{Var}(X) = \mathsf{E}(X^2) - (\mathsf{E}(X))^2$ implies convergence of the variance (the reader may check). Proceeding in this way implies weak convergence with respect to all moments.

Since the properties of the integrals on which expectation is based lead to

$$|\mathsf{E}(X) - \mathsf{E}(Y)| = |\mathsf{E}(X - Y)| \le \mathsf{E}(|X - Y|),$$

we confirm that strong convergence implies weak convergence with respect to $g(x) = x$.

When weakly convergent methods are evaluated, the increments ΔW can be replaced by other random variables $\widehat{\Delta W}$ that have the same expectation and variance. If the replacing random variables are easier to evaluate, costs can be saved significantly.

3.2 Stochastic Taylor Expansion

The derivation of algorithms for the integration of SDEs is based on stochastic Taylor expansions. To facilitate the understanding of stochastic Taylor expansions we confine ourselves to the scalar and autonomous[1] case, and first introduce the terminology by means of the deterministic case. That is, we begin with $\frac{d}{dt}X_t = a(X_t)$. The chain rule for arbitrary $f \in C^1(\mathbb{R})$ is

$$\frac{d}{dt}f(X_t) = a(X_t)\frac{\partial}{\partial x}f(X_t) =: Lf(X_t).$$

With the linear operator L this rule in integral form is

$$f(X_t) = f(X_{t_0}) + \int_{t_0}^{t} Lf(X_s)ds. \tag{3.3}$$

This version is resubstituted for the integrand $\tilde{f}(X_s) := Lf(X_s)$, which requires at least $f \in C^2$, and gives the term in braces:

$$\begin{aligned}
f(X_t) =& f(X_{t_0}) + \int_{t_0}^{t} \left\{ \tilde{f}(X_{t_0}) + \int_{t_0}^{s} L\tilde{f}(X_z)dz \right\} ds \\
=& f(X_{t_0}) + \tilde{f}(X_{t_0}) \int_{t_0}^{t} ds + \int_{t_0}^{t} \int_{t_0}^{s} L\tilde{f}(X_z)dzds \\
=& f(X_{t_0}) + Lf(X_{t_0})(t - t_0) + \int_{t_0}^{t} \int_{t_0}^{s} L^2 f(X_z)dzds
\end{aligned}$$

This version of the Taylor expansion consists of two terms and the remainder as double integral. To get the next term of the second-order derivative, apply (3.3) for $L^2 f(X_z)$, and split off the term

---

[1] An *autonomous* differential equation does not explicitly depend on the independent variable, here $a(X_t)$ rather than $a(X_t, t)$. The standard Model 1.13 of the stock market is autonomous for constant μ and σ.

$$L^2 f(X_{t_0}) \int_{t_0}^{t} \int_{t_0}^{s} dz\,ds = L^2 f(X_{t_0}) \frac{1}{2}(t - t_0)^2$$

from the remainder double integral. At this stage, the remainder is a triple integral. This procedure is repeated to obtain the Taylor formula in integral form. Each further step requires more differentiability of f.

We now devote our attention to the stochastic case and investigate the *Itô-Taylor expansion* of the autonomous scalar SDE $dX_t = a(X_t)dt + b(X_t)dW_t$. Itô's Lemma for $g(x,t) := f(x)$ is

$$df(X_t) = \underbrace{\left\{ a\frac{\partial}{\partial x}f(X_t) + \frac{1}{2}b^2\frac{\partial^2}{\partial x^2}f(X_t) \right\}}_{=:L^0 f(X_t)} dt + \underbrace{b\frac{\partial}{\partial x}f(X_t)}_{=:L^1 f(X_t)} dW_t,$$

or in integral form

$$f(X_t) = f(X_{t_0}) + \int_{t_0}^{t} L^0 f(X_s)ds + \int_{t_0}^{t} L^1 f(X_s)dW_s. \qquad (3.4)$$

This SDE will be applied for different choices of f. Specifically for $f(x) \equiv x$ the SDE (3.4) recovers the original SDE

$$X_t = X_{t_0} + \int_{t_0}^{t} a(X_s)ds + \int_{t_0}^{t} b(X_s)dW_s . \qquad (3.5)$$

As first applications of (3.4) we substitute $f = a$ and $f = b$. The resulting versions of (3.4) are substituted in (3.5) leading to

$$X_t = X_{t_0} + \int_{t_0}^{t} \left\{ a(X_{t_0}) + \int_{t_0}^{s} L^0 a(X_z)dz + \int_{t_0}^{s} L^1 a(X_z)dW_z \right\} ds$$
$$+ \int_{t_0}^{t} \left\{ b(X_{t_0}) + \int_{t_0}^{s} L^0 b(X_z)dz + \int_{t_0}^{s} L^1 b(X_z)dW_z \right\} dW_s$$

with

$$L^0 a = aa' + \frac{1}{2}b^2 a'' \qquad L^0 b = ab' + \frac{1}{2}b^2 b''$$
$$L^1 a = ba' \qquad\qquad L^1 b = bb'. \qquad (3.6)$$

Summarizing the four double integrals into one remainder expression R, we have

$$X_t = X_{t_0} + a(X_{t_0}) \int_{t_0}^{t} ds + b(X_{t_0}) \int_{t_0}^{t} dW_s + R , \qquad (3.7a)$$

with

$$R = \int_{t_0}^{t} \int_{t_0}^{s} L^0 a(X_z) dz ds + \int_{t_0}^{t} \int_{t_0}^{s} L^1 a(X_z) dW_z ds$$
$$+ \int_{t_0}^{t} \int_{t_0}^{s} L^0 b(X_z) dz dW_s + \int_{t_0}^{t} \int_{t_0}^{s} L^1 b(X_z) dW_z dW_s. \qquad (3.7b)$$

The order of the terms is limited by the number of repeated integrations. In view of (1.28), $dW^2 = dt$, we expect the last of the integrals in (3.7b) to be of first order (and show this below).

In an analogous fashion the integrands in (3.7b) can be replaced using (3.4) with appropriately chosen f. In this way triple integrals occur. We illustrate this for the integral on $f = L^1 b$, which is the double integral of lowest order. The non-integral term of (3.4) allows to split off another "ground integral" with constant integrand,

$$R = L^1 b(X_{t_0}) \int_{t_0}^{t} \int_{t_0}^{s} dW_z dW_s + \tilde{R} .$$

In view of (3.6) and (3.7a) this result can be summarized as

$$X_t = X_{t_0} + a(X_{t_0}) \int_{t_0}^{t} ds + b(X_{t_0}) \int_{t_0}^{t} dW_s + b(X_{t_0}) b'(X_{t_0}) \int_{t_0}^{t} \int_{t_0}^{s} dW_z dW_s + \tilde{R}.$$
$$(3.8)$$

A general treatment of the Itô-Taylor expansion with an appropriate formalism is found in [KP92].

The next step is to formulate numerical algorithms out of the equations derived by the stochastic Taylor expansion. To this end the integrals must be solved. For (3.8) we need a solution of the double integral. For $X_t = W_t$ the Itô Lemma with $a = 0$, $b = 1$ and $y = g(x) := x^2$ leads to the equation $d(W_t^2) = dt + 2W_t dW_t$. Specifically for $t_0 = 0$ this is the equation

$$\int_{0}^{t} \int_{0}^{s} dW_z dW_s = \int_{0}^{t} W_s dW_s = \tfrac{1}{2} W_t^2 - \tfrac{1}{2} t. \qquad (3.9)$$

Another derivation of (3.9) uses

$$\sum_{j=1}^{n} W_{t_j} (W_{t_{j+1}} - W_{t_j}) = \tfrac{1}{2} W_t^2 - \tfrac{1}{2} \sum_{j=1}^{n} (W_{t_{j+1}} - W_{t_j})^2$$

for $t = t_{n+1}$ and $t_1 = 0$, and takes the limit in the mean on both sides ($\longrightarrow$ Exercise 3.2). The general version of (3.9) needed for (3.8) is

$$\int_{t_0}^{t} W_s dW_s = \tfrac{1}{2} (W_t - W_{t_0})^2 - \tfrac{1}{2} (t - t_0).$$

With $\Delta t := t - t_0$ and the random variable $\Delta W_t := W_t - W_{t_0}$ this is rewritten as

$$\int_{t_0}^{t} \int_{t_0}^{s} dW_z \, dW_s = \tfrac{1}{2} (\Delta W_t)^2 - \tfrac{1}{2} \Delta t. \tag{3.10}$$

Since this double integral is of order Δt, it completes the list of first-order terms. Also the three other double integrals

$$\int_{t_0}^{t} \int_{t_0}^{s} dz \, ds \quad , \quad \int_{t_0}^{t} \int_{t_0}^{s} dW_z \, ds \quad , \quad \int_{t_0}^{t} \int_{t_0}^{s} dz \, dW_s$$

are needed for the construction of higher-order numerical methods. The first integral is elementary, second order and not stochastic. The two others depend on each other via the equation

$$\int_{t_0}^{t} \int_{t_0}^{s} dz \, dW_s + \int_{t_0}^{t} \int_{t_0}^{s} dW_z \, ds = \int_{t_0}^{t} dW_s \int_{t_0}^{t} ds \tag{3.11}$$

($\longrightarrow$ Exercise 3.3). This indicates that the two remaining double integrals are of order 1.5. We will return to these integrals in the following section.

3.3 Examples of Numerical Methods

Now we apply the stochastic Taylor expansion to construct numerical methods for SDEs. First we check how Euler's method (3.1) evolves. Here we evaluate the integrals in (3.7a) and substitute

$$t_0 \to t_j, \quad t \to t_{j+1} = t_j + \Delta t.$$

This leads to

$$X_{t_{j+1}} = X_{t_j} + a(X_{t_j}) \Delta t + b(X_{t_j}) \Delta W_j + R.$$

After neglecting the remainder R the Euler scheme of (3.1) results, here for autonomous SDEs.

To obtain higher-order methods, further terms of the stochastic Taylor expansions are added. We may expect a "repair" of the half-order $O(\sqrt{\Delta t})$ by including the lowest-order double integral of (3.8), which is calculated in (3.10). The resulting correction term, after multiplying with bb', is added to the Euler scheme. Discarding the remainder $\widetilde{R}$, an algorithm results, which is due to Milstein (1974).

Algorithm 3.5   (Milstein)

> *Start:* $t_0 = 0,\ y_0 = X_0,\ W_0 = 0,\ \Delta t = T/m$
>
> *loop* $j = 0, 1, 2, ..., m - 1 :$
>
> $t_{j+1} = t_j + \Delta t$
>
> Calculate the values $a(y_j),\ b(y_j),\ b'(y_j)$
>
> $\Delta W = Z\sqrt{\Delta t}$ with $Z \sim \mathcal{N}(0, 1)$
>
> $y_{j+1} = y_j + a\Delta t + b\Delta W + \dfrac{1}{2}bb' \cdot ((\Delta W)^2 - \Delta t)$

This integration method by Milstein is strongly convergent with order one ($\longrightarrow$ Exercise 3.8). Adding the correction term has raised the strong convergence order of Euler's method to 1. The derivation of this order is based on the stochastic Taylor expansion. For the advanced calculus with stochastic integrals we refer to [KP92].

Runge-Kutta Methods

A disadvantage of the Taylor-expansion methods is the use of the derivatives $a',\ b',\ ...$ Analogously as with deterministic differential equations there is the alternative of Runge–Kutta–type methods, which only evaluate a or b for appropriate arguments.

As an example we discuss the factor bb' of Algorithm 3.5, and see how to replace it by an approximation. Starting from

$$b(y + \Delta y) - b(y) = b'(y)\Delta y + O((\Delta y)^2)$$

and using $\Delta y = a\Delta t + b\Delta W$ we deduce in view of (1.28) that

$$b(y + \Delta y) - b(y) = b'(y)(a\Delta t + b\Delta W) + O(\Delta t)$$
$$= b'(y)b(y)\Delta W + O(\Delta t).$$

Applying (1.28) again, we substitute $\Delta W = \sqrt{\Delta t}$ and arrive at an $O(\sqrt{\Delta t})$-approximation of the product bb', namely

$$\frac{1}{\sqrt{\Delta t}} \left(b[y_j + a(y_j)\Delta t + b(y_j)\sqrt{\Delta t}] - b(y_j) \right).$$

This expression is used in the Milstein scheme of Algorithm 3.5. The resulting variant

$$\hat{y} := y_j + a\Delta t + b\sqrt{\Delta t}$$
$$y_{j+1} = y_j + a\Delta t + b\Delta W + \frac{1}{2\sqrt{\Delta t}}(\Delta W^2 - \Delta t)[b(\hat{y}) - b(y_j)] \tag{3.12}$$

is a Runge-Kutta method, which also converges strongly with order one. Versions of these schemes for nonautonomous SDEs read analogously.

Taylor Scheme with Weak Second-Order Convergence.

Next we investigate the method that results when in the remainder term (3.7b) of all double integrals the ground integrals are split off. This is done by applying (3.4) for $f = L^0 a$, $f = L^1 a$, $f = L^0 b$, $f = L^1 b$. Then the new remainder $\tilde{R}$ consists of triple integrals. For $f = L^1 b$ this analysis was carried out at the end of Section 3.2. With (3.6) and (3.10) the correction term

$$bb' \frac{1}{2} \left((\Delta W)^2 - \Delta t \right)$$

resulted, which has lead to the strong convergence order one of the Milstein scheme. For $f = L^0 a$ the integral is not stochastic and the term

$$\left(aa' + \frac{1}{2} b^2 a'' \right) \frac{1}{2} \Delta t^2$$

is an immediate consequence. For $f = L^1 a$ and $f = L^0 b$ the integrals are again stochastic, namely

$$I_{(1,0)} := \int_{t_0}^{t} \int_{t_0}^{s} dW_z ds = \int_{t_0}^{t} (W_s - W_{t_0}) ds,$$

$$I_{(0,1)} := \int_{t_0}^{t} \int_{t_0}^{s} dz dW_s = \int_{t_0}^{t} (s - t_0) dW_s.$$

Summarizing all terms, the preliminary numerical scheme is

$$\begin{aligned} y_{j+1} = y_j &+ a\Delta t + b\Delta W + \frac{1}{2} bb' \left((\Delta W)^2 - \Delta t \right) \\ &+ \frac{1}{2} \left(aa' + \frac{1}{2} b^2 a'' \right) \Delta t^2 + ba' I_{(1,0)} + \left(ab' + \frac{1}{2} b^2 b'' \right) I_{(0,1)}. \end{aligned} \qquad (3.13)$$

It remains to approximate the two stochastic integrals $I_{(0,1)}$ and $I_{(1,0)}$. Setting $\Delta Y := I_{(1,0)}$ we have in view of (3.11)

$$I_{(0,1)} = \Delta W \Delta t - \Delta Y.$$

At this state the two stochastic double integrals $I_{(0,1)}$ and $I_{(1,0)}$ are expressed in terms of only one random variable ΔY, in addition to the variable ΔW used before. Since for weak convergence only the correct moments are needed, all occuring random variables (here ΔW and ΔY) can be replaced by other random variables with the same moments. The normally distributed random variable ΔY has expectation, variance and covariance

$$\mathsf{E}(\Delta Y) = 0, \;\; \mathsf{E}(\Delta Y^2) = \frac{1}{3} (\Delta t)^3, \;\; \mathsf{E}(\Delta Y \Delta W) = \frac{1}{2} (\Delta t)^2 \qquad (3.14)$$

($\longrightarrow$ Exercise 3.4). Such a random variable can be realized by two independent normally distributed variates Z_1 and Z_2,

$$\Delta Y = \frac{1}{2}(\Delta t)^{3/2}\left(Z_1 + \frac{1}{\sqrt{3}}Z_2\right)$$

$$\text{with } Z_i \sim \mathcal{N}(0,1), \quad i = 1,2 \tag{3.15}$$

($\longrightarrow$ Exercise 3.5). With this realization of ΔY we have approximations of $I_{(0,1)}$ and $I_{(1,0)}$, which are substituted into (3.13).

Next the random variable ΔW is replaced by other variates having the same moments. ΔW_j can be replaced by the simple approximation $\widehat{\Delta W}_j = \pm\sqrt{\Delta t}$, where both values have probability 1/2. Expectation and variance of $\widehat{\Delta W}$ and ΔW are the same: $\mathsf{E}(\widehat{\Delta W}) = 0$, $\mathsf{E}(\widehat{\Delta W}^2) = \Delta t$. For the numerical scheme (3.13) there is an even better approximation: Choosing $\widetilde{\Delta W}$ trivalued such that the two values $\pm\sqrt{3\Delta t}$ occur with probability 1/6, and the value 0 with probability 2/3, then the random variable $\widetilde{\Delta Y} := \frac{1}{2}\Delta t\,\widetilde{\Delta W}$ has up to terms of order $O(\Delta t^3)$ the moments in (3.14) ($\longrightarrow$ Exercise 3.6). As a consequence, the variant of (3.13)

$$\begin{aligned}
y_{j+1} = {}& y_j + a\Delta t + b\widetilde{\Delta W} + \frac{1}{2}bb'\left((\widetilde{\Delta W})^2 - \Delta t\right) \\
& + \frac{1}{2}\left(aa' + \frac{1}{2}b^2a''\right)\Delta t^2 + \frac{1}{2}\left(a'b + ab' + \frac{1}{2}b^2b''\right)\widetilde{\Delta W}\Delta t
\end{aligned} \tag{3.16}$$

is second-order weakly convergent.

Higher–Dimensional Cases

In higher-dimensional cases there are mixed terms. We distinguish two kinds of "higher–dimensional":

1.) $y \in \mathbb{R}^n$, a, $b \in \mathbb{R}^n$. Then, for instance, replace bb' by $\frac{\partial b}{\partial y}b$, where $\frac{\partial b}{\partial y}$ is the Jacobian matrix of all first-order partial derivatives.
2.) For multiple Wiener processes the situation is more complicated, because then simple explicit integrals as in (3.9) do not exist. Only the Euler scheme remains simple: for m Wiener processes the Euler scheme is

$$y_{j+1} = y_j + a\Delta t + b^{(1)}\Delta W^{(1)} + ... + b^{(m)}\Delta W^{(m)}.$$

The Figure 3.1 depicts two components of the system of Example 1.15.

3.4 Intermediate Values

Integration methods as discussed in the previous section calculate approximations y_j only at the grid points t_j. This leaves the question how to obtain intermediate values, namely approximations $y(t)$ for $t \neq t_j$. This situation is simple for deterministic ODEs. There we have in general smooth solutions, which suggests to construct an interpolation curve joining the calculated points (y_j, t_j).

A smooth interpolation is at variance with the stochastic nature of solutions of SDEs. When Δt is small, a linear interpolation matches the appearance of a stochastic process, and is easy to be carried out. Such an interpolating continuous polygon was used for the Figures 1.15 and 1.16. Another easily executable alternative would be to construct an interpolating step function with step length Δt.

The situation is different when the gaps between two calculated y_j and y_{j+1} are large. Then the *Brownian bridge* is a proper means to fill the gap [KS91], [RY91], [KP92], [Øk98], [Mo98]. For illustration assume that y_0 (for $t = 0$) and y_T (for $t = T$) are to be connected. Then the Brownian bridge is defined by

$$B_t = y_0 \left(1 - \frac{t}{T} \right) + y_T \frac{t}{T} + \left\{ W_t - \frac{t}{T} W_T \right\}.$$

The first two terms represent a straight-line connection between y_0 and y_T. This straight line stands for the trend. The term $W_t - \frac{t}{T} W_T$ describes the stochastic fluctuation. For its realization an appropriate volatility can be prescribed ($\longrightarrow$ Exercise 3.7).

Another alternative to fill large gaps is to apply fractal interpolation [Man99].

3.5 Monte Carlo Simulation

As pointed out in Section 2.4 in the context of calculating integrals, Monte Carlo is attractive in high-dimensional spaces. The same characterization holds when Monte Carlo is applied to the valuation of options. For sake of clarity we describe the approach in the one-dimensional context.

From Section 1.7.2 we take the one-factor model of a geometric Brownian motion of the asset price S_t,

$$\frac{dS}{S} = \mu \, dt + \sigma \, dW.$$

Here μ is the expected growth rate. When options are to be priced we assume a risk-neutral world and set $\mu = r$ (compare Section 1.7.3 and Remark 1.14). Recall the lognormal distribution of GBM, with density (1.47).

The Monte Carlo simulation of options can be seen in two ways. Either dynamically as a process of simulating numerous prices with subsequent appropriate valuation, or as the MC approximation of integrals. For the latter view we discuss the integral representation of options.

3.5.1 Integral Representation

In the one-period model of Section 1.5 the valuation of an option was summarized in (1.19) as the discounted values of a probable payoff,

$$V_0 = e^{-rT} \mathsf{E}_{\mathsf{Q}}(V_T) \ .$$

For the binomial model we proved for European options in Exercise 1.8 that this method produces

$$V_0^{(M)} = e^{-rT} \mathsf{E}(V_T) \ ,$$

where E reflects expectation with respect to the risk-free probability of the binomial method. And for the continuous-time Black-Scholes model, the result in (A4.11b) for a put is

$$V_0 = e^{-rT} \left[K \, F(-d_2) - e^{(r-\delta)T} S \, F(-d_1) \right] \ , \tag{3.17}$$

similarly for a call. Since F is an integral ($\longrightarrow$ Appendix D2), equation (3.17) is a first version of an integral representation. Its origin is either the analytic solution of the Black-Scholes PDE, or the representation

$$V_0 = e^{-rT} \int_0^\infty (K - S_T)^+ \, f(S_T; T, S_0, r, \sigma) \, dS_T \ . \tag{3.18}$$

Here $f(S_T; T, S_0, \mu, \sigma)$ is the density (1.47) of the lognormal distribution, with $\mu = r$, or μ replaced by $r - \delta$ to match a continuous dividend yield. It is not difficult to prove that (3.18) and (3.17) are equivalent ($\longrightarrow$ Exercise 3.9). We summarize the integral representation as

$$\boxed{\quad V(S_0, 0) = e^{-rT} \widetilde{\mathsf{E}}(V(S_T, T)) \quad} \tag{3.19}$$

The risk-neutral expectation $\widetilde{\mathsf{E}}$ corresponds to E_{Q} in Section 1.5; μ is replaced by r.

The integral representation (3.18) offers another way to calculate V_0, namely via a numerical approximation of (3.18). Of course, in this one-dimensional situation, the approximation of the closed-form solution (3.17) is more efficient. But in higher-dimensional spaces integrals corresponding to (3.18) can be become highly attractive for computational purposes. Note that the integrand is smooth because the zero branch of $(K - S_T)^+$ needs not be

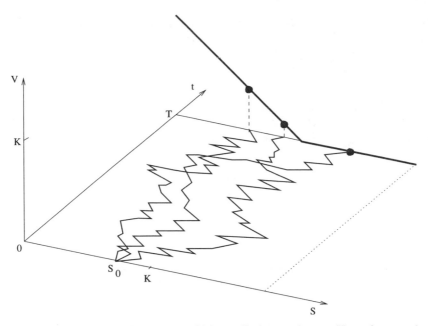

Fig. 3.2. Schematic illustration of Monte Carlo simulation. Here three paths start at S_0; at $t = T$ two paths end at a value of S for which the put option is in the money, and one out of the money. (See the website for an illustration with realistic simulations.)

integrated; in (3.18) the integration is cut to the interval $0 \leq S_T \leq K$. Any numerical method can be applied, such as sparse-grid quadrature [GeG98], and —of course— Monte Carlo methods.

3.5.2 The Basic Version for European Options

The basic idea of Monte Carlo simulation is to calculate a large number of trajectories of the SDE, always starting from S_0, and then average over these results in order to obtain information on the probable behavior of the process. Specifically when European options are to be valued, a Monte Carlo simulation calculates the expectation of the terminal payoff of the option ($t = T$), which is then discounted to obtain the value for $t = 0$. This simulation approach can again be summarized by (3.19). Figure 3.2 illustrates the procedure. One obtains $\widetilde{\mathsf{E}}$ by simulation of N paths of the asset with $\mu = r$, each starting from S_0:

Algorithm 3.6   (Monte Carlo simulation of European options)

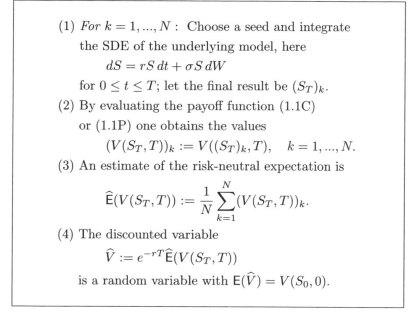

(1) *For $k = 1, ..., N$:* Choose a seed and integrate
the SDE of the underlying model, here
$$dS = rS\,dt + \sigma S\,dW$$
for $0 \leq t \leq T$; let the final result be $(S_T)_k$.

(2) By evaluating the payoff function (1.1C)
or (1.1P) one obtains the values
$$(V(S_T, T))_k := V((S_T)_k, T), \quad k = 1, ..., N.$$

(3) An estimate of the risk-neutral expectation is
$$\widehat{\mathsf{E}}(V(S_T, T)) := \frac{1}{N} \sum_{k=1}^{N} (V(S_T, T))_k.$$

(4) The discounted variable
$$\widehat{V} := e^{-rT}\widehat{\mathsf{E}}(V(S_T, T))$$
is a random variable with $\mathsf{E}(\widehat{V}) = V(S_0, 0)$.

The resulting $\widehat{V}$ is the desired approximation $\widehat{V} \approx V(S_0, 0)$. In this simple form, the Monte Carlo simulation can only be applied to European options where the exercise date is fixed. Only the value $V(S_0, 0)$ is obtained, and the lack of other information on $V(S, t)$ does not allow to check whether the early-exercise constraint of an American option is violated. For American options a greater effort in simulation is necessary, see Section 3.5.5. In practice the number N must be chosen large, for example, $N = 10000$. The convergence behavior corresponds to that discussed for Monte Carlo integration, see Section 2.4. This explains why Monte Carlo simulation in general is expensive. For standard European options with univariate underlying that satisfies the Assumption 1.2, the alternative of evaluating the Black-Scholes formula is by far cheaper. But in principle both approaches provide the same result, where we neglect that accuracies and costs are different.

Example 3.7   (European put)

We consider a European put with the parameters $S_0 = 5$, $K = 10$, $r = 0.06$, $\sigma = 0.3$, $T = 1$. For the linear SDE with constant coefficients $dS = rSdt + \sigma SdW$ the theoretical solution is known, see equation (1.49). For the chosen parameters we have

$$S_1 = 5\exp(0.015 + 0.3W_1),$$

which requires "the" value of the Wiener process at $t = 1$. Related values W_1 of the Wiener process can be obtained from (1.22) with $\Delta t = T$ as

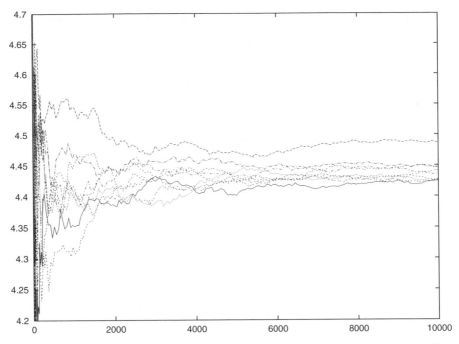

Fig. 3.3. Ten sequences of Monte Carlo simulations on Example 3.7, each with a maximum of 10000 paths. horizontal axis: N, vertical axis: mean value $\widehat{V}$

$W_1 = Z\sqrt{T}$, $Z \sim \mathcal{N}(0,1)$. But for this illustration we do not take advantage of the analytic solution formula, because the Monte Carlo approach is not limited to linear SDEs with constant coefficients. To demonstrate the general procedure we integrate the SDE numerically with step length $\Delta t < T$, in order to calculate an approximation to S_1. Any of the methods derived in Section 3.3 can be applied. For simplicity we use Euler's method. Since the chosen value of r is small, the discretization error of the drift term is small compared to the standard deviation of W_1. As a consequence, the accuracy of the integration for small values of Δt is hardly better than for larger values of the step size. Artificially we choose $\Delta t = 0.02$ for the time step. Hence each trajectory requires to calculate 50 normal variates $\sim \mathcal{N}(0,1)$. The Figure 3.3 shows the resulting values $\widehat{V} \approx V(S_0, 0)$ for 10 sequences of simulations, each with a maximum of $N = 10000$ trajectories. Each sequence has started with a different seed for the calculation of the random numbers from Section 2.3.

The Example 3.7 is a European put with the same parameters as Example 1.5. This allows to compare the results of the simulation with the more accurate results from Table 1.2, where we had obtained $V(5,0) \approx 4.43$. The simulations reported in Figure 3.3 have difficulties to come close to this value. Since Figure 3.3 depicts all intermediate results for sample sizes $N < 10000$, the convergence behavior of Monte Carlo can be observed. For this example and $N < 2000$ the accuracy is bad; for $N \approx 6000$ it reaches acceptable values, and hardly improves for $6000 < N \leq 10000$. Note that the "convergence" is not monotonous, and one of the simulations delivers a frustratingly inaccurate result.

3.5.3 Bias

The sampling error of Monte Carlo was already discussed in Section 2.4. Recall the size of this error is proportional to $N^{-1/2}$. The same error is encountered when Monte Carlo is applied to option valuation. In case the closed-form solution (1.49) of the SDE is used in step (1) of Algorithm 3.6, the sampling error is basically the only error. But for more general options, approximations are often based on discretizations, and some bias is encountered.

This typically occurs, when the option is path-dependent— that is, its value depends on S_t for possibly all $t \leq T$. For example, the volatility may be *local*, which means that it depends on S_t, $\sigma = \sigma(S)$. Another example is furnished by the *lookback* option, where the valuation depends on

$$x := \mathsf{E}\left(\max_{0\leq t\leq T} S_t\right) .$$

In both examples, a time discretization is required with a finite number m of values $S(t_j)$, with the notation as used in (3.1). Even if the underlying SDE is such that a closed-form solution is available, the estimator provided by the discretely sampled maximum

$$\hat{x} := \max_{0\leq j\leq m} S(t_j)$$

almost surely underestimates x. That is, the estimator $\hat{x}$ of x is biased, with

$$\mathrm{bias}(\hat{x}) := \mathsf{E}(\hat{x}) - x \neq 0 .$$

This and other occurrences of a bias are discussed in [Gla04].

Fortunately, this bias can be made arbitrarily small by taking sufficiently large values of m. In practice there is a tradeoff between making the variance small ($N \to \infty$), and making the bias small ($m \to \infty$). The mean square error

$$\mathrm{MSE}(\hat{x}) := \mathsf{E}[(\hat{x} - x)^2] \tag{3.20}$$

measures both errors: A straightforward calculation (which the reader may check) shows

$$\mathrm{MSE}(\hat{x}) = (\mathsf{E}(\hat{x}) - x)^2 + \mathsf{E}[(\hat{x} - \mathsf{E}(\hat{x}))^2]$$
$$= (\mathrm{bias}(\hat{x}))^2 + \mathsf{Var}(\hat{x})$$

3.5.4 Variance Reduction

To improve the accuracy of simulation and thus the efficiency, it is essential to apply methods of variance reduction. We explain the methods of the *antithetic variates* and the *control variates*. In many cases these methods decrease the variances.

Antithetic Variates

If a random variable satisfies $Z \sim \mathcal{N}(0,1)$, then $-Z \sim \mathcal{N}(0,1)$. Let us denote by $\widehat{V}$ the approximation obtained by Monte Carlo simulation. With little extra effort during the original Monte Carlo simulation we can run in parallel a side calculation which uses $-Z$ instead of Z. We denote the resulting *antithetic variate* by V^-. By taking the average

$$V_{\mathrm{AV}} := \tfrac{1}{2}\left(\widehat{V} + V^-\right) \tag{3.21}$$

(AV for *antithetic variate*) we obtain a new approximation, which in many cases is more accurate than $\widehat{V}$. Since $\widehat{V}$ and V_{AV} are random variables we can only aim at

$$\mathsf{Var}(V_{\mathrm{AV}}) < \mathsf{Var}(\widehat{V}) .$$

In view of the properties of variance and covariance (equation (B1.7) in Appendix B1) we have

$$\begin{aligned}
\mathsf{Var}(V_{\mathrm{AV}}) &= \tfrac{1}{4}\mathsf{Var}(\widehat{V} + V^-) \\
&= \tfrac{1}{4}\mathsf{Var}(\widehat{V}) + \tfrac{1}{4}\mathsf{Var}(V^-) + \tfrac{1}{2}\mathsf{Cov}(\widehat{V}, V^-).
\end{aligned} \tag{3.22}$$

From

$$|\mathsf{Cov}(X,Y)| \leq \frac{1}{2}[\mathsf{Var}(X) + \mathsf{Var}(Y)]$$

(follows from (B1.7)) we deduce

$$\mathsf{Var}(V_{\mathrm{AV}}) \leq \frac{1}{2}(\mathsf{Var}(\widehat{V}) + \mathsf{Var}(V^-)).$$

This shows that in the worst case only the efficiency is slightly deteriorated by the additional calculation of V^-. The favorable situation is when the covariance is negative. Then (3.22) shows that the variance of V_{AV} can become significantly smaller than that of $\widehat{V}$ and V^-. Since we have chosen the random numbers $-Z$ for the calculation of V^- the chances are high that $\widehat{V}$ and V^- are negatively correlated and hence $\mathsf{Cov}(\widehat{V}, V^-) < 0$. In this situation V_{AV} is a better approximation than $\widehat{V}$.

In Figure 3.4 we simulate Example 3.7 again, now with antithetic variates. With this example and the chosen random number generator the variance reaches small values already for small N. Compared to Figure 3.3 the convergence is somewhat smoother. The accuracy the experiment shown in Figure 3.3 reaches with $N = 6000$ is achieved already with $N = 2000$ in Figure 3.4.

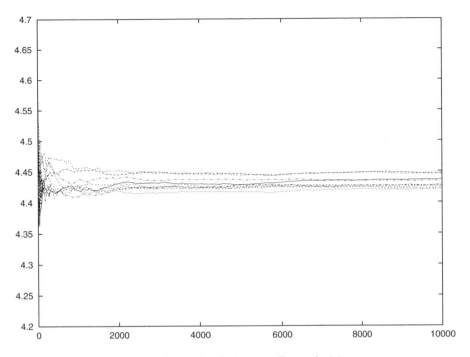

Fig. 3.4. Ten series of antithetic simulations on Example 3.7

But in the end, the error has not become really small. The main reason for the remaining significant error in the experiment reported by Figure 3.4 is the discretization error of the Euler scheme. To remove this source of error, we repeat the above experiments with the analytical solution of (1.49). The result is shown in Figure 3.5 for standard Monte Carlo, and in Figure 3.6 for antithetic variates. These figures better reflect the convergence behavior of Monte Carlo simulation. By the way, applying the Milstein scheme of Algorithm 3.5 does not improve the picture: No qualitative change is visible if we replace the Euler-generated simulations of Figures 3.3/3.4 by their Milstein counterparts. This may be explained by the fact that the weak convergence order of Milstein's method equals that of the Euler method. — Recall that Example 3.7 is chosen merely for illustration; here other methods are by far more efficient than Monte Carlo approaches.

Control Variates

Again V denotes the exact value of the option and $\widehat{V}$ a Monte Carlo approximation. For comparison we calculate in parallel another option, which is closely related to the original option, and for which we know the exact value V^*. Let the Monte Carlo approximation of V^* be denoted $\widehat{V}^*$. This variate serves as *control variate* with which we wish to "control" the error. The additional effort to calculate the control variate $\widehat{V}^*$ is small in case the simulations

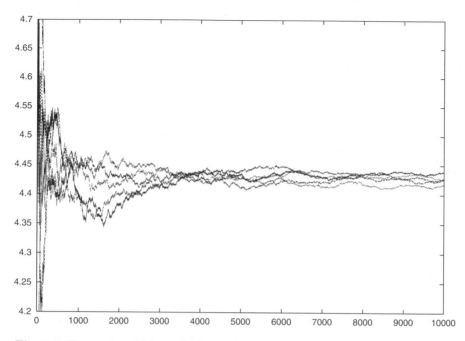

Fig. 3.5. Five series of Monte Carlo simulations on Example 3.7, using the analytic solution of the SDE (compare to Fig. 3.3)

of the asset S are identical for both options. This situation arises when S_0, μ and σ are identical and only the payoff differs. When the two options are similar enough one may expect a strong positive correlation between them. So we expect relatively large values of $\mathsf{Cov}(V, V^*)$ or $\mathsf{Cov}(\widehat{V}, \widehat{V}^*)$, close to its upper bound,

$$\mathsf{Cov}(\widehat{V}, \widehat{V}^*) \approx \frac{1}{2}\mathsf{Var}(\widehat{V}) + \frac{1}{2}\mathsf{Var}(\widehat{V}^*).$$

This leads us to define "closeness" between the options as sufficiently large covariance in the sense

$$\mathsf{Cov}(\widehat{V}, \widehat{V}^*) > \frac{1}{2}\mathsf{Var}(\widehat{V}^*). \tag{3.23}$$

The method is motivated by the assumption that the unknown error $V - \widehat{V}$ has the same order of magnitude as the known error $V^* - \widehat{V}^*$. This expectation can be written $V \approx \widehat{V} + (V^* - \widehat{V}^*)$, which leads to define as another approximation

$$V_{\mathrm{CV}} := \widehat{V} + V^* - \widehat{V}^* \tag{3.24}$$

(CV for *control variate*). We see from (B1.6) (with $\beta = V^*$) and (B1.7) that

$$\mathsf{Var}(V_{\mathrm{CV}}) = \mathsf{Var}(\widehat{V} - \widehat{V}^*) = \mathsf{Var}(\widehat{V}) + \mathsf{Var}(\widehat{V}^*) - 2\mathsf{Cov}(\widehat{V}, \widehat{V}^*).$$

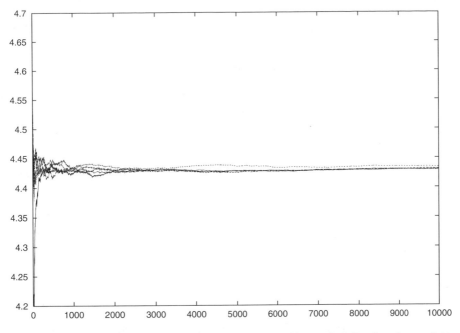

Fig. 3.6. Five series of Monte Carlo simulations on Example 3.7 using the analytic solution of the SDE and antithetic variates (3.21) (compare to Fig. 3.4)

If (3.23) holds, then $\mathsf{Var}(V_{\mathrm{CV}}) < \mathsf{Var}(\widehat{V})$. In this sense $\mathsf{Var}(V_{\mathrm{CV}})$ is a better approximation than $\widehat{V}$.

3.5.5 American Options

The equation (3.19) can be generalized to American options. Monte Carlo applied to both European and American options requires simulating paths S_t by integrating $dS_t = rS_t dt + \sigma S_t dW_t$ for $t \geq 0$. Whereas for European options it is clear to integrate until expiration, $t = T$, the American option requires to continuously investigate whether early exercise is advisable. To mimic reality, one must take care that for any t the decision on early exercise is *only based on the information that is known so far*. Recall that the filtration $\mathcal{F}_t$ is interpreted as a model of the available information at time t. This situation suggests to require a *stopping time* be defined accordingly:

Definition 3.8   (stopping time)
    A stopping time τ with respect to a filtration $\mathcal{F}_t$ is a random variable that is $\mathcal{F}_t$-measurable for all $t \geq 0$.

That is, $\{\tau \leq t\} \in \mathcal{F}_t$ for all $t \geq 0$. Typically a decision is made when τ is reached, such as exercising early. Two examples should make the concept of a stopping time clearer.

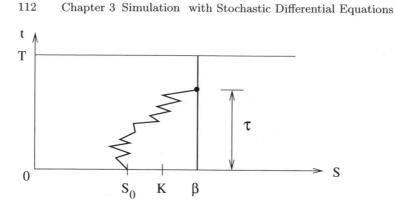

Fig. 3.7. The strategy of Example 3.9 to define a stopping time τ

Example 3.9   (hitting time)
    For a value β, which fixes a level of S, define

$$\tau := \inf\{t > 0 \mid S_t \geq \beta\} \quad,$$
$$\text{and } \tau := \infty \quad \text{if such a } t \text{ does not exist.}$$

This example, illustrated in Figure 3.7, fulfils the requirements of a stopping time ([HuK00], p.42). The example

$$\tau := \text{ moment when } S_t \text{ reaches its maximum over } 0 \leq t \leq T$$

is no stopping time, because for each $t < T$ knowledge of the future states of S is needed.

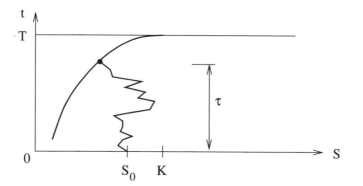

Fig. 3.8. The optimal stopping time τ of a put. The heavy curve is the early-exercise curve.

    Of all possible stopping times, the stopping at the early-exercise curve is optimal. This optimal stopping is illustrated in Figure 3.8, it gives the American option its optimal value. From a practical point of view, the stopping at the early-exercise curve can not be established as in Example 3.9, because the curve is not known initially. But the following characterization of the value $V(S,0)$ of an American option holds true:

$$V(S,0) = \sup_{0 \le \tau \le T} \mathsf{E}_\mathsf{Q}(e^{-r\tau}\,\mathrm{payoff}\,(S_\tau))\;,$$

(3.25)

where τ is a stopping time.

This result is a special case for $t = 0$ of a more general formula for $V(S,t)$, which is proved in [Ben84]. Clearly, (3.25) includes the case of a European option for $\tau := T$, in which case taking the supremum is not effective.

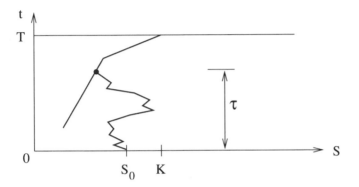

Fig. 3.9. An approximation of the optimal stopping time

Bounds

A practical realization of (3.25) can be based on calculating lower bounds $V^{\mathrm{low}}(S,0)$ and upper bounds $V^{\mathrm{up}}(S,0)$ such that

$$V^{\mathrm{low}}(S,0) \le V(S,0) \le V^{\mathrm{up}}(S,0)\;.$$

Since by (3.25) $V(S,0)$ is given by taking the supremum over *all* stopping times, a lower bound is obtained by taking a *specific* stopping strategy. For example, choose a level β and the stopping strategy of Example 3.9, see Figure 3.7. If we denote for each calculated path the resulting stopping time by $\tilde{\tau}$, a lower bound to $V(S,0)$ is given by

$$V^{\mathrm{low}(\beta)}(S,0) := \mathsf{E}_\mathsf{Q}(e^{-r\tilde{\tau}}\,\mathrm{payoff}\,(S_{\tilde{\tau}}))\;.$$

(3.26)

This value depends on the level β, which is indicated by writing $V^{\mathrm{low}(\beta)}$. The bound is calculated by Monte Carlo simulation over a sample of N paths, similar as in Algorithm 3.6. A related experiment can be found in [Hig04].

There are many examples how to obtain better lower bounds. For instance, the early-exercise curve might be approximated by pieces of curves or pieces of straight lines, which are defined by *parameters*. This is illustrated by Figure

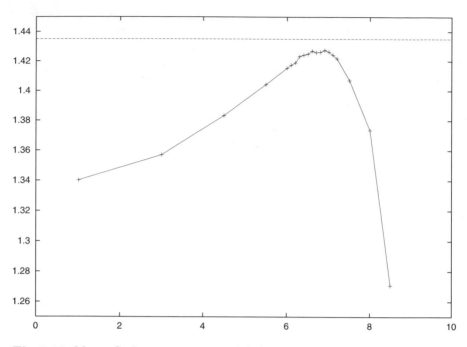

Fig. 3.10. Monte Carlo approximations (+) for several values of β (Exercise 3.10, random numbers from [MaN98]). The dashed line represents the exact value.

3.9, where three parameters $\beta_1, \beta_2, \beta_3$ are needed to define the two straight lines providing a crude approximation of the unknown early-exercise curve shown in Figure 3.8. The resulting lower bound $V^{\mathrm{low}(\beta_1, \beta_2, \beta_3)}$ would depend on the choice of the parameters $\beta_1, \beta_2, \beta_3$. A simpler and nicely working approximation uses a parabola in the (S,t)-domain with horizontal tangent at $t = T$. This approach requires only one parameter β ($\longrightarrow$ Exercise 3.10). The approach is illustrated in Figure 3.10.

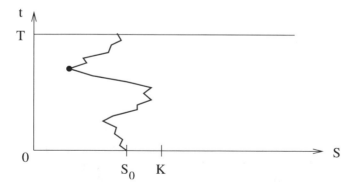

Fig. 3.11. No stopping time; maximizing the payoff of a *given* path

Within the class of chosen parametric approximations, one may try to optimize over the β. But what kind of parametric approximation, and what choice of the parameters can be considered "good" when $V(S,t)$ is still unknown? To this end, upper bounds V^{up} can be constructed, and one attempts to push the difference $V^{\mathrm{up}} - V^{\mathrm{low}}$ close to zero. An upper bound can be obtained, for example, when one peers into the future. As a crude example, the entire path S_t for $0 \le t \le T$ may be simulated, and the option "exercised" in retrospect when

$$e^{-rt}\, \mathrm{payoff}\,(S_t)$$

is maximal. This is illustrated in Figure 3.11. Pushing the lower bounds towards upper bounds amounts to search in the β-parameter space for a better combination of β-values. As a by-product of approximating $V(S,0)$, the corresponding parameters β provide an approximation of the early-exercise curve.

The above is just a crude strategy how Monte Carlo can be applied to approximate American options. In particular, the described simple approach to obtain upper bounds is not satisfactory.

Refined Methods

Recently, many refined Monte Carlo methods for the calculation of American options have been suggested. These include the use of stochastic grids [BrG04], and least-squares approximations of conditional expectation [LonS01]. For an overview on related methods, see Chapter 8 in [Gla04]. Here we outline some ideas on which many of these methods rely.

One essential step is to approximate the American-style option by a *Bermuda*-style option. A Bermudan option restricts early exercise to specified dates during its life. In our context, the "dates" are created artificially by a finite set of discrete time instances t_i. This resembles the time discretization of the binomial method of Section 1.4, see Figure 1.8. In this semi-discretized setting the value of the dynamic programming procedure of equation (1.14) generalizes to

$$V_i(S) = \max\{\, \mathrm{payoff}\,(S)\,,\ V_i^{\mathrm{cont}}(S)\} \tag{3.27}$$

where the *continuation function* V^{cont} is defined by the conditional expectation

$$V_i^{\mathrm{cont}}(S) = e^{-r\Delta t}\,\tilde{\mathsf{E}}(V_{i+1}(S_{i+1})\,|\,S_i = S) \tag{3.28}$$

(On the binomial tree, this is equation (1.13).) $\tilde{\mathsf{E}}$ (or E_{Q}) is calculated as before under the assumption of risk neutrality. In (3.28) the period between two time levels t_i and t_{i+1} can be interpreted as a local one-period model. For each value S along the time level t_i, the $V_i^{\mathrm{cont}}(S)$ in (3.28) represents the expected value of continuing (holding the option). In (3.27) this is compared to the immediate exercise value payoff(S), the intrinsic value of the option. The approach of [LonS01] estimates the conditional expectation in (3.28) by a least-squares approximation, which is based on simulating paths of the underlying SDE.

3.5.6 Further Hints

In summary we emphasize that Monte Carlo simulation is of great importance for general models where no specific assumptions (as those of Black, Merton and Scholes) have lead to efficient approaches. For example, in case the interest rate r cannot be regarded as constant but is modeled by some SDE (such as equation (1.40)), then a system of SDEs must be integrated. An example of a stochastic volatility is provided by Example 1.15, compare Figure 3.1. In such cases the Black-Scholes equation may not help and a Monte Carlo simulation can be the method of choice. Then the Algorithm 3.6 is adapted appropriately. In this situation, in case options are to be priced, the question arises whether the risk-neutral valuation principle is obeyed. An important application of Monte Carlo methods is the calculation of risk indices such as *value at risk*, see the notes on Section 1.7/1.8. Monte Carlo methods are especially attractive for multifactor models with high dimension.

The demands for **accuracy** of Monte Carlo simulation should be kept on a low level. In many cases an error of 1% must suffice. Note that it does not make sense to decrease the Monte Carlo error significantly below the error of the time discretization of the underlying SDE (and vice versa). When the amount of available random numbers is too small or its quality poor, then no improvement of the error can be expected. The methods of variance reduction can save a significant amount of costs [BBG97], [SH97], [Pl99]. Note that different variance-reduction techniques can be combined with each other.

When results are required for slightly changed parameter values, it may be necessary to rerun Monte Carlo. But sometimes this can be avoided. For example, options are often priced for different maturities. When Monte Carlo is combined with a bridging technique, several such options can be priced effectively in a single run [RiW02]. Another example occurs when Greeks are calculated by Monte Carlo. Here we comment on approximating delta$=\frac{\partial V}{\partial S}$. Applying two runs of Monte Carlo simulation, one for S_0 and one for a close value $S_0 - \Delta S$, an approximation of delta is obtained by the difference quotient

$$\frac{V(S_0) - V(S_0 - \Delta S)}{\Delta S} .$$

As an alternative, Malliavin calculus allows to shift the differencing to the density function, which leads via a kind of differentiation by parts to a different integral to be approximated by Monte Carlo. For references on this technique, see [FoLLLT99].

Finally we test the Monte Carlo simulation in a fully deterministic variant. To this end we insert the quasi-random two-dimensional Halton points into Algorithm 2.13 and use the resulting quasi normal deviates to calculate solutions of the SDE. In this way, for Example 3.7 acceptable accuracy is reached already for about 2000 paths, much better than what is shown in Figure 3.2.

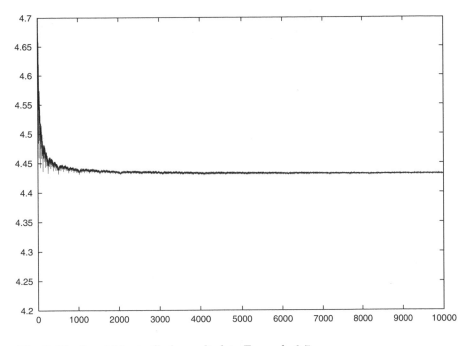

Fig. 3.12. Quasi Monte Carlo applied to Example 3.7

A closer investigation reveals that normal deviates based on Box-Muller-Marsaglia (Algorithm 2.13) with two-dimensional Halton points lose the equi-distributedness; the low discrepancy is not preserved. Apparently the quasi-random method does not simulate independence [Ge98]. A related visual inspection resembles Figure 2.6. This sets the stage for the slightly faster inversion method [Moro95] ($\longrightarrow$ Appendix D2), based on one-dimensional low-discrepancy sequences. Figure 3.12 shows the result. The scaling of the figure is the same as before.

Notes and Comments

on Section 3.1:

Under suitable assumptions it is possible to prove for strong solutions existence and uniqueness, see [KP92]. Usually the discretization error dominates other sources of error. We have neglected the sampling error (the difference between $\widehat{\epsilon}$ and ϵ), imperfections in the random number generator, and rounding errors. Typically these errors are likely to be less significant.

on Section 3.2:

This section closely follows Section 5.1 of [KP92].

on Section 3.3:

[KP92] discusses many methods for the approximation of paths of SDEs, and proves their convergence. An introduction is given in [Pl99]. Possible orders of strongly converging schemes are integer multiples of $\frac{1}{2}$ whereas the orders of weakly converging methods are whole numbers. Simple adaptions of deterministic schemes do not converge for SDEs. For the integration of *random ODEs* we refer to [GK01]. Maple routines for SDEs can be found in [CKO01], and MATLAB routines in [Hig01].

For ODEs and SDEs linear stability is investigated. This is concerned with the long-time behavior of solutions of the test equation $dX_t = \alpha X_t dt + \beta X_t dW_t$, where α is a complex number with negative real part. This situation does not appear relevant for applications in finance. The numerical stability in the case $Re(\alpha) < 0$ depends on the step size h and the relation among the three parameters α, β, h. For this topic and further references we refer to [SM96], [Hig01], [Pl99].

Also SDEs may suffer from stiffness. Then semi-implicit methods are used. The implicit term is the drift term but not the diffusion term, such that Itô's choice of intermediate values is obeyed. The Euler scheme modifies to

$$y_{j+1} = y_j + [\alpha a(y_{j+1}, t_{j+1}) + (1 - \alpha)a(y_j, t_j)]\Delta t + b(y_j, t_j)\Delta W_j$$

with, for example $\alpha = \frac{1}{2}$.

on Section 3.4:

Other bridges than Brownian bridges are possible. For a Gamma process and a Gaussian bridge this is shown in [RiW02], [RiW03]. Bridges play a role when the effectiveness of Monte Carlo integration is improved [CaMO97].

on Section 3.5:

In the literature the basic idea of the approach summarized by equation (3.19) is analyzed using martingale theory, compare the references in Chapter 1 and Appendix B2. The assumption $\mu = r$ reflects our standard scenario of an asset that pays no dividend. For Monte Carlo simulation on American options see also [BrG97], [BBG97], [Kwok98], [Ro00], [Fu01], [LonS01], [Gla04]. The equivalence of the Monte Carlo simulation with the solution of the Black-Scholes equation is guaranteed by the theorem of Feynman and Kac [Ne96], [Re96], [Øk98], [KS91], [TR00], [Shr04].

Monte Carlo simulations can be parallelized in a trivial way: The single simulations can be distributed among the processors in a straightforward fashion because they are independent of each other. If M processors are available, the speed reduces by a factor of $1/M$. But the streams of random numbers in each processor must be independent. For related generators see

[Mas99]. In doubtful and sensitive cases Monte Carlo simulation should be repeated with other random-number generators and with low-discrepancy numbers.

The method of control variates can be modified with a parameter α,

$$V_{\mathrm{CV}}^{\alpha} := \widehat{V} + \alpha(V^* - \widehat{V}^*),$$

where one tries to find a value of α such that the variance is minimized. For a discussion of variance reduction and examples see Chapter 4 in [Gla04]. For the variance-reduction method of *importance sampling*, see [New97].

Exercises

Exercise 3.1   Implementing Euler's Method

Implement Algorithm 1.11. Start with a test version for one scalar SDE, then develop a version for a system of SDEs. Test examples:

a) Perform the experiment of Figure 1.17.
b) Integrate the system of Example 1.15 for $\alpha = 0.3$, $\beta = 10$ and the initial values $S_0 = 50$, $\sigma_0 = 0.2$, $\xi_0 = 0.2$ for $0 \leq t \leq 1$.

We recommend to plot the calculated trajectories.

Exercise 3.2   Itô Integral in Equation (3.9)

Let the interval $0 \leq s \leq t$ be partitioned into n subintervals, $0 = t_1 < t_2 < \dots < t_{n+1} = t$. For a Wiener process W_t assume $W_{t_1} = 0$.

a)   Show $\displaystyle\sum_{j=1}^{n} W_{t_j} \left(W_{t_{j+1}} - W_{t_j}\right) = \frac{1}{2}W_t^2 - \frac{1}{2}\sum_{j=1}^{n} \left(W_{t_{j+1}} - W_{t_j}\right)^2$

b)   Use Lemma 1.9 to deduce Equation (3.9).

Exercise 3.3   Integration by Parts for Itô Integrals
a) Show

$$\int_{t_0}^{t} s\,dW_s = tW_t - t_0 W_{t_0} - \int_{t_0}^{t} W_s\,ds$$

Hint: Start with the Wiener process $X_t = W_t$ and apply the Itô Lemma with the transformation $y = g(x,t) := tx$.

b) Denote $\Delta Y := \int_{t_0}^{t} \int_{t_0}^{s} dW_z\,ds$. Show by using a) that

$$\int_{t_0}^{t} \int_{t_0}^{s} dz\,dW_s = \Delta W \Delta t - \Delta Y.$$

Exercise 3.4   Moments of Itô Integrals for Weak Solutions
a) Use the Itô isometry

$$\mathsf{E}\left[\left(\int_a^b f(t,\omega)dW_t\right)^2\right] = \int_a^b \mathsf{E}\left[f^2(t,\omega)\right]dt$$

to show its generalization

$$\mathsf{E}\left[I(f)I(g)\right] = \int_a^b \mathsf{E}[fg]dt, \qquad \text{where} \quad I(f) = \int_a^b f(t,\omega)dW_t.$$

Hint: $4fg = (f+g)^2 - (f-g)^2$.
b) For $\Delta Y := \int_{t_0}^t \int_{t_0}^s dW_z ds$ the moments are

$$\mathsf{E}[\Delta Y] = 0, \quad \mathsf{E}[\Delta Y^2] = \frac{\Delta t^3}{3}, \quad \mathsf{E}[\Delta Y \Delta W] = \frac{\Delta t^2}{2} \quad \text{and} \quad \mathsf{E}[\Delta Y \Delta W^2] = 0.$$

Show this by using a) and $\mathsf{E}\left[\int_a^b f(t,\omega)dW_t\right] = 0$.

Exercise 3.5
By transformation of two independent standard normally distributed random varables $Z_i \sim \mathcal{N}(0,1)$, $i = 1, 2$, two new random variables are obtained by

$$\Delta\widehat{W} := Z_1\sqrt{\Delta t}, \qquad \Delta\widehat{Y} := \frac{1}{2}(\Delta t)^{3/2}\left(Z_1 + \frac{1}{\sqrt{3}}Z_2\right).$$

Show that $\Delta\widehat{W}$ and $\Delta\widehat{Y}$ have the moments of (3.14).

Exercise 3.6
In addition to (3.14) further moments are

$$\mathsf{E}(\Delta W) = \mathsf{E}(\Delta W^3) = \mathsf{E}(\Delta W^5) = 0, \quad \mathsf{E}(\Delta W^2) = \Delta t, \quad \mathsf{E}(\Delta W^4) = 3\Delta t^2.$$

Assume a new random variable $\Delta\widetilde{W}$ satisfying

$$P\left(\Delta\widetilde{W} = \pm\sqrt{3\Delta t}\right) = \frac{1}{6}, \qquad P\left(\Delta\widetilde{W} = 0\right) = \frac{2}{3}$$

and the additional random variable

$$\Delta\widetilde{Y} := \frac{1}{2}\Delta\widetilde{W}\Delta t.$$

Show that the random variables $\Delta\widetilde{W}$ and $\Delta\widetilde{Y}$ have up to terms of order $O(\Delta t^3)$ the same moments as ΔW and ΔY.

Exercise 3.7   Brownian Bridge

For a Wiener process W_t consider

$$X_t := W_t - \frac{t}{T} W_T \quad \text{for } 0 \le t \le T.$$

Calculate $\mathsf{Var}(X_t)$ and show that

$$\sqrt{t \left(1 - \frac{t}{T} \right)} Z \quad \text{with } Z \sim \mathcal{N}(0,1)$$

is a realization of X_t.

Exercise 3.8   Error of the Milstein Scheme

To which formula does the Milstein scheme reduce for linear SDEs? Perform the experiment outlined in Example 3.2 using the Milstein scheme of Algorithm 3.5. Set up a table similar as in Table 3.1 to show

$$\widehat{\varepsilon}(h) \approx h$$

for Example 3.2.

Exercise 3.9   Monte Carlo and European Option

For a European put with time to maturity $\tau := T - t$ prove that

$$V(S_t, t) = e^{-r\tau} \int_0^\infty (K - S_T)^+ \frac{1}{S_T \sigma \sqrt{2\pi\tau}} \exp\left\{ -\frac{[\ln(S_T/S_t) - (r - \frac{\sigma^2}{2})\tau]^2}{2\sigma^2 \tau} \right\} dS_T$$

$$= e^{-r\tau} K F(-d_2) - S_t F(-d_1) ,$$

where d_1 and d_2 are defined in (A4.10).

Hints: The second equation is to be shown, the first only collects the terms of (3.18). Use $(K - S_T)^+ = 0$ for $S_T > K$, and get two integrals.

Exercise 3.10   Project: Monte Carlo Experiment

Construct as hitting curve a parabola with horizontal tangent at $(S, t) = (K, T)$, similar as in Figure 3.8, compare Figure 3.10. The parabola is defined by the intersection with the S-axis, $(S, t) = (\beta, 0)$. Choose $K = 10$, $r = 0.006$, $\sigma = 0.3$, and $S_0 = 9$ and simulate for several values of β the GBM $dS = rS dt + \sigma S dW$ several thousand times, and calculate the hitting time for each trajectory. Estimate a lower bound to $V(S_0, 0)$ using (3.25). Decide whether an exact calculation of the hitting point makes sense. (Run experiments comparing such a strategy to implementing the hitting time restricted to the discrete time grid.) Think about how to implement upper bounds.

Chapter 4 Standard Methods for Standard Options

We now enter the part of the book that is devoted to the numerical solution of equations of the Black-Scholes type. Here we discuss "standard" options in the sense as introduced in Section 1.1. Accordingly, let us assume the scenario characterized by the Assumptions 1.2. In case of European options the function $V(S, t)$ solves the Black-Scholes equation (1.2). It is not really our aim to solve this partial differential equation because it possesses an analytic solution ($\longrightarrow$ Appendix A4). Ultimately it is our intention to solve more general equations and inequalities. In particular, American options will be calculated numerically. The goal is not only to calculate single values $V(S_0, 0)$ —for this purpose binomial methods can be applied— but also to approximate the curve $V(S, 0)$, or even the surface defined by $V(S, t)$ on the half strip $S > 0$, $0 \leq t \leq T$.

American options obey *inequalities* of the type of the Black-Scholes equation (1.2). To allow for early exercise, the Assumptions 1.2 must be weakened. As a further generalization, the payment of dividends must be taken into account because otherwise early exercise does not make sense for American calls.

The main part of this chapter outlines an approach based on finite differences. We begin with unrealistically simplified boundary conditions in order to keep the explanation of the discretization schemes transparent. Later sections will discuss the full boundary conditions, which turn out to be tricky in the case of American options. At the end of this chapter we will be able to implement a finite-difference algorithm that can calculate standard American (and European) options. If we work carefully, the resulting finite-difference computer program will yield correct approximations. But the finite-difference approach is not necessarily the most efficient one. Hints on other methods will be given at the end of this chapter. For nonstandard options we refer to Chapter 6.

The finite-difference methods will be explained in some detail because they are the most elementary approaches to approximate differential equations. As a side-effect, this chapter serves as introduction into several fundamental concepts of numerical mathematics. A trained reader may like to skip Sections 4.2 and 4.3. The aim of this chapter is to introduce concepts, as well as a characterization of the free boundary (early-exercise curve), and of linear complementarity.

In addition to the classical finite-difference approach, "standard methods" include analytic methods, which to a significant part are based on nonnumerical methods. The final Section 4.8 will give an introduction.

4.1 Preparations

We assume that dividends are paid with a continuous yield of constant level. In case of a discrete payment of, for example, one payment per year, the payment can be converted into a continuous yield ($\longrightarrow$ Exercise 4.1). To this end one has to take into consideration that at the instant of a discrete payment the price $S(t)$ of the asset instantaneously drops by the amount of the payment. This holds true because of the no-arbitrage principle. The continuous flow of dividends is modeled by a decrease of S in each time interval dt by the amount

$$\delta S \; dt,$$

with a constant $\delta \geq 0$. This continuous dividend model can be easily built into the Black-Scholes framework. To this end the standard model of a geometric Brownian motion represented by the SDE (1.33) is generalized to

$$\frac{dS}{S} = (\mu - \delta)dt + \sigma dW.$$

The corresponding Black-Scholes equation for $V(S,t)$ is

$$\frac{\partial V}{\partial t} + \frac{\sigma^2}{2} S^2 \frac{\partial^2 V}{\partial S^2} + (r - \delta)S \frac{\partial V}{\partial S} - rV = 0. \tag{4.1}$$

This equation is equivalent to the equation

$$\frac{\partial y}{\partial \tau} = \frac{\partial^2 y}{\partial x^2} \tag{4.2}$$

for $y(x,\tau)$ with $0 \leq \tau$, $x \in \mathbb{R}$. This equivalence can be proved by means of the transformations

$$
\boxed{
\begin{aligned}
&S = Ke^x, \quad t = T - \frac{2\tau}{\sigma^2}, \quad q := \frac{2r}{\sigma^2}, \quad q_\delta := \frac{2(r-\delta)}{\sigma^2}, \\
&V(S,t) = V\left(Ke^x, T - \tfrac{2\tau}{\sigma^2}\right) =: v(x,\tau) \quad \text{and} \\
&v(x,\tau) =: K \exp\left\{-\tfrac{1}{2}(q_\delta - 1)x - \left(\tfrac{1}{4}(q_\delta - 1)^2 + q\right)\tau\right\} y(x,\tau).
\end{aligned}
}
\tag{4.3}
$$

For the slightly simpler case of no dividend payments ($\delta = 0$) the derivation was carried out earlier ($\longrightarrow$ Exercise 1.2). Intrinsic to the transformation (4.3) is that the constants r, σ, δ must be constants. The Black-Scholes equation in the version (4.1) has variable coefficients S^j with powers matching the order of the derivative with respect to S. That is, the relevant terms in (4.1) are of the type

$$S^j \frac{\partial^j V}{\partial S^j}, \quad \text{for } j = 0, 1, 2.$$

Linear differential equations with such terms are known as Euler's differential equations; their analysis motivates the transformation $S = K e^x$. The transformed version in equation (4.2) has constant coefficients ($=1$), which simplifies implementing numerical algorithms.

In view of the time transformation in (4.3) the expiration time $t = T$ is determined in the "new" time by $\tau = 0$, and $t = 0$ is transformed to $\tau = \frac{1}{2}\sigma^2 T$. Up to the scaling by $\frac{1}{2}\sigma^2$ the new time variable τ represents the remaining life time of the option. And the original domain of the half strip $S > 0, \, 0 \le t \le T$ belonging to (4.1) becomes the strip

$$-\infty < x < \infty, \quad 0 \le \tau \le \tfrac{1}{2}\sigma^2 T,$$

on which we are going to approximate a solution $y(x, \tau)$ to (4.2). After that calculation we again apply the transformations of (4.3) to derive out of $y(x, \tau)$ the value of the option $V(S, t)$ in the original variables.

Under the transformations (4.3) the terminal conditions (1.1C) and (1.1P) become **initial conditions** for $y(x, 0)$. A call, for example, satisfies

$$V(S, T) = \max\{S - K, 0\} = K \cdot \max\{e^x - 1, 0\}.$$

We see from (4.3) that

$$V(S, T) = K \exp\left\{-\frac{x}{2}(q_\delta - 1)\right\} y(x, 0),$$

and thus

$$y(x, 0) = \exp\left\{\frac{x}{2}(q_\delta - 1)\right\} \max\{e^x - 1, 0\}$$

$$= \begin{cases} \exp\left\{\frac{x}{2}(q_\delta - 1)\right\}(e^x - 1) & \text{for } x > 0 \\ 0 & \text{for } x \le 0. \end{cases}$$

Using

$$\exp\left\{\frac{x}{2}(q_\delta - 1)\right\}(e^x - 1) = \exp\left\{\frac{x}{2}(q_\delta + 1)\right\} - \exp\left\{\frac{x}{2}(q_\delta - 1)\right\}$$

we write the initial conditions $y(x, 0)$ in the new variables

$$\text{call:} \quad y(x, 0) = \max\left\{e^{\frac{x}{2}(q_\delta + 1)} - e^{\frac{x}{2}(q_\delta - 1)}, \; 0\right\} \tag{4.4C}$$

$$\text{put:} \quad y(x, 0) = \max\left\{e^{\frac{x}{2}(q_\delta - 1)} - e^{\frac{x}{2}(q_\delta + 1)}, \; 0\right\} \tag{4.4P}$$

The boundary-value problem is completed by boundary conditions for $x \to -\infty$ and $x \to +\infty$ (in Section 4.4).

The equation (4.2) is of the type of a parabolic partial differential equation and is the simplest diffusion or heat-conducting equation. Both equations (4.1) and (4.2) are linear in the dependent variables V or y. The differential equation (4.2) is also written $y_\tau = y_{xx}$ or $\dot{y} = y''$. The diffusion term is y_{xx}.

In principle, the methods of this chapter can be applied directly to (4.1). But the equations and algorithms are easier to derive for the algebraically equivalent version (4.2). Note that numerically the two equations are *not* equivalent. A direct application of this chapter's methods to version (4.1) can cause severe difficulties. This will be discussed in Chapter 6 in the context of Asian options. These difficulties will not occur for equation (4.2), which is well-suited for standard options with constant coefficients. The equation (4.2) is integrated in forward time —that is, for increasing τ starting from $\tau = 0$. This fact is important for stability investigations. For increasing τ the version (4.2) makes sense; this is equivalent to the well-posedness of (4.1) for *decreasing* t.

4.2 Foundations of Finite-Difference Methods

In this section we describe the basic ideas of finite differences as they are applied to the PDE (4.2).

4.2.1 Difference Approximation

Each two times continuously differentiable function f satisfies

$$f'(x) = \frac{f(x + h) - f(x)}{h} + \frac{h}{2} f''(\xi);$$

where ξ is an intermediate number between x and $x+h$. The accurate position of ξ is usually unknown. Such expressions are derived by Taylor expansions. We discretize $x \in \mathbb{R}$ by introducing a one-dimensional grid of discrete points x_i with

$$... < x_{i-1} < x_i < x_{i+1} < ...$$

For example, choose an equidistant grid with mesh size $h := x_{i+1} - x_i$. The x is discretized, but the function values $f_i := f(x_i)$ are not discrete, $f_i \in \mathbb{R}$. For $f \in \mathcal{C}^2$ the derivative f'' is bounded, and the term $\frac{h}{2} f''(\zeta)$ can be conveniently written as $O(h)$. This leads to the practical notation

$$f'(x_i) = \frac{f_{i+1} - f_i}{h} + O(h). \qquad (4.5)$$

Analogous expressions hold for the partial derivatives of $y(x, \tau)$, which includes a discretization in τ. This suggests to replace the neutral notation h by either Δx or $\Delta \tau$, respectively. The fraction in (4.5) is the difference quotient that approximates the differential quotient f' of the left-hand side; the $O(h^p)$-term is the error. The one-sided (i.e. nonsymmetric) difference quotient of (4.5) is of the order $p = 1$. Error orders of $p = 2$ are obtained by central differences

$$f'(x_i) = \frac{f_{i+1} - f_{i-1}}{2h} + O(h^2) \qquad \text{(for } f \in \mathcal{C}^3)$$

$$f''(x_i) = \frac{f_{i+1} - 2f_i + f_{i-1}}{h^2} + O(h^2) \qquad \text{(for } f \in \mathcal{C}^4)$$

or by one-sided differences that involve more terms, such as

$$f'(x_i) = \frac{-f_{i+2} + 4f_{i+1} - 3f_i}{2h} + O(h^2) \qquad \text{(for } f \in \mathcal{C}^3).$$

Rearranging terms and indices provides the approximation formula

$$f_i \approx \frac{4}{3}f_{i-1} - \frac{1}{3}f_{i-2} + \frac{2}{3}hf'(x_i) \;, \qquad \text{(BDF2)}$$

which is of second order. The latter difference quotient leads to one example of a *backward differentiation formula* (BDF). Equidistant grids are advantagous in that algorithms are easy to implement, and error terms are easily derivated by Taylor's expansion. This explains why this chapter works with equidistant grids.

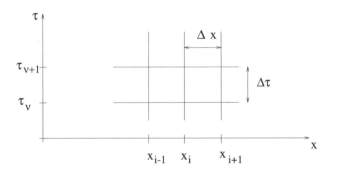

Fig. 4.1. Detail and notations of the grid

4.2.2 The Grid

Either the x-axis, or the τ-axis, or both can be discretized. If only one of the two independent variables x or τ is discretized, one obtains a semi-discretization consisting of parallel lines. This is used in Exercise 4.10 and in Section 4.8.3. Here we perform a full discretization leading to a two-dimensional grid.

Let $\Delta\tau$ and Δx be the mesh sizes of the discretizations of τ and x. The step in τ is $\Delta\tau := \tau_{\max}/\nu_{\max}$ for $\tau_{\max} := \frac{1}{2}\sigma^2 T$ and a suitable integer $\nu_{\max}$. The choice of the x-discretization is more complicated. The infinite interval $-\infty < x < \infty$ must be replaced by a finite interval $a \le x \le b$. Here the end values $a = x_{\min} < 0$ and $b = x_{\max} > 0$ must be chosen such that for the corresponding $S_{\min} = Ke^a$ and $S_{\max} = Ke^b$ and the interval $S_{\min} \le S \le S_{\max}$ a sufficient quality of approximation is obtained. For a suitable integer m the step length in x is defined by $\Delta x := (b-a)/m$. Additional notations for the grid are

$\tau_\nu := \nu \cdot \Delta\tau$ for $\nu = 0, 1, ..., \nu_{\max}$

$x_i := a + i\Delta x$ for $i = 0, 1, ..., m$

$y_{i\nu} := y(x_i, \tau_\nu)$,

$w_{i\nu}$ approximation to $y_{i\nu}$.

This defines a two-dimensional uniform grid as illustrated in Figure 4.1. (Writing the indices as in $y_{i\nu}$ is meant in the sense $y_{i,\nu}$.) Note that the equidistant grid in this chapter is defined in terms of x and τ, and not for S and t. Transforming the (x, τ)-grid via the transformation in (4.3) back to the (S, t)-plane, leads to a nonuniform grid with unequal distances of the grid lines $S = S_i = Ke^{x_i}$: The grid is increasingly dense close to $S_{\min}$. (This is not advantagous for the accuracy of the approximations of $V(S, t)$. We will come back to this in Section 5.2.) The Figure 4.1 illustrates only a small part of the entire grid in the (x, τ)-strip. The grid lines $x = x_i$ and $\tau = \tau_\nu$ can be indicated by their indices (Figure 4.2).

The points where the grid lines $\tau = \tau_\nu$ and $x = x_i$ intersect, are called *nodes*. In contrast to the theoretical solution $y(x, \tau)$, which is defined on a continuum, the $w_{i\nu}$ are only defined for the nodes. The error $w_{i\nu} - y_{i\nu}$ depends on the choice of parameters $\nu_{\max}$, m, $x_{\min}$, $x_{\max}$. A priori we do not know which choice of parameters matches a prespecified error tolerance. An example of the order of magnitude of these parameters is given by $x_{\min} = -5$, $x_{\max} = 5$, $\nu_{\max} = 100$, $m = 100$. This choice of $x_{\min}$, $x_{\max}$ has shown to be reasonable for a wide range of r, σ-values and accuracies. The actual error is then controlled via the numbers $\nu_{\max}$ und m of grid lines.

4.2.3 Explicit Method

Substituting

$$\frac{\partial y_{i\nu}}{\partial \tau} = \frac{y_{i,\nu+1} - y_{i\nu}}{\Delta\tau} + O(\Delta\tau)$$

$$\frac{\partial^2 y_{i\nu}}{\partial x^2} = \frac{y_{i+1,\nu} - 2y_{i\nu} + y_{i-1,\nu}}{\Delta x^2} + O(\Delta x^2)$$

into (4.2) and discarding the error terms leads to the equation

$$\frac{w_{i,\nu+1} - w_{i\nu}}{\Delta\tau} = \frac{w_{i+1,\nu} - 2w_{i\nu} + w_{i-1,\nu}}{\Delta x^2}$$

for the approximation w. Solving for $w_{i,\nu+1}$ we obtain

$$w_{i,\nu+1} = w_{i,\nu} + \frac{\Delta\tau}{\Delta x^2}(w_{i+1,\nu} - 2w_{i\nu} + w_{i-1,\nu}).$$

With the abbreviation

$$\lambda := \frac{\Delta\tau}{\Delta x^2}$$

the result is written compactly

$$w_{i,\nu+1} = \lambda w_{i-1,\nu} + (1 - 2\lambda)w_{i\nu} + \lambda w_{i+1,\nu} \qquad (4.6)$$

The Figure 4.2 accentuates the nodes that are connected by this formula. Such a graphical scheme that illustrates the structure of the equation, is called *molecule*.

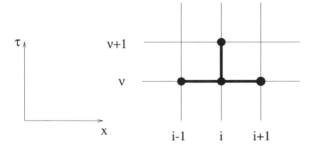

Fig. 4.2. Connection scheme of the explicit method

The equation (4.6) and the Figure 4.2 suggest an evaluation organized by time levels. All nodes with the same index ν form the ν-th time level. For a fixed ν the values $w_{i,\nu+1}$ for all i of the time level $\nu + 1$ are calculated. Then we advance to the next time level. The formula (4.6) is an explicit expression for each of the $w_{i,\nu+1}$; the values w at level $\nu + 1$ are not coupled. Since (4.6) provides an explicit formula for all $w_{i,\nu+1}$ $(i = 0, 1, ..., m)$, this method is called *explicit method* or *forward-difference method*.

Start: For $\nu = 0$ the values of w_{i0} are given by the initial conditions

$$w_{i0} = y(x_i, 0) \quad \text{for } y \text{ from (4.4)}, 0 \leq i \leq m.$$

The $w_{0\nu}$ and $w_{m\nu}$ for $1 \leq \nu \leq \nu_{\max}$ are fixed by boundary conditions. Provisonally we take $w_{0\nu} = w_{m\nu} = 0$. The correct boundary conditions are deferred to Section 4.4.

For the following analysis it is useful to collect all values w of the time level ν into a vector,

$$w^{(\nu)} := (w_{1\nu}, ..., w_{m-1,\nu})^{tr}.$$

The next step towards a vector notation of the explicit method is to introduce the constant $(m-1) \times (m-1)$ tridiagonal matrix

$$A := A_{\text{expl}} := \begin{pmatrix} 1-2\lambda & \lambda & 0 & \cdots & 0 \\ \lambda & 1-2\lambda & \ddots & \ddots & \vdots \\ 0 & \ddots & \ddots & \ddots & 0 \\ \vdots & \ddots & \ddots & \ddots & \lambda \\ 0 & \cdots & 0 & \lambda & \ddots \end{pmatrix}. \tag{4.7a}$$

Now the explicit method in matrix-vector notation is written

$$w^{(\nu+1)} = Aw^{(\nu)} \quad \text{for } \nu = 0, 1, 2, ... \tag{4.7b}$$

The formulation of (4.7) with the matrix A and the iteration (4.7b) is needed only for theoretical investigations. An actual computer program would rather use the version (4.6). The inner-loop index i does not occur explicitly in the vector notation of (4.7).

To illustrate the behavior of the explicit method, we perform an experiment with an artificial example, where initial conditions and boundary conditions are not related to finance.

Example 4.1

$y_\tau = y_{xx}$, $y(x,0) = \sin \pi x$, $x_0 = 0$, $x_m = 1$, boundary conditions $y(0,\tau) = y(1,\tau) = 0$ (that is, $w_{0\nu} = w_{m\nu} = 0$).

The aim is to calculate an approximation w for one (x,τ), for example, for $x = 0.2$, $\tau = 0.5$. The exact solution is $y(x,\tau) = e^{-\pi^2 \tau} \sin \pi x$, such that $y(0.2, 0.5) = 0.004227....$ We carry out two calculations with the same $\Delta x = 0.1$ (hence $0.2 = x_2$), and two different $\Delta \tau$:

(a) $\Delta\tau = 0.0005 \implies \lambda = 0.05$
  $0.5 = \tau_{1000}$, $\quad w_{2,1000} \doteq 0.00435$
(b) $\Delta\tau = 0.01 \implies \lambda = 1$,
  $0.5 = \tau_{50}$, $\quad w_{2,50} \doteq -1.5 * 10^8$ (the actual numbers depend on the computer)

It turns out that the choice of $\Delta\tau$ in (a) has led to a reasonable approximation, whereas the choice in (b) has caused a disaster. Here we have a stability problem!

4.2.4 Stability

Let us perform an error analysis of the iteration $w^{(\nu+1)} = Aw^{(\nu)}$. In general we use the same notation w for the theoretical definition of w and for the values of w that are obtained by numerical calculations in a computer. Since we now discuss rounding errors, we must distinguish between the two meanings. Let $w^{(\nu)}$ denote the vectors theoretically defined by (4.7). Hence, by definition, the $w^{(\nu)}$ are free of rounding errors. But in computational reality, rounding errors are inevitable. We denote the computer-calculated vector by $\bar{w}^{(\nu)}$ and the error vectors by

$$e^{(\nu)} := \bar{w}^{(\nu)} - w^{(\nu)},$$

for $\nu \geq 0$. The result in a computer can be written

$$\bar{w}^{(\nu+1)} = A\bar{w}^{(\nu)} + r^{(\nu+1)},$$

where $r^{(\nu+1)}$ amounts to rounding errors that occur during the calculation of $A\bar{w}^{(\nu)}$. Let us concentrate on the effect of the rounding errors that occur for an arbitrary ν, say for $\nu = \nu^*$. We ask for the propagation of this error for increasing $\nu > \nu^*$. Without loss of generality we set $\nu^* = 0$, and for simplicity take $r^{(\nu)} = 0$ for $\nu > 1$. That is, we investigate the effect the initial rounding error $e^{(0)}$ has on the iteration. The initial error $e^{(0)}$ represents the rounding error during the evaluation of the initial condition (4.4), when $\bar{w}^{(0)}$ is calculated. According to this scenario we have $\bar{w}^{(\nu+1)} = A\bar{w}^{(\nu)}$. The relation

$$Ae^{(\nu)} = A\bar{w}^{(\nu)} - Aw^{(\nu)} = \bar{w}^{(\nu+1)} - w^{(\nu+1)} = e^{(\nu+1)}$$

between consecutive errors is applied repeatedly and results in

$$e^{(\nu)} = A^\nu e^{(0)}. \tag{4.8}$$

For the method to be *stable*, previous errors must be damped. This leads to require $A^\nu e^{(0)} \to 0$ for $\nu \to \infty$. Elementwise this means $\lim_{\nu \to \infty}\{(A^\nu)_{ij}\} = 0$ for $\nu \to \infty$ and for any pair of indices (i, j). The following lemma provides a criterion for this requirement.

Lemma 4.2

$$\rho(A) < 1 \iff A^\nu z \to 0 \text{ for all } z \text{ and } \nu \to \infty$$
$$\iff \lim_{\nu \to \infty}\{(A^\nu)_{ij}\} = 0$$

Here $\rho(A)$ is the *spectral radius* of A,

$$\rho(A) := \max_i |\mu_i^A|,$$

where $\mu_1^A, ..., \mu_{m-1}^A$ are the eigenvalues of A. The proof can be found in text books of numerical analysis, for example, in [IK66]. As a consequence of Lemma 4.2 we require for stable behavior that $|\mu_i^A| < 1$ for all eigenvalues, here for $i = 1, ..., m - 1$. To check the criterion of Lemma 4.2, the eigenvalues μ_i^A of A are needed. To this end we split the matrix A into

$$\Lambda = I - \lambda \cdot \underbrace{\begin{pmatrix} 2 & -1 & & & & 0 \\ -1 & \ddots & & \ddots & & \\ & & \ddots & \ddots & \ddots & \\ & & & \ddots & \ddots & \ddots \\ 0 & & & & \ddots & \ddots \end{pmatrix}}_{=:G}.$$

It remains to investigate the eigenvalues μ^G of the tridiagonal matrix G. These eigenvalues are provided by

Lemma 4.3

Let $\quad G = \begin{pmatrix} \alpha & \beta & & & 0 \\ \gamma & \ddots & \ddots & & \\ & \ddots & \ddots & \ddots & \\ & & \ddots & \ddots & \beta \\ 0 & & & \gamma & \alpha \end{pmatrix} \quad$ be an N^2-matrix.

The eigenvalues μ_k^G and the eigenvectors $v^{(k)}$ of G are

$$\mu_k^G = \alpha + 2\beta \sqrt{\frac{\gamma}{\beta}} \cos \frac{k\pi}{N+1} \ , \quad k = 1, ..., N,$$

$$v^{(k)} = \left(\sqrt{\frac{\gamma}{\beta}} \sin \frac{k\pi}{N+1}, \left(\sqrt{\frac{\gamma}{\beta}}\right)^2 \sin \frac{2k\pi}{N+1}, ..., \left(\sqrt{\frac{\gamma}{\beta}}\right)^N \sin \frac{Nk\pi}{N+1} \right)^{tr}.$$

Proof: Substitute into $Gv = \mu^G v$.

To apply the lemma observe $N = m - 1$, $\alpha = 2$, $\beta = \gamma = -1$, and obtain the eigenvalues μ^G and finally the eigenvalues μ^A of A:

$$\mu_k^G = 2 - 2\cos \frac{k\pi}{m} = 4\sin^2 \left(\frac{k\pi}{2m}\right)$$

$$\mu_k^A = 1 - 4\lambda \sin^2 \frac{k\pi}{2m}$$

Now we can state the stability requirement $|\mu_k^A| < 1$ as

$$\left| 1 - 4\lambda \sin^2 \frac{k\pi}{2m} \right| < 1, \quad k = 1, ..., m - 1.$$

This implies the two inequalities $\lambda > 0$ and

$$-1 < 1 - 4\lambda \sin^2 \frac{k\pi}{2m}, \quad \text{rewritten as} \quad \frac{1}{2} > \lambda \sin^2 \frac{k\pi}{2m}.$$

The largest sin-term is $\sin \frac{(m-1)\pi}{2m}$; for increasing m this term grows monotonically approaching 1.

In summary we have shown

For $0 < \lambda \le \dfrac{1}{2}$ the explicit method $w^{(\nu+1)} = Aw^{(\nu)}$ is stable.

In view of $\lambda = \Delta\tau/\Delta x^2$ this stability criterion amounts to bounding the $\Delta\tau$ step size,

$$0 < \Delta\tau \le \frac{\Delta x^2}{2} \tag{4.9}$$

This explains what happened with Example 4.1. The values of λ in the two cases of this example are

$$\text{(a)} \quad \lambda = 0.05 \le \frac{1}{2}$$

$$\text{(b)} \quad \lambda = 1 > \frac{1}{2}$$

In case (b) the chosen $\Delta\tau$ and hence λ were too large, which led to an amplification of rounding errors resulting eventually in the "explosion" of the w-values.

The explicit method is stable only as long as (4.9) is satisfied. As a consequence, the parameters m and $\nu_{\max}$ of the grid resolution can not be chosen independent of each other. If the demands for accuracy are high, the step size Δx will be small, which in view of (4.9) bounds $\Delta\tau$ quadratically. This situation suggests searching for a method that is unconditionally stable.

4.2.5 An Implicit Method

When we introduced the explicit method in Subsection 4.2.3, we approximated the time derivative with a forward difference, "forward" as seen from the ν-th time level. Now we try the backward difference

$$\frac{\partial y_{i\nu}}{\partial \tau} = \frac{y_{i\nu} - y_{i,\nu-1}}{\Delta\tau} + O(\Delta\tau),$$

which yields the alternative to (4.6)

$$-\lambda w_{i+1,\nu} + (2\lambda+1)w_{i\nu} - \lambda w_{i-1,\nu} = w_{i,\nu-1} \tag{4.10}$$

The equation (4.10) relates the time level ν to the time level $\nu - 1$. For the transition from $\nu-1$ to ν only the value $w_{i,\nu-1}$ on the right-hand side of (4.10) is known, whereas on the left-hand side three unknown values of w wait to be computed. Equation (4.10) couples three unknowns. The corresponding molecule is shown in Figure 4.3. There is no simple explicit formula with which the unknown can be obtained one after the other. Rather a system must be considered, all equations simultaneously. A vector notation reveals the structure of (4.10): With the matrix

$$A := A_{\text{impl}} := \begin{pmatrix} 2\lambda+1 & -\lambda & & & 0 \\ -\lambda & \ddots & \ddots & & \\ & \ddots & \ddots & \ddots & \\ & & \ddots & \ddots & \ddots \\ 0 & & & \ddots & \ddots \end{pmatrix} \qquad (4.11a)$$

the vector $w^{(\nu)}$ is implicitly defined as solution of the system of linear equations

$$Aw^{(\nu)} = w^{(\nu-1)} \quad \text{for } \nu = 1, ..., \nu_{\max} \qquad (4.11b)$$

Here we again have assumed $w_{0\nu} = w_{m\nu} = 0$. For each time level ν such a system of equations must be solved. This method is sometimes called *implicit method*. But to distinguish it from other implicit methods, we call it *fully implicit*, or *backward-difference method*, or more accurately *backward time centered space* scheme (BTCS). The method is unconditionally stable for all $\Delta\tau > 0$. This is shown analogously as in the explicit case ($\longrightarrow$ Exercise 4.2). The costs of this implicit method are low, because the matrix A is constant and tridiagonal. Initially, for $\nu = 0$, the LR-decomposition ($\longrightarrow$ Appendix C1) is calculated once. Then the costs for each ν are only of the order $O(m)$.

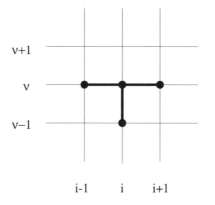

Fig. 4.3. Molecule of the backward-difference method (4.10)

4.3 Crank-Nicolson Method

For the methods of the previous section the discretizations of $\frac{\partial y}{\partial \tau}$ are of the order $O(\Delta\tau)$. It seems preferable to use a method where the time discretization of $\frac{\partial y}{\partial \tau}$ has the better order $O(\Delta\tau^2)$ and the stability is unconditional. Let us again consider equation (4.2), the equivalent to the Black-Scholes equation,

$$\frac{\partial y}{\partial \tau} = \frac{\partial^2 y}{\partial x^2}.$$

Crank and Nicolson suggested to average the forward- and the backward difference method. For easy reference, let us sum up the underlying approaches:

forward for ν:

$$\frac{w_{i,\nu+1} - w_{i\nu}}{\Delta\tau} = \frac{w_{i+1,\nu} - 2w_{i\nu} + w_{i-1,\nu}}{\Delta x^2}$$

backward for $\nu + 1$:

$$\frac{w_{i,\nu+1} - w_{i\nu}}{\Delta\tau} = \frac{w_{i+1,\nu+1} - 2w_{i,\nu+1} + w_{i-1,\nu+1}}{\Delta x^2}$$

Addition yields

$$\frac{w_{i,\nu+1} - w_{i\nu}}{\Delta\tau} = \frac{1}{2\Delta x^2}(w_{i+1,\nu} - 2w_{i\nu} + w_{i-1,\nu} + w_{i+1,\nu+1} - 2w_{i,\nu+1} + w_{i-1,\nu+1})$$

(4.12)

The equation (4.12) involves in each of the time levels ν and $\nu + 1$ three values w (Figure 4.4). This is the basis of an efficient method. Its features are as follows:

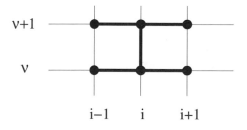

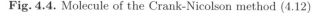

Fig. 4.4. Molecule of the Crank-Nicolson method (4.12)

Theorem 4.4   (Crank-Nicolson)
  Suppose y is smooth in the sense $y \in \mathcal{C}^4$. Then:
  1.) The order of the method is $O(\Delta\tau^2) + O(\Delta x^2)$.
  2.) For each ν a linear system of a simple tridiagonal structure must be solved.
  3.) Stability holds for all $\Delta\tau > 0$.

Proof:

1.) order: A practical notation for the symmetric difference quotient of second order for y_{xx} is

$$\delta_x^2 w_{i\nu} := \frac{w_{i+1,\nu} - 2w_{i\nu} + w_{i-1,\nu}}{\Delta x^2}. \tag{4.13}$$

Apply the operator δ_x^2 to the exact solution y. Then by Taylor expansion for $y \in \mathcal{C}^4$ one can show

$$\delta_x^2 y_{i\nu} = \frac{\partial^2}{\partial x^2} y_{i\nu} + \frac{\Delta x^2}{12} \frac{\partial^4}{\partial x^4} y_{i\nu} + O(\Delta x^4).$$

The *local discretization error* ϵ describes how well the exact solution y of (4.2) satisfies the difference scheme,

$$\epsilon := \frac{y_{i,\nu+1} - y_{i\nu}}{\Delta \tau} - \frac{1}{2}(\delta_x^2 y_{i\nu} + \delta_x^2 y_{i,\nu+1}).$$

Applying the operator δ_x^2 of (4.13) to the expansion of $y_{i,\nu+1}$ at τ_ν and observing $y_\tau = y_{xx}$ leads to

$$\epsilon = O(\Delta \tau^2) + O(\Delta x^2)$$

($\longrightarrow$ Exercise 4.3)

2.) system of equations: With $\lambda := \frac{\Delta \tau}{\Delta x^2}$ the equation (4.12) is rewritten

$$\boxed{\begin{aligned} &-\frac{\lambda}{2} w_{i-1,\nu+1} + (1+\lambda)w_{i,\nu+1} - \frac{\lambda}{2} w_{i+1,\nu+1} \\ &= \frac{\lambda}{2} w_{i-1,\nu} + (1-\lambda)w_{i\nu} + \frac{\lambda}{2} w_{i+1,\nu} \end{aligned}} \tag{4.14}$$

The values of the new time level $\nu + 1$ are implicitly given by the system of equations (4.14). For the simplest boundary conditions $w_{0\nu} = w_{m\nu} = 0$ equation (4.14) is a system of $m - 1$ equations. With matrices

$$A := A_{\mathrm{CN}} := \begin{pmatrix} 1+\lambda & -\frac{\lambda}{2} & & & 0 \\ -\frac{\lambda}{2} & \ddots & \ddots & & \\ & \ddots & \ddots & \ddots & \\ 0 & & & \ddots & \ddots \end{pmatrix},$$

$$B := B_{\mathrm{CN}} := \begin{pmatrix} 1-\lambda & \frac{\lambda}{2} & & & 0 \\ \frac{\lambda}{2} & \ddots & \ddots & & \\ & \ddots & \ddots & \ddots & \\ 0 & & & \ddots & \ddots \end{pmatrix} \tag{4.15a}$$

the system (4.14) is rewritten

$$Aw^{(\nu+1)} = Bw^{(\nu)}. \tag{4.15b}$$

The eigenvalues of A are real and lie between 1 and $1+2\lambda$. (This follows from the Theorem of Gerschgorin, see Appendix C1). This rules out a zero eigenvalue, and so A must be nonsingular and the solution of (4.15b) is uniquely defined.

3.) stability: The matrices A and B can be rewritten in terms of a constant tridiagonal matrix,

$$A = I + \tfrac{\lambda}{2}G, \quad G := \begin{pmatrix} 2 & -1 & & & 0 \\ -1 & \ddots & \ddots & & \\ & \ddots & \ddots & \ddots & \\ & & \ddots & \ddots & \ddots \\ 0 & & & \ddots & \ddots \end{pmatrix}, \quad B = I - \tfrac{\lambda}{2}G.$$

Now the equation (4.15b) reads

$$\underbrace{(2I + \lambda G)}_{=:C} w^{(\nu+1)} = (2I - \lambda G)w^{(\nu)}$$

$$= (4I - 2I - \lambda G)w^{(\nu)}$$

$$= (4I - C)w^{(\nu)},$$

which leads to the formally explicit iteration

$$w^{(\nu+1)} = (4C^{-1} - I)w^{(\nu)}. \tag{4.16}$$

The eigenvalues μ_k^C of C for $k = 1, ..., m-1$ are known from Lemma 4.3,

$$\mu_k^C = 2 + \lambda\mu_k^G = 2 + \lambda(2 - 2\cos\frac{k\pi}{m}) = 2 + 4\lambda\sin^2\frac{k\pi}{2m}.$$

In view of (4.16) we require for a stable method that for all k

$$\left| \frac{4}{\mu_k^C} - 1 \right| < 1.$$

This is guaranteed because of $\mu_k^C > 2$. Consequently, the Crank-Nicolson method (4.15) is unconditionally stable for all $\lambda > 0$ $(\Delta\tau > 0)$.

Although the correct boundary conditions are still lacking, it makes sense to formulate the basic version of the Crank-Nicolson algorithm for the PDE (4.2).

Algorithm 4.5   (Crank-Nicolson)

Start: Choose m, $\nu_{\max}$; calculate $\Delta x, \Delta \tau$

$w_i^{(0)} = y(x_i, 0)$ with y from (4.4), $0 \le i \le m$

Calculate the LR-decomposition of A

loop: *for* $\nu = 0, 1, ..., \nu_{\max} - 1$:

Calculate $c := Bw^{(\nu)}$ (preliminary)

Solve $Ax = c$ using the LR-decomposition—

that is, solve $Lz = Bw^{(\nu)}$ and $Rx = z$

$w^{(\nu+1)} := x$

It is obvious that the matrices A and B are not stored in the computer. —
Next we show how the vector c in Algorithm 4.5 is modified to realize the
correct boundary conditions.

4.4 Boundary Conditions

The Black-Scholes equation (4.1), the transformed version (4.2), and the discretized versions of the previous sections, they all need boundary conditions.
In particular, the values

$V(S, t)$ for $S = 0$ and $S \to \infty$, or

$y(x, \tau)$ for $x_{\min}$ and $x_{\max}$, or

$w_{0\nu}$ and $w_{m\nu}$ for $\nu = 1, ..., \nu_{\max}$,

respectively, must be prescribed by boundary conditions. The preliminary
homogenous boundary conditions $w_{0\nu} = w_{m\nu} = 0$ of the previous sections do
not match the scenario of Black, Merton and Scholes. In order to complete
and adapt the Algorithm 4.5 we must define realistic boundary conditions.

The boundary conditions for the expiration time $t = T$ are obvious. They
give rise to the simplest cases of boundary conditions for $t < T$: As motivated
by the Figures 1.1 and 1.2 and the equations (1.1C), (1.1P), the value V_C of
a call and the value V_P of a put must satisfy

$$
\begin{aligned}
V_C(S, t) = 0 \quad \text{for } S = 0, \text{ and} \\
V_P(S, t) = 0 \quad \text{for } S \to \infty
\end{aligned}
\tag{4.17}
$$

also for all $t < T$. This follows from the integral representation (3.19), because
discounting does not affect the value 0 of the payoff. And $S(0) = 0$ implies
because of $dS = S(\mu dt + \sigma dW)$ that $S(t) = 0$ for all $t > 0$; hence the value
$V_C(0, t) = 0$ can be predicted safely. The same holds true for $S(0) \to \infty$ and

V of (1.1P). This holds for European as well as for American options, with or without dividend payments.

The boundary conditions on each of the "other sides" of S, where $V \neq 0$, are more difficult. We postpone the boundary conditions for the American option to the next section, and investigate European options in this section.

From the put-call parity ($\longrightarrow$ Exercise 1.1) we deduce the additional boundary conditions for European options without dividend payment ($\delta = 0$). The result is

$$V_C(S,t) = S - Ke^{-r(T-t)} \quad \text{for } S \to \infty$$
$$V_P(S,t) = Ke^{-r(T-t)} - S \quad \text{for } S \approx 0. \tag{4.18}$$

The lower bounds for European options ($\longrightarrow$ Appendix D1) are attained at the boundaries. In (4.18) for $S \approx 0$ we do not discard the term S, because the realization of the transformation (4.3) requires $S_{\min} > 0$, see Section 4.2.2. Boundary conditions analogous as in (4.18) hold for the case of a continuous flow of dividend payments ($\delta \neq 0$). We skip the derivation, which can be based on transformation (4.3) and the additional transformation $S = \overline{S}e^{\delta(T-t)}$ ($\longrightarrow$ Exercise 4.4). In summary, the boundary conditions for European options in the (x, τ)-world are as follows:

Boundary Conditions 4.6   (European options)

$$y(x,\tau) = r_1(x,\tau) \text{ for } x \to -\infty,$$
$$y(x,\tau) = r_2(x,\tau) \text{ for } x \to \infty, \quad \text{with}$$
$$\text{call: } r_1(x,\tau) := 0,$$
$$r_2(x,\tau) := \exp\left(\tfrac{1}{2}(q_\delta + 1)x + \tfrac{1}{4}(q_\delta + 1)^2\tau\right) \tag{4.19}$$
$$\text{put: } r_1(x,\tau) := \exp\left(\tfrac{1}{2}(q_\delta - 1)x + \tfrac{1}{4}(q_\delta - 1)^2\tau\right),$$
$$r_2(x,\tau) := 0$$

Truncation: Instead of the theoretical domain $-\infty < x < \infty$ the practical realization truncates the infinite interval to the finite interval

$$a := x_{\min} \leq x \leq x_{\max} =: b,$$

see Section 4.2.2. This suggests the boundary conditions

$$w_{0\nu} = r_1(a, \tau_\nu)$$
$$w_{m\nu} = r_2(b, \tau_\nu)$$

for all ν. These are explicit formulas and easy to implement. To this end return to the Crank-Nicolson equation (4.14), in which some of the terms on both sides of the equations are known by the boundary conditions. For the equation with $i = 1$ these are terms

$$\text{from the left-hand side: } -\frac{\lambda}{2}w_{0,\nu+1} = -\frac{\lambda}{2}r_1(a,\tau_{\nu+1})$$

$$\text{from the right-hand side: } \frac{\lambda}{2}w_{0\nu} = \frac{\lambda}{2}r_1(a,\tau_\nu)$$

and for $i = m - 1$:

$$\text{from the left-hand side: } -\frac{\lambda}{2}w_{m,\nu+1} = -\frac{\lambda}{2}r_2(b,\tau_{\nu+1})$$

$$\text{from the right-hand side: } \frac{\lambda}{2}w_{m\nu} = \frac{\lambda}{2}r_2(b,\tau_\nu)$$

These known boundary values are collected on the right-hand side of system (4.14). So we finally arrive at

$$Aw^{(\nu+1)} = Bw^{(\nu)} + d^{(\nu)}$$

$$d^{(\nu)} := \frac{\lambda}{2} \cdot \begin{pmatrix} r_1(a,\tau_{\nu+1}) + r_1(a,\tau_\nu) \\ 0 \\ \vdots \\ 0 \\ r_2(b,\tau_{\nu+1}) + r_2(b,\tau_\nu) \end{pmatrix} \tag{4.20}$$

The previous version (4.15b) is included as special case, with $d^{(\nu)} = 0$. The statement in Algorithm 4.5 that defines c is modified to the statement

$$\text{Calculate } c := Bw^{(\nu)} + d^{(\nu)}.$$

The methods of Section 4.2 can be adapted by analogous formulas. The stability is not affected by adding the vector d, which is constant with respect to w.

4.5 American Options as Free Boundary Problems

In Sections 4.1 through 4.3 we so far have considered tools for the Black-Scholes differential equation —that is, we have investigated European options. Now we turn our attention to American options. Recall that the value of an American option can never be smaller than the value of a European option,

$$V^{\text{am}} \geq V^{\text{eur}}.$$

In addition, an American option has at least the value of the payoff. So we have elementary lower bounds for the value of American options, but —as we will see— additional numerical problems to cope with.

4.5.1 Early-Exercise Curve

A European option can have a value that is smaller than the payoff (compare, for example, Figure 1.6). This can not happen with American options. Recall the arbitrage strategy: if for instance an American put would have a value $V_P^{am} < (K - S)^+$, one would simultaneously purchase the asset and the put, and exercise immediately. An analogous arbitrage argument implies that for an American call the situation $V_C^{am} < (S - K)^+$ can not prevail. Therefore the inequalities

$$V_P^{am}(S, t) \geq (K - S)^+ \quad \text{for all } (S, t)$$
$$V_C^{am}(S, t) \geq (S - K)^+ \quad \text{for all } (S, t) \tag{4.21}$$

hold. This result is illustrated schematically for a put in Figure 4.5.

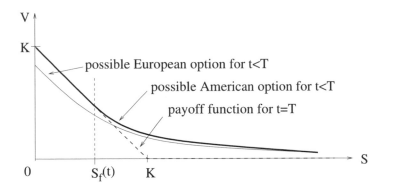

Fig. 4.5. $V(S, t)$ for a put and a $t < T$, schematically

For American options we have noted in (4.17) the boundary conditions that prescribe $V = 0$. The boundary conditions at each of the other "ends" of the S-axis are still needed. In view of the inequalities (4.21) it is clear that the missing boundary conditions will be of a different kind than those for European options, which are listed in (4.18). Let us investigate the situation of an **American put**, which is illustrated in Figure 4.5. First discuss the left-end part of the *curve* $V_P(S, t)$, for small $S < 0$, and some $t < T$. Without the possibility of early exercise the inequality $V_P(S, t) < K - S$ holds for sufficiently small S. But in view of (4.21) the American put should satisfy $V_P(S, t) \equiv K - S$ for small S. To understand what happens for "medium" values of S, imagine to approach from the right, where $V_P^{am}(S, t) > (K - S)^+$. Continuity and monotony of V_P suggest the curve $V_P^{am}(S, t)$ hits the straight line of the payoff at some value S_f with $0 < S_f < K$, see Figure 4.5. This **contact point** S_f is defined by

$$V_P^{am}(S, t) > (K - S)^+ \quad \text{for } S > S_f(t),$$
$$V_P^{am}(S, t) = K - S \quad \text{for } S \leq S_f(t). \tag{4.22}$$

For $S < S_f$ the value V_P^{am} equals the straight line of the payoff and nothing needs to be calculated. For each t, the *curve* $V_P^{am}(S, t)$ reaches its left boundary at $S_f(t)$.

The above situation holds for any $t < T$, and the contact point S_f varies with t, $S_f = S_f(t)$. For all $0 \le t < T$, the contact points form a curve in the (S, t)-half strip. The curve S_f is the boundary separating the area with $V >$ payoff and the area with $V =$ payoff. The curve S_f of a put is illustrated in the left-hand diagram of Figure 4.6. A priori the location of the boundary S_f is unknown, the curve is "free." This explains why the problem of calculating $V_P^{am}(S, t)$ for $S > S_f(t)$ is called **free boundary problem**.

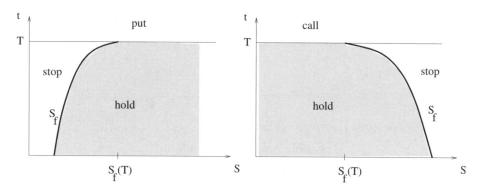

Fig. 4.6. Continuation region (shaded) and stopping region for American options

For **American calls** the situation is similar, except that the contact only occurs for dividend-paying assets, $\delta \neq 0$. This is seen from

$$V_C^{am} \ge V_C^{eur} \ge S - Ke^{-r(T-t)} > S - K$$

for $\delta = 0$, $r \neq 0$, $t < T$, compare Exercise 1.1. $V_C^{am} > S - K$ for $\delta = 0$ implies that early-exercise does not pay. American and European calls on assets that pay no dividends are identical, $V_C^{am} = V_C^{eur}$. A typical curve $V_C^{am}(S, t)$ for $\delta \neq 0$ contacting the payoff is shown in Figure 4.9. And the free boundary S_f may look like the right-hand diagram of Figure 4.6.

Our notation $S_f(t)$ for the free boundary is motivated by the process of solving PDEs. But the primary meaning of the curve S_f is economical. The free boundary S_f is the **early-exercise curve**. The time instance t_s when a price process S_t reaches the early-exercise curve is the optimal stopping time, compare also the illustration of Figure 3.8. Let us explain this for the case of a put; for a call with $\delta \neq 0$ the argument is similar.

In case $S > S_f$, early-exercise causes an immediate loss, because (4.22) implies $-V + K - S < 0$. Receiving the strike price K does not compensate the loss of S and V. Accordingly, the holder of the option does not exercise when $S > S_f$. This explains why the area $S > S_f$ is called **continuation region** (shaded in Figure 4.6). On the other side of the boundary curve

S_f, characterized by $V = K - S$, each change of S is compensated by a corresponding move of V. Here the only way to create a profit is to exercise and invest the proceeds K at the risk-free rate for the remaining time period $T - t$. The resulting profit will be

$$Ke^{r(T-t)} - K .$$

To maximize the profit, the holder of the option will maximize $T - t$, and accordingly exercise as soon as $V \equiv K - S$ is reached. Hence, the boundary curve S_f is the early-exercise curve. And the area $S \leq S_f$ is called **stopping region**. — So much for the basic principle. Of course, the profit depends on $r > 0$, and the holder must watch the market, see [Hull00].

Now that the curve S_f is recognized as having such a distinguished importance as early-exercise curve, we should make sure that the properties of S_f are as suggested by Figures 4.5 and 4.6. In fact, the curves $S_f(t)$ are continuously differentiable in t, and monotonous not decreasing / not increasing as illustrated. There are both upper and lower bounds to $S_f(t)$. For more details and proofs see Appendix A5. Here we confine ourselves to the bounds given by the limit $t \to T$ ($t < T$, $\delta > 0$):

$$\text{put:} \quad \lim_{t \to T} S_f(t) = \min(K, \frac{r}{\delta}K) \qquad (4.23\text{P})$$

$$\text{call:} \quad \lim_{t \to T} S_f(t) = \max(K, \frac{r}{\delta}K) \qquad (4.23\text{C})$$

These bounds express a qualitatively different behavior of the early-exercise curve in the two situations $0 < \delta < r$ and $\delta > r$. This is illustrated in Figure 4.7 for a put. For the chosen numbers for all $\delta \geq 0.06$ the limit of (4.23P) is the strike K (lower diagram). Compare to Figures 1.4 and 1.5, and to the title figure of this book to get a feeling for the geometrical importance of the curve as contact line where two surfaces merge. For larger values of S the surface $V(S,t)$ approaches 0 in a way illustrated by Figure 4.8.

4.5.2 Free Boundary Problem

Again we start with a put. For the European option, the left-end boundary condition is formulated for $S = 0$. For the American option, the left-end boundary is given along the curve S_f. In order to calculate the free boundary $S_f(t)$ we need an additional condition. To this end consider the slope $\frac{\partial V}{\partial S}$ with which $V_P^{am}(S,t)$ touches at $S_f(t)$ the straight line $K - S$, which has the constant slope -1. By geometrical reasons we can rule out for V_P^{am} the case $\frac{\partial V(S_f(t),t)}{\partial S} < -1$, because otherwise (4.21) and (4.22) would be violated. Using arbitrage arguments, the case $\frac{\partial V(S_f(t),t)}{\partial S} > -1$ can also be ruled out ($\longrightarrow$ Exercise 4.9). It remains the condition $\partial V_P^{am}(S_f(t),t)/\partial S = -1$. That is, $V(S,t)$ touches the payoff function *tangentially*. This tangency condition

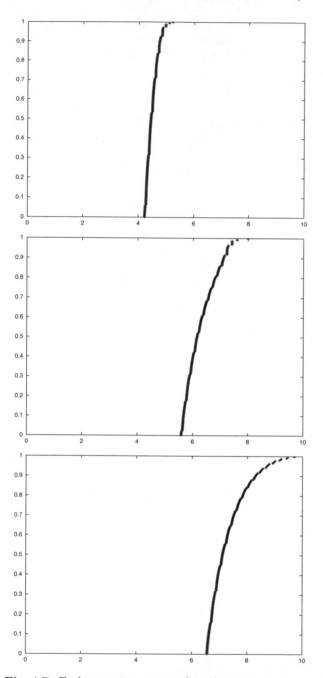

Fig. 4.7. Early-exercise curves of an American put, $r = 0.06$, $\sigma = 0.3$, $K = 10$, and dividend rates $\delta = 0.12$ (top figure), $\delta = 0.08$ (middle), $\delta = 0.04$ (bottom); raw data of a finite-difference calculation without interpolation or smoothing

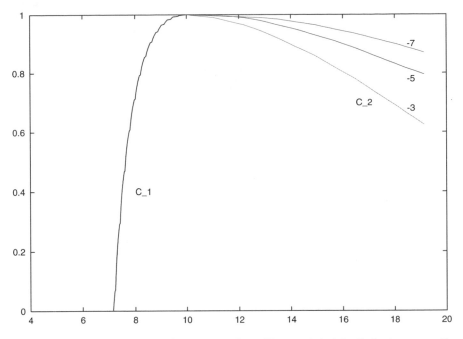

Fig. 4.8. Calculated curves of a put matching Figures 1.4, 1.5. C_1 is the curve S_f. The three curves C_2 have the meaning $V < 10^{-k}$ for $k = 3, 5, 7$.

is commonly called the *high-contact condition*, or *smooth pasting*. For the somewhat hypothetical case of a *perpetual option* ($T = \infty$) the tangential touching can be calculated analytically ($\longrightarrow$ Exercise 4.8). In summary, *two boundary conditions* must hold at the contact point $S_f(t)$:

$$
\begin{aligned}
V_P^{am}(S_f(t), t) &= K - S_f(t) \\
\frac{\partial V_P^{am}(S_f(t), t)}{\partial S} &= -1
\end{aligned}
\tag{4.24P}
$$

As before, the right-end boundary condition $V_P(S, t) \to 0$ must be observed for $S \to \infty$.

For **American calls** analogous boundary conditions can be formulated. For a call in case $\delta > 0$, $r > 0$ the free boundary conditions

$$
\begin{aligned}
V_C^{am}(S_f(t), t) &= S_f(t) - K \\
\frac{\partial V_C^{am}(S_f(t), t)}{\partial S} &= 1
\end{aligned}
\tag{4.24C}
$$

must hold along the right-end boundary for $S_f(t) > K$. The left-end boundary condition at $S = 0$ remains unchanged. Figure 4.9 shows an American call on a dividend-paying asset. The high contact on the payoff is visible.

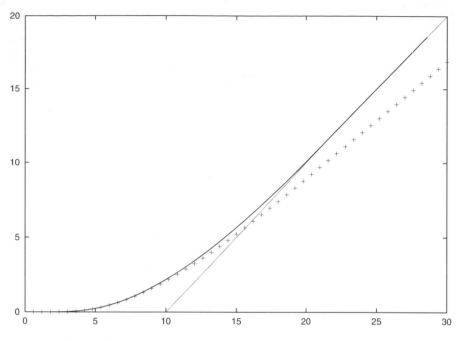

Fig. 4.9. Value $V(S,0)$ of an American call with $K = 10$, $r = 0.25$, $\sigma = 0.6$, $T = 1$ and dividend flow $\delta = 0.2$. Crosses indicate the corresponding curve of a European call; the payoff is shown. A special value is $V(K,0) = 2.18728$.

We note in passing that the transformation $\zeta := S/S_f(t)$, $y(\zeta,t) := V(S,t)$ allows to set up a Black-Scholes-type PDE on a rectangle. In this way, the unknown front $S_f(t)$ is fixed at $\zeta = 1$, and is given implicitly by an ordinary differential equation as part of a nonlinear PDE. This *front-fixing* approach is numerically relevant ($\longrightarrow$ Exercise 4.11).

4.5.3 Black-Scholes Inequality

The Black-Scholes equation (4.1) is valid on the continuation region (shaded areas in Figure 4.6). For the numerical approach of the following Section 4.6 the computational domain will be the entire half strip with $S > 0$, including the stopping areas. This will allow locating the early-exercise curve S_f. The approach requires to adapt the Black-Scholes equation in some way to the stopping areas.

To this end, define the Black-Scholes operator as

$$\mathcal{L}_{\mathrm{BS}}(V) := \frac{1}{2}\sigma^2 S^2 \frac{\partial^2 V}{\partial S^2} + (r - \delta)S\frac{\partial V}{\partial S} - rV \ .$$

With this notation the Black-Scholes equation reads

$$\frac{\partial V}{\partial t} + \mathcal{L}_{\mathrm{BS}}(V) = 0 .$$

What happens with this operator on the stopping regions? To this end we substitute the payoff into $\frac{\partial V}{\partial t} + \mathcal{L}_{\mathrm{BS}}(V)$ for the case of a put. (The reader may carry out the analysis for the case of a call.) For the put, for $S \leq S_{\mathrm{f}}$,

$$V = K - S , \quad \frac{\partial V}{\partial t} = 0 , \quad \frac{\partial V}{\partial S} = -1 , \quad \frac{\partial^2 V}{\partial S^2} = 0 .$$

Hence

$$\frac{\partial V}{\partial t} + \mathcal{L}_{\mathrm{BS}}(V) = -(r - \delta)S - r(K - S) = \delta S - rK .$$

From (4.23P) we have the bound $\delta S < rK$, which leads to conclude

$$\frac{\partial V}{\partial t} + \mathcal{L}_{\mathrm{BS}}(V) < 0 .$$

The Black-Scholes equation changes to an *inequality* on the stopping region. The same inequality holds for the call. In summary, on the entire half strip American options must satisfy an *inequality* of the Black-Scholes type,

$$\frac{\partial V}{\partial t} + \frac{1}{2}\sigma^2 S^2 \frac{\partial^2 V}{\partial S^2} + (r - \delta)S \frac{\partial V}{\partial S} - rV \leq 0. \tag{4.25}$$

The inequalities (4.21) and (4.25) hold for all (S, t). In case the strict inequality ">" holds in (4.21), equality holds in (4.25). The contact boundary S_{f} divides the half strip into the stopping region and the continuation region, each with appropriate version of V:

put: $V_{\mathrm{P}}^{\mathrm{am}} = K - S$ for $S \leq S_{\mathrm{f}}$ (stop)

$V_{\mathrm{P}}^{\mathrm{am}}$ solves (4.1)   for $S > S_{\mathrm{f}}$ (hold)

and

call: $V_{\mathrm{C}}^{\mathrm{am}} = S - K$ for $S \geq S_{\mathrm{f}}$ (stop)

$V_{\mathrm{C}}^{\mathrm{am}}$ solves (4.1)   for $S < S_{\mathrm{f}}$ (hold)

This shows that also for American options the Black-Scholes equation (4.1) must be solved, however, with special arrangements because of the free boundary. We have to look for methods that simultaneously calculate V along with the unknown S_{f}.

Note that $\frac{\partial V}{\partial S}$ is continuous when S_{f} is crossed, but $\frac{\partial^2 V}{\partial S^2}$ and $\frac{\partial V}{\partial t}$ are not continuous. It must be expected that this lack of smoothness affects the accuracy of numerical approximations.

4.5.4 Obstacle Problems

A brief digression into obstacle problems will motivate the procedure. We assume an "obstacle" $f(x)$, say with $f(x) > 0$ for $\alpha < x < \beta$, $f \in \mathcal{C}^2$, $f'' < 0$ and $f(-1) < 0$, $f(1) < 0$, compare Figure 4.10. Across the obstacle a function u with minimal length is stretched like a rubber thread. Between $x = \alpha$ and $x = \beta$ the curve u clings to the boundary of the obstacle. For α and β we encounter high-contact conditions, where the curve of u touches the obstacle tangentially. These two values $x = \alpha$ and $x = \beta$ are unknown initially. This obstacle problem is a simple free boundary problem.

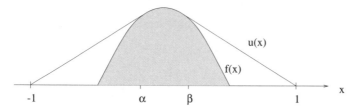

Fig. 4.10. Function $u(x)$ across an obstacle $f(x)$

The aim is to reformulate the obstacle problem such that the free boundary conditions do not show up explicitly. This may promise computational advantages. The function u shown in Figure 4.10 is defined by the requirement $u \in \mathcal{C}^1[-1,1]$, and by:

$$
\begin{array}{lll}
\text{for } -1 < x < \alpha : & u'' = 0 & (\text{then } u > f) \\
\text{for } \alpha < x < \beta : & u = f & (\text{then } u'' = f'' < 0) \\
\text{for } \beta < x < 1 : & u'' = 0 & (\text{then } u > f)
\end{array}
$$

The characterization of the two outer intervals is identical. The situation manifests a complementarity in the sense

$$
\begin{aligned}
\text{if } u > f, & \text{ then } u'' = 0; \\
\text{if } u = f, & \text{ then } u'' < 0.
\end{aligned}
$$

In retrospect it is clear that American options are complementary in an analogous way:

$$\text{if } V > \text{payoff, then Black-Scholes } \textit{equation } \frac{\partial V}{\partial t} + \mathcal{L}_{\text{BS}}(V) = 0$$

$$\text{if } V = \text{payoff, then Black-Scholes } \textit{inequality } \frac{\partial V}{\partial t} + \mathcal{L}_{\text{BS}}(V) < 0$$

This analogy motivates searching for a solution of the obstacle problem. The obstacle problem can be reformulated as

$$
\begin{cases}
\text{find a function } u \text{ such that} \\
u''(u - f) = 0, \quad -u'' \geq 0, \quad u - f \geq 0, \\
u(-1) = u(1) = 0, \; u \in \mathcal{C}^1[-1,1].
\end{cases}
\tag{4.26}
$$

The key line (4.26) is a **linear complementarity problem**. This formulation does not mention the free boundary conditions at $x = \alpha$ and $x = \beta$ explicitly. This will be advantageous because α and β are unknown. If a solution to (4.26) is known, then α and β are read off from the solution. So we will construct a numerical solution procedure for the complementarity version (4.26) of the obstacle problem. But first we derive an additional formulation of the obstacle problem.

Formulation as Variational Inequality

The function u can be characterized by comparing it to functions v out of a set $\mathcal{K}$ of *competing functions*

$$\mathcal{K} := \{v \in \mathcal{C}^0[-1,1] : \; v(-1) = v(1) = 0,$$
$$v(x) \geq f(x) \;\; \text{for} \; -1 \leq x \leq 1, \; v \; \text{piecewise} \in \mathcal{C}^1\}.$$

The requirements on u imply $u \in \mathcal{K}$. For $v \in \mathcal{K}$ we have $v - f \geq 0$ and in view of $-u'' \geq 0$ also $-u''(v - f) \geq 0$. Hence for all $v \in \mathcal{K}$ the inequality

$$\int_{-1}^{1} -u''(v - f)dx \geq 0$$

must hold. By (4.26) the integral

$$\int_{-1}^{1} -u''(u - f)dx = 0$$

vanishes. Subtracting yields

$$\int_{-1}^{1} -u''(v - u)dx \geq 0 \;\; \text{for any} \; v \in \mathcal{K}.$$

The obstacle function f does not occur explicitly in this formulation; the obstacle is implicitly defined in $\mathcal{K}$. Integration by parts leads to

$$\underbrace{[-u'(v - u)]_{-1}^{1}}_{=0} + \int_{-1}^{1} u'(v - u)'dx \geq 0.$$

The integral-free term vanishes because of $u(-1) = v(-1)$, $u(1) = v(1)$. In summary, we have derived the statement:

> If u solves the obstacle problem (4.26), then
>
> $$\int_{-1}^{1} u'(v - u)'dx \geq 0 \quad \text{for all} \; v \in \mathcal{K}.$$

(4.27)

Since v varies in the set $\mathcal{K}$ of competing functions, an inequality such as in (4.27) is called *variational inequality*. The characterization of u by (4.27) can be used to construct an approximation w: Instead of u, find a $w \in \mathcal{K}$ such that the inequality (4.27) is satisfied for all $v \in \mathcal{K}$,

$$\int_{-1}^{1} w'(v-w)'dx \geq 0 \quad \text{for all } v \in \mathcal{K}$$

The characterization (4.27) is related to a minimum problem, because the integral in (4.27) vanishes for $v = u$. This chapter will not elaborate further the formulation as a variational problem. But we will return to this version of the obstacle problem in Chapter 5.

Discretization of the Obstacle Problem

For a numerical solution, let us return to the complementarity version (4.26) of the obstacle problem. A finite-difference approximation for u'' on the grid $x_i = -1 + i\Delta x$, with $\Delta x = \frac{2}{m}$, $f_i := f(x_i)$ leads to

$$\left\{ \begin{array}{l} (w_{i-1} - 2w_i + w_{i+1})(w_i - f_i) = 0, \\ -w_{i-1} + 2w_i - w_{i+1} \geq 0, \quad w_i \geq f_i \end{array} \right\} \quad 0 < i < m,$$

and $w_0 = w_m = 0$. The w_i are approximations to $u(x_i)$. In view of the signs of the factors in the first line in this discretization scheme it can be written using a scalar product. To this end we define a vector notation using

$$B := \begin{pmatrix} 2 & -1 & & & 0 \\ -1 & \ddots & \ddots & & \\ & \ddots & \ddots & \ddots & \\ & & \ddots & \ddots & -1 \\ 0 & & & -1 & 2 \end{pmatrix} \quad \text{and} \quad w := \begin{pmatrix} w_1 \\ \vdots \\ w_{m-1} \end{pmatrix}, \quad f := \begin{pmatrix} f_1 \\ \vdots \\ f_{m-1} \end{pmatrix}.$$

Then the discretized complementarity problem is rewritten in the form

$$\left\{ \begin{array}{l} (w-f)^{tr}Bw = 0 , \\ Bw \geq 0 , \quad w \geq f \end{array} \right. \tag{4.28}$$

To calculate (4.28) one solves $Bw = 0$ under the side condition $w \geq f$. This will be explained in Section 4.6.2.

4.5.5 Linear Complementarity for American Put Options

In analogy to the simple obstacle problem described above we now derive a linear complementarity problem for American options. Here we confine ourselves to American puts without dividends ($\delta = 0$); the general case will be listed in Section 4.6. The transformations (4.3) lead to

$$\frac{\partial y}{\partial \tau} = \frac{\partial^2 y}{\partial x^2} \quad \text{as long as} \quad V_{\mathrm{P}}^{\mathrm{am}} > (K - S)^+.$$

Also the side condition (4.21) is transformed: The relation

$$V_{\mathrm{P}}^{\mathrm{am}}(S,t) \geq (K - S)^+ = K \max\{1 - e^x, 0\}$$

leads to the inequality

$$y(x,\tau) \geq \exp\{\tfrac{1}{2}(q - 1)x + \tfrac{1}{4}(q + 1)^2 \tau\} \max\{1 - e^x, 0\}$$
$$= \exp\{\tfrac{1}{4}(q + 1)^2 \tau\} \max\{(1 - e^x)e^{\frac{1}{2}(q-1)x}, 0\}$$
$$= \exp\{\tfrac{1}{4}(q + 1)^2 \tau\} \max\{e^{\frac{1}{2}(q-1)x} - e^{\frac{1}{2}(q+1)x}, 0\}$$
$$=: g(x, \tau)$$

This function g allows to write the initial condition (4.4) as $y(x,0) = g(x,0)$. In summary, we require $y_\tau = y_{xx}$ as well as

$$y(x,0) = g(x,0) \quad \text{and} \quad y(x,\tau) \geq g(x,\tau),$$

and, in addition, the boundary conditions, and $y \in \mathcal{C}^1$. For $x \to \infty$ the function g vanishes, $g(x,\tau) = 0$, so the boundary condition $y(x,\tau) \to 0$ for $x \to \infty$ can be written

$$y(x,\tau) = g(x,\tau) \quad \text{for} \quad x \to \infty .$$

The same holds for $x \to -\infty$ ($\longrightarrow$ Exercise 4.5). In practice, the boundary conditions are formulated for $x_{\min}$ and $x_{\max}$. Collecting all expressions, the American put is formulated as linear complementarity problem:

$$\begin{cases} \left(\dfrac{\partial y}{\partial \tau} - \dfrac{\partial^2 y}{\partial x^2}\right)(y - g) = 0, \\[2mm] \dfrac{\partial y}{\partial \tau} - \dfrac{\partial^2 y}{\partial x^2} \geq 0, \quad y - g \geq 0 \\[2mm] y(x,0) = g(x,0), \quad y(x_{\min}, \tau) = g(x_{\min}, \tau), \\[2mm] y(x_{\max}, \tau) = g(x_{\max}, \tau), \ y \in \mathcal{C}^1 . \end{cases}$$

The exercise boundary is automatically captured by this formulation. An analogous formulation holds for the American call. Both of the formulations are listed in the beginning of the following section. We will return to the versions as variational problems in Section 5.3.

4.6 Computation of American Options

Let us first summarize the results of the previous sections, for both put and call:

Problem 4.7   (linear complementarity problem)

$$q = \frac{2r}{\sigma^2}; \quad q_\delta = \frac{2(r-\delta)}{\sigma^2}$$

put: $g(x,\tau) := \exp\{\frac{1}{4}((q_\delta - 1)^2 + 4q)\tau\} \max\{e^{\frac{1}{2}(q_\delta-1)x} - e^{\frac{1}{2}(q_\delta+1)x}, 0\}$

call $(\delta > 0)$:

$$g(x,\tau) := \exp\{\tfrac{1}{4}((q_\delta - 1)^2 + 4q)\tau\} \max\{e^{\frac{1}{2}(q_\delta+1)x} - e^{\frac{1}{2}(q_\delta-1)x}, 0\}$$

$$\left(\frac{\partial y}{\partial \tau} - \frac{\partial^2 y}{\partial x^2}\right)(y - g) = 0$$

$$\frac{\partial y}{\partial \tau} - \frac{\partial^2 y}{\partial x^2} \geq 0, \quad y - g \geq 0$$

$$y(x,0) = g(x,0), \quad 0 \leq \tau \leq \frac{1}{2}\sigma^2 T$$

$$\lim_{x \to \pm\infty} y(x,\tau) = \lim_{x \to \pm\infty} g(x,\tau)$$

As outlined in Section 4.5, the free boundary problem of American options is described in Problem 4.7 such that the free boundary condition does not show up explicitly. We now enter the discussion of the numerical solution of Problem 4.7.

4.6.1 Discretization with Finite Differences

We use the same grid as in Section 4.2.2, with $w_{i\nu}$ denoting an approximation to $y(x_i, \tau_\nu)$, and $g_{i\nu} := g(x_i, \tau_\nu)$. The backward difference, the explicit, and the Crank-Nicolson method can be combined into one formula,

$$\frac{w_{i,\nu+1} - w_{i\nu}}{\Delta\tau} = \theta\frac{w_{i+1,\nu+1} - 2w_{i,\nu+1} + w_{i-1,\nu+1}}{\Delta x^2} +$$
$$(1-\theta)\frac{w_{i+1,\nu} - 2w_{i\nu} + w_{i-1,\nu}}{\Delta x^2},$$

with the choices $\theta = 0$ (explicit), $\theta = \frac{1}{2}$ (Crank–Nicolson), $\theta = 1$ (backward-difference method). This family of numerical schemes parameterized by θ is often called θ-method.

The differential inequality $\frac{\partial y}{\partial \tau} - \frac{\partial^2 y}{\partial x^2} \geq 0$ becomes the discrete version

$$w_{i,\nu+1} - \lambda\theta(w_{i+1,\nu+1} - 2w_{i,\nu+1} + w_{i-1,\nu+1})$$
$$- w_{i\nu} - \lambda(1-\theta)(w_{i+1,\nu} - 2w_{i\nu} + w_{i-1,\nu}) \geq 0, \tag{4.29}$$

where we use again the abbreviation $\lambda := \frac{\Delta \tau}{\Delta x^2}$. With the notations

$$b_{i\nu} := w_{i\nu} + \lambda(1-\theta)(w_{i+1,\nu} - 2w_{i\nu} + w_{i-1,\nu}) \,, \ i = 2, \ldots, m-2$$

$b_{1,\nu}$ and $b_{m-1,\nu}$ incorporate the boundary conditions

$$b^{(\nu)} := (b_{1\nu}, ..., b_{m-1,\nu})^{tr}$$

$$w^{(\nu)} := (w_{1\nu}, ..., w_{m-1,\nu})^{tr}$$

$$g^{(\nu)} := (g_{1\nu}, ..., g_{m-1,\nu})^{tr}$$

and

$$A := \begin{pmatrix} 1+2\lambda\theta & -\lambda\theta & & & 0 \\ -\lambda\theta & \ddots & \ddots & & \\ & \ddots & \ddots & \ddots & \\ & & \ddots & \ddots & \ddots \\ 0 & & & \ddots & \ddots \end{pmatrix} \in \mathbb{R}^{(m-1)\times(m-1)} \tag{4.30}$$

(4.29) is rewritten in vector form as

$$Aw^{(\nu+1)} \geq b^{(\nu)} \quad \text{for all } \nu.$$

Such inequalities for vectors are understood componentwise. The inequality $y - g \geq 0$ leads to

$$w^{(\nu)} \geq g^{(\nu)},$$

and $\left(\frac{\partial y}{\partial \tau} - \frac{\partial^2 y}{\partial x^2}\right)(y-g) = 0$ becomes

$$\left(Aw^{(\nu+1)} - b^{(\nu)}\right)^{tr} \left(w^{(\nu+1)} - g^{(\nu+1)}\right) = 0.$$

The initial and boundary conditions are

$$w_{i0} = g_{i0}, \quad i = 1, ..., m-1, \quad \text{also } w^{(0)} = g^{(0)};$$

$$w_{0\nu} = g_{0\nu}, \quad w_{m\nu} = g_{m\nu}, \quad \nu \geq 1$$

The boundary conditions are realized in the vectors $b^{(\nu)}$ as follows:

$b_{2\nu}, ..., b_{m-2,\nu}$ as defined above,

$$b_{1\nu} = w_{1\nu} + \lambda(1-\theta)(w_{2\nu} - 2w_{1\nu} + g_{0\nu}) + \lambda\theta g_{0,\nu+1} \tag{4.31}$$

$$b_{m-1,\nu} = w_{m-1,\nu} + \lambda(1-\theta)(g_{m\nu} - 2w_{m-1,\nu} + w_{m-2,\nu}) + \lambda\theta g_{m,\nu+1}$$

We summarize the discrete version of the Problem 4.7 into an Algorithm:

Algorithm 4.8   (computation of American options)

> *For* $\nu = 0, 1, ..., \nu_{\max} - 1$:
>
> Calculate the vectors $g := g^{(\nu+1)}$,
>
> $\qquad b := b^{(\nu)}$ from (4.30), (4.31).
>
> Calculate the vector w as solution of the problem
>
> $\qquad Aw - b \geq 0, \quad w \geq g, \quad (Aw - b)^{tr}(w - g) = 0.$ (4.32)
>
> $w^{(\nu+1)} := w$

This completes the chosen finite-difference discretization.

The remaining problem is to solve the complementarity problem in matrix-vector form (4.32). In principle, how to solve (4.32) is a new topic independent of the discretization background. But accuracy and efficiency will depend on the context of selected methods. We pause for a moment to become aware how broad the range of possible finite-difference methods is.

Recall from Subsection 4.5.3 that $V(S, t)$ is not C^2-smooth over the free boundary S_f. This is a source of possible inaccuracies. The order two of the basic Crank-Nicolson must be expected to be deteriorated. The effect caused by lacking smoothness depends on the choice of several items, namely the

(1) kind of transformation/PDE (from no transformation over a mere $\tau := T - t$ to the transformation (4.3)),
(2) kind of discretization (from backward-difference over Crank-Nicolson to more refined schemes like BDF2),
(3) method of solution for (4.32).

The latter can be an efficient direct elimination method, as suggested in [BrS77], see also [DeHR98], [IkT04]. Large systems as they occur in PDE context are frequently solved iteratively, in particular in high-dimensional spaces. Such approaches sometimes benefit from smoothing properties. For this exposition, we select as third item a simple iterative procedure, following [WDH96]. But note that in the one-dimensional scenario of this chapter (one underlying asset), the direct approach is faster ($\longrightarrow$ Exercise 4.12).

4.6.2 Iterative Solution

In each time level ν in Algorithm 4.8 a linear complementarity problem (4.32) must be solved. This is the bulk of work in Algorithm 4.8. In what follows, the solution of the problems (4.32) is done iteratively. To this end we will use the projected SOR method following Cryer. Since this subsection is general numerical analysis independent of the finance framework, we momentarily use vectors x, y, r freely in other context.[1] For an introduction into iterative methods for the solution of systems of linear equations $Ax = b$ we refer to Appendix C2. Note that (4.32) is not in the easy form of equation $Ax = b$ discussed in Appendix C2; a modification of the standard SOR will be necessary. To derive the algorithm we transform problem (4.32) from the w-world into an x-world with

$$x := w - g.$$

Then with $y := Aw - b$ it is easy to see (the reader may check) that the task of calculating a solution w for (4.32) is equivalent to the following problem:

Problem 4.9   (Cryer)

> Find vectors x and y such that for $\hat{b} := b - Ag$
>
> $Ax - y = \hat{b}, \quad x \geq 0, \quad y \geq 0, \quad x^t y = 0.$

$\hfill (4.33)$

For this problem an algorithm has been based on the SOR method [Cr71]. The iteration of the SOR method for $Ax = \hat{b} = b - Ag$ is written componentwise ($\longrightarrow$ Exercise 4.6) as iteration for the correction vector $x^{(k)} - x^{(k-1)}$:

$$r_i^{(k)} := \hat{b}_i - \sum_{j=1}^{i-1} a_{ij} x_j^{(k)} - a_{ii} x_i^{(k-1)} - \sum_{j=i+1}^{n} a_{ij} x_j^{(k-1)} \qquad (4.34a)$$

$$x_i^{(k)} = x_i^{(k-1)} + \omega_R \frac{r_i^{(k)}}{a_{ii}}. \qquad (4.34b)$$

Here k denotes the number of the iteration, $n = m - 1$, and in the cases $i = 1$, $i = m - 1$ one of the sums in (4.34a) is empty. The *relaxation parameter* ω_R is a factor chosen in a way that should improve the convergence of the iteration. The projected SOR method for solving (4.33) starts from a vector $x^{(0)} \geq 0$ and is identical to the SOR method up to a modification on (4.34b) serving for $x_i^{(k)} \geq 0$.

---

[1] Notation: In this Subsection 4.6.2 , x does not have the meaning of transformation (4.3), and r not that of an interest rate, and y is no PDE solution.

Algorithm 4.10   (projected SOR for Problem 4.9)

> *outer loop: $k = 1, 2, \ldots$*
> *inner loop: $i = 1, \ldots, m - 1$*
>
> $r_i^{(k)}$ as in (4.34a)
>
> $$x_i^{(k)} = \max \left\{ 0, \; x_i^{(k-1)} + \omega_R \frac{r_i^{(k)}}{a_{ii}} \right\},$$
>
> $$y_i^{(k)} = -r_i^{(k)} + a_{ii} \left(x_i^{(k)} - x_i^{(k-1)} \right)$$

(4.35)

We see that the method solves $Ax = \hat{b}$ for $\hat{b} = b - Ag$ iteratively by *component-wise* considering $x^{(k)} \geq 0$. The vector y or the components $y_i^{(k)}$ converging against it, are not used explicitly for the algorithm. As we will see, y serves an important role in the proof of convergence. Transformed back into the w-world of problem (4.32), the Algorithm 4.10 is a solution to (4.32). It solves

Problem 4.11   (Cryer's problem restated)

> Solve $Aw = b$ such that the side condition
> $w \geq g$ is obeyed componentwise.

Adapting the formula (4.35) for $x \geq 0$ to $w \geq g$ is easy.

It remains to prove that the projected SOR method converges toward a unique solution. One first shows the equivalence of Problem 4.9 with a minimization problem.

Lemma 4.12

The Problem 4.9 is equivalent to the minimization problem

$$\min_{x \geq 0} G(x), \quad \text{where } G(x) := \frac{1}{2}(x^{tr} Ax) - \hat{b}^{tr} x \quad \text{is strictly convex.} \quad (4.36)$$

Proof. The derivatives of G are $G_x = Ax - \hat{b}$ and $G_{xx} = A$. The Lemma 4.3 implies that A has positive eigenvalues. Hence the Hessian matrix G_{xx} is symmetric and positive definite. So G is strictly convex, and has a unique minimum on each convex set in $\mathbb{R}^n$, for example on $x \geq 0$. The Theorem of Kuhn and Tucker minimizes G under $H_i(x) \leq 0$, $i = 1, \ldots, m$. According to this theorem, a vector x_0 to be a minimum is equivalent to the existence of a Lagrange multiplier $y \geq 0$ with

$$\text{grad } G(x_0) + \left(\frac{\partial H(x_0)}{\partial x} \right)^{tr} y = 0, \quad y^{tr} H(x_0) = 0.$$

The set $x \geq 0$ leads to define $H(x) := -x$. Hence the Kuhn-Tucker condition is $Ax - \widehat{b} + (-I)^{tr}y = 0$, $y^{tr}x = 0$, and we have reached equation (4.33). (For the Kuhn-Tucker theory we refer to [SW70], [St86]).

A proof of the convergence of Algorithm 4.10 is based on Lemma 4.12. One shows that the sequence defined in Algorithm 4.10 minimizes G. The main steps of the argumentation are sketched as follows:

For $0 < \omega_R < 2$ the sequence $G(x^{(k)})$ is decreasing monotonically;
Show $x^{(k+1)} - x^{(k)} \to 0$ for $k \to \infty$;
The limit exists because $x^{(k)}$ moves in a compact set $\{x|G(x) \leq G(x^{(0)})\}$;
The vector r from (4.34) converges toward $-y$;
Assuming $r \geq 0$ and $r^{tr}x \neq 0$ leads to a contradiction to $x^{(k+1)} - x^{(k)} \to 0$.
(For the proof see [Cr71].)

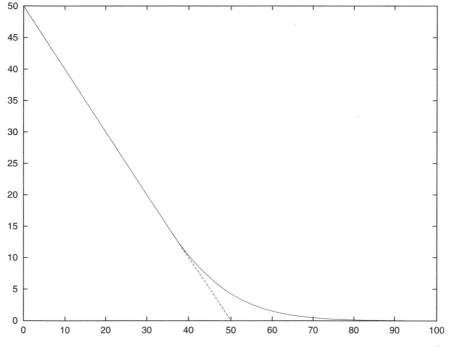

Fig. 4.11. (Example 1.6) American put, $K = 50$, $S = 50$, $r = 0.1$, $\sigma = 0.4$, $T = \frac{5}{12}$. $V(S,0)$ (solid curve) and payoff $V(S,T)$ (dashed). Special value: $V(K,0) = 4.2842$

4.6.3 An Algorithm for Calculating American Options

We return to the original meaning of the variables x, y, r, as used for instance in (4.2), (4.3). It remains to adapt Algorithm 4.10 for the w-vectors, use the resulting algorithm for (4.32) and substitute it into Algorithm 4.8 ($\longrightarrow$ Exercise 4.7). The algorithm is formulated in Algorithm 4.13. Recall

$g_{i\nu} := g(x_i, \tau_\nu)$ and $g^{(\nu)} := (g_{1\nu}, \dots, g_{m-1,\nu})^{tr}$. The Figure 4.11 depicts a result of Algorithm 4.13 for Example 1.6. Here we obtained for contact point the value $S_f(0) = 36.3$. Figure 4.13 shows the American put that corresponds to the call in Figure 4.9.

Algorithm 4.13   (prototype core algorithm)

Set up the function $g(x, \tau)$ listed in Problem 4.7.

(The x is again that of transformation (4.3).)

Choose θ ($\theta = 1/2$ for Crank-Nicolson);

choose $1 \leq \omega_R < 2$ (for example, $\omega_R = 1$).

Fix the discretization by choosing x_{min}, x_{max}, m, ν_{max}

(for example, $x_{min} = -5$, $x_{max} = 5$, $\nu_{max} = m = 100$).

Fix an error bound ε (for example, $\varepsilon = 10^{-5}$).

Calculate $\Delta x := (x_{max} - x_{min})/m$,

$$\Delta\tau := \tfrac{1}{2}\sigma^2 T/\nu_{max}$$

$$x_i := x_{min} + i\Delta x \text{ for } i = 1, \dots, m$$

Initialize the iteration vector w with

$g^{(0)} = (g(x_1, 0), \dots, g(x_{m-1}, 0))$.

Calculate $\lambda := \Delta\tau/\Delta x^2$ and $\alpha := \lambda\theta$.

τ-loop: for $\nu = 0, 1, \dots, \nu_{max} - 1$:

$\tau_\nu := \nu\Delta\tau$

$b_i := w_i + \lambda(1 - \theta)(w_{i+1} - 2w_i + w_{i-1})$ for $2 \leq i \leq m - 2$

$b_1 := w_1 + \lambda(1 - \theta)(w_2 - 2w_1 + g_{0\nu}) + \alpha g_{0,\nu+1}$

$b_{m-1} := w_{m-1} + \lambda(1 - \theta)(g_{m\nu} - 2w_{m-1} + w_{m-2}) + \alpha g_{m,\nu+1}$

Set componentwise $v = \max(w, g^{(\nu+1)})$

(v is the iteration vector of the projected SOR.)

SOR-loop: for $k = 1, 2, \dots$:

as long as $\|v^{new} - v\|_2 > \epsilon$:

for $i = 1, 2, \dots, m - 1$:

$\rho := (b_i + \alpha(v_{i-1}^{new} + v_{i+1}))/(1 + 2\alpha)$

(with $v_0^{new} = v_m = 0$)

$v_i^{new} = \max\{g_{i,\nu+1}, \ v_i + \omega_R(\rho - v_i)\}$

$v := v^{new}$ (after testing for convergence)

$w^{(\nu+1)} = w = v$

European options:

For completeness we mention that it is possible to calculate European options with Algorithm 4.13 after some modifications. Replacing the line

$$v_i^{\text{new}} = \max\{g_{i,\nu+1},\ v_i + \omega_R(\rho - v_i)\}$$

by the line

$$v_i^{\text{new}} = v_i + \omega_R(\rho - v_i)$$

recovers the standard SOR for iteratively solving $Aw = b$ (without $w \geq g$). If in addition the boundary conditions are adapted, then the program resulting from Algorithm 4.13 can be applied to European options. But note again that the iterative SOR procedure is less efficient than the direct solution by means of the LR-decomposition of A. And applying the analytic solution formula should be most economical, when the entire surface is not required. But for the purpose of testing Algorithm 4.13 it may be recommendable to compare its results to something "known."

Back to American options, we may wonder how a concrete financial task is solved with the core Algorithm 4.13, which is formulated in artificial variables such as $x_i, g_{i\nu}, w_i$ and not in financial variables. We require an interface between the real world and the core algorithm. The interface is provided by the transformations in (4.3). This important ingredient must be included for completeness. So we formulate the required transition between the real world and the numerical machinery of Algorithm 4.13 as another algorithm:

Algorithm 4.14   (American options)

Input: strike K, time to expiration T, spot price S_0, r, δ, σ

Perform the core Algorithm 4.13.

(The τ-loop ends at $\tau_{\text{end}} = \frac{1}{2}\sigma^2 T$.)

For $i = 1, \ldots, m - 1$:

w_i approximates $y(x_i, \frac{1}{2}\sigma^2 T)$,

$S_i = K \exp\{x_i\}$

$V(S_i, 0) = K w_i \exp\{-\frac{x_i}{2}(q_\delta - 1)\} \exp\{-\tau_{\text{end}}(\frac{1}{4}(q_\delta - 1)^2 + q)\}$

Test for early exercise: Approximate $S_f(0)$:

Choose $\varepsilon^* = K \cdot 10^{-5}$ (for example)

For a put:

$i_f := \max\{i\ :\ |V(S_i, 0) + S_i - K| < \varepsilon^*\}$

$S_0 < S_{i_f}$: stopping region!

For a call:

$i_f := \min\{i\ :\ |K - S_i + V(S_i, 0)| < \varepsilon^*\}$

$S_0 > S_{i_f}$: stopping region!

The Algorithm 4.14 evaluates the data at the final time level τ_{end}, which corresponds to $t = 0$. The computed information for the intermediate time levels can be evaluated analogously. In this way, the locations of $S_{i_{\text{f}}}$ can be put together to form an approximation of the free-boundary or stopping-time curve $S_{\text{f}}(t)$. But note that this approximation will be a crude step function. It requires some effort to calculate the curve $S_{\text{f}}(t)$ with reasonable accuracy, see the illustration of curve C_1 in Figure 4.8.

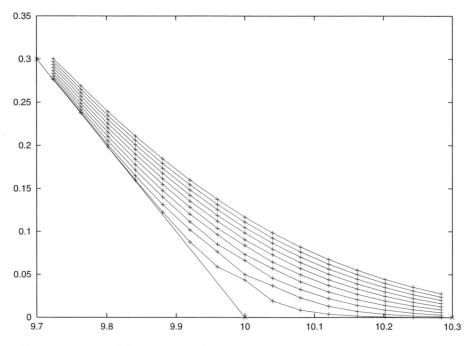

Fig. 4.12. Finite differences, Crank-Nicolson, PSOR; American put with $r = 0.06$, $\sigma = 0.3$, $T = 1$, $K = 10$; $M = 1000$, $x_{\min} = -2$, $x_{\max} = 2$, $\Delta x = 1/250$, $\Delta t = 1/1000$, payoff and $V(S, t_\nu)$ for $t_\nu = 1 - \nu \Delta t$, $\nu = 1, \ldots, 10$.

Modifications

The above Algorithm 4.13 (along with Algorithm 4.14) is the prototype of a finite-difference algorithm. Improvements are possible. For example, the equidistant time step $\Delta \tau$ can be given up in favor of a variable time stepping. A few very small time steps initially will help to quickly damp the influence of the nonsmooth payoff. The effect of the lack in smoothness is illustrated by Figure 4.12. The turmoil at the corner is seen, but also the relatively rapid smoothing within a few time steps. In this context it may be advisable to start with a few fully implicit backward time steps ($\theta = 1$) before switching to Crank-Nicolson ($\theta = 1/2$), see [Ran84] and the Notes on Section 4.2. After one run of the algorithm it is advisable to refine the initial grid to have a possibility to control the error. This simple strategy will be discussed in some more detail in Section 4.7.

And, of course, Problem 4.11 is easily solved by a direct method ($\longrightarrow$ Exercise 4.12). Then Algorithm 4.13 is adapted by replacing the SOR-module.

4.7 On the Accuracy

Necessarily, each result obtained with the means of this chapter is subjected to errors in several ways. The most important errors have been mentioned earlier; in this section we collect them. Let us emphasize again that in general the *existence* of errors must be accepted, but not their magnitude. By investing sufficient effort, many of the errors can be kept at a tolerable level.

(a) modeling error

The assumptions defining the underlying financial model are restrictive. The Assumption 1.2, for example, will not exactly match the reality of a financial market. And the parameters of the equations (such as volatility σ) are unknown and must be estimated. Hence the equations of the model are only approximations of the "reality."

(b) discretization errors

Under the heading "discretization error" we summarize several errors that are introduced when the continuous PDE is replaced by a set of approximating equations defined on a grid. An essential portion of the discretization error is the error between differential quotients and difference quotients. For example, a Crank-Nicolson discretization is of the order $O(\Delta^2)$, if Δ is a measure of the grid size and the solution function is sufficiently smooth. Other discretization errors with mostly smaller influence are the error caused by truncating the infinite interval $-\infty < x < \infty$ to a finite interval, the implementation of the boundary conditions, or a quantification error when the strike ($x = 0$) is not part of the grid. In passing we recommend that the strike be one of the grid points, $x_k = 0$ for one k.

(c) error from solving the linear equation

An iterative solution of the linear systems of equation $Aw = b$ means that the error approaches 0 when $k \to \infty$, where k counts the number of iterations. By practical reasons the iteration must be terminated at a finite $k_{\max}$ such that the effort is bounded. So an error remains from the linear equations. The error tends to be small for direct elimination methods.

(d) rounding error

The finite number of digits l of the mantissa is the reason for rounding errors.

In general one has no *accurate* informations on the size of these errors. Typically the modeling errors are larger than the discretization errors. For a stable method, the rounding errors are the least problem. The numerical analyst, as a rule, has limited potential in manipulating the modeling error. So the numerical analyst concentrates on the other errors, especially on discretization errors. To this end we may use the qualitative assertion of Theorem 4.4. But such an a priori result is only a basic step toward our ultimate goal.

4.7.1 Elementary Error Control

We neglect modeling errors and try to solve the a posteriori error problem:

Problem 4.15   (principle of an error control)
 Let the exact result of a solution of the continuous equations be denoted η^*. The approximation η calculated by a given algorithm depends on a representative grid size Δ, on $k_{\max}$, on the wordlength l of the computer, and maybe on several additional parameters, symbolically written

$$\eta = \eta(\Delta, k_{\max}, l).$$

Choose $\Delta, k_{\max}, l$ such that the absolute error of η does not exceed a prescribed error tolerance ϵ,

$$|\eta - \eta^*| < \epsilon.$$

This problem is difficult to solve, because we implicitly assume an *efficient* approximation avoiding an overkill with extremely small values of Δ or large values of $k_{\max}$ or l. Time counts in real-time application. So we try to avoid unnecessary effort of achieving a tiny error $|\eta - \eta^*| \ll \epsilon$. The exact size of the error is unknown. But its order of magnitude can be estimated as follows.

Let us assume the method is of order p. We simplify this statement to

$$\eta(\Delta) - \eta^* = \gamma \Delta^p. \tag{4.37}$$

Here γ is a priori unknown. By calculating two approximations, say for Δ_1 and Δ_2, the constant γ can be calculated. To this end subtract the two calculated approximations η_1 and η_2,

$$\eta_1 := \eta(\Delta_1) = \gamma \Delta_1^p + \eta^*$$
$$\eta_2 := \eta(\Delta_2) = \gamma \Delta_2^p + \eta^*$$

to obtain

$$\gamma = \frac{\eta_1 - \eta_2}{\Delta_1^p - \Delta_2^p} \, .$$

A simple choice of the grid size Δ_2 for the second approximation is the refinement $\Delta_2 = \frac{1}{2}\Delta_1$. This leads to

Table 4.1. Results reported in Figure 4.13

$m = \nu_{\max}$	$V(10,0)$
50	1.8562637
100	1.8752110
200	1.8800368
400	1.8812676
800	1.8815842
1600	1.8816652

$$\gamma \left(\frac{\Delta_1}{2} \right)^p = \frac{\eta_1 - \eta_2}{2^p - 1}. \tag{4.38}$$

Especially for $p = 2$ the relation

$$\gamma \Delta_1^2 = \tfrac{4}{3}(\eta_1 - \eta_2)$$

results. In view of the scenario (4.37) the absolute error of the approximation η_1 is given by

$$\tfrac{4}{3}|\eta_1 - \eta_2|$$

and the error of η_2 by (4.38).

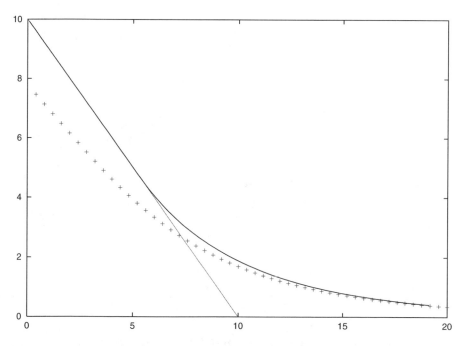

Fig. 4.13. Value $V(S,0)$ of an American put with $K = 10$, $r = 0.25$, $\sigma = 0.6$, $T = 1$ and dividend flow $\delta = 0.2$. For special values see Table 4.1. Crosses mark the corresponding curve of a European option.

The above procedure does not guarantee that the error η is bounded by ϵ. This flaw is explained by the simplification in (4.37), and by neglecting the other type of errors of the above list (b)–(c). Here we have assumed γ constant, which in reality depends on the parameters of the model, for example, on the volatility σ. But testing the above rule of thumb (4.37)/(4.38) on European options shows that it works reasonably well. Here we compare the finite-difference results to the analytic solution formula (A4.10), the numerical errors of which are comparatively negligible. The procedure works similar well for American options, although here the function $V(S,t)$ is not $\mathcal{C}^2$-smooth at $S_f(t)$. (The effect of the lack in smoothness is similar as in Figure 4.12.) In practical applications of Crank-Nicolson's method one can observe quite well that doubling of m and $\nu_{\max}$ decreases the absolute error approximately by a factor of four. To obtain a minimum of information on the error, the core Algorithm 4.13 should be applied at least for two grids following the lines outlined above. The information on the error can be used to match the grid size Δ to the desired accuracy.

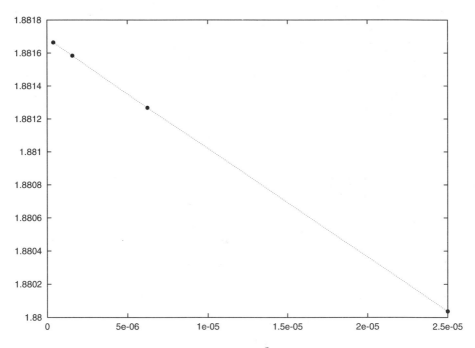

Fig. 4.14. Approximations depending on Δ^2, with $\Delta = (x_{\max} - x_{\min})/m = 1/\nu_{\max}$; results of Figure 4.13 and Table 4.1.

Let us illustrate the above considerations with an example, compare Figures 4.13 and 4.14, and Table 4.1. For an American put and $x_{\max} = -x_{\min} = 5$ we calculate four approximations, and test equation (4.37) in the form $\eta(\Delta) = \eta^* + \gamma\Delta^2$. We illustrate the approximations as points in the (Δ^2, η)-

plane. The better the assumption (4.37) is satisfied, the closer the calculated points lie on a straight line. Figure 4.14 shows that this error-control model can be expected to work well.

4.7.2 Extrapolation

The obviously reasonable error model suggests applying (4.37) to obtain an improved approximation η at practically zero cost. Such a procedure is called *extrapolation*. In a graphical illustration η over Δ^2 as in Figure 4.14, extrapolation amounts to construct a straight line through two of the calculated points. The value of the straight line for $\Delta^2 = 0$ gives the extrapolated value from

$$\eta^* \approx \frac{4\eta_2 - \eta_1}{3} \ . \tag{4.39}$$

In our example, this procedure allows to estimate the correct value to be close to 1.8817. Combining, for example, two approximations of rather low quality, namely $m = 50$ with $m = 100$, gives already an extrapolated approximation of 1.8815. And based on the two best approximations of Table 4.1, the extrapolated value is 1.881690.

Typically the extrapolation formula provided by (4.39) is significantly more accurate than η_2. But we have no further information on the accuracy from the calculated η_1, η_2. Calculating a third approximation η_3 reveals more information. For example, a higher-order extrapolation can be constructed ($\longrightarrow$ Exercise 4.13). Figure 4.15 reports on the accuracies.

The convergence rate in Theorem 4.4 was derived under the assumptions of a structured equidistant grid and a $\mathcal{C}^4$-smooth solution. Practical experiments with unstructured grids and not so smooth data suggest that the convergence rate may still behave reasonably. The finite-difference truncation error is not the whole story. The more flexible finite-element approaches in Chapter 5 will shed light on convergence under more general conditions.

4.8 Analytic Methods

Numerical methods typically are designed such that they achieve convergence. So, in principle, every accuracy can be reached, only limited by the available computer time and by hardware restrictions. In several cases this high potential of numerical methods is not needed. Rather, some analytic formula may be sufficient that delivers medium accuracy at low cost. Such "analytic methods" have been developed. This Section discusses related approaches. A calculator that applies analytic methods can be found on the website www.compfin.de. This calulator may be used for tests, for example, using the data of Figures 4.11 (Example 1.6), and of Figure 4.13 (Table 4.1).

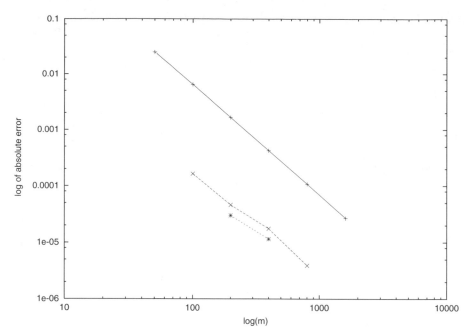

Fig. 4.15. Finite difference methods, log of absolute error in $V(K,0)$ over $\log(m)$, where $m = \nu_{\max}$, and the basis of the logarithm is 10. Solid line: plain algorithm, results in Table 4.1; dashed line: extrapolation (4.39) based on two approximations; dotted line: higher-order extrapolation of Exercise 4.13

In reality there is hardly a clear-cut between numerical and analytic methods. On the one hand, numerical methods require analysis for their derivation. And on the other hand, analytic methods involve numerical algorithms. These may be elementary evaluations of functions like the logarithm or the square root such as in the Black-Scholes formula, or may consist of a subalgorithm like Newton's method for zero finding. There is hardly a purely analytic method.

The finite-difference approach, which approximates the surface $V(S,t)$, requires intermediate values for $0 < t < T$ for the purpose of approximating $V(S,0)$. In the financial practice one is basically interested in $V(S,0)$ only, intermediate values are rarely asked for. So the only temporal input parameter is the time to maturity $T - t$ (or T in case the current time is set to zero, $t = 0$). Recall that also in the Black-Scholes formula, time only enters in the form $T - t$ ($\longrightarrow$ Appendix A4). So it makes sense to write the formula in terms of the time to maturity τ,

$$\tau := T - t \, ,$$

which leads to the compact version of the Black-Scholes formulas (A4.10),

$$d_1(S, \tau; K, r, \sigma) := \frac{1}{\sigma\sqrt{\tau}} \left\{ \log \frac{S}{K} + \left(r + \frac{\sigma^2}{2} \right) \tau \right\}$$

$$d_2(S, \tau; K, r, \sigma) := \frac{1}{\sigma\sqrt{\tau}} \left\{ \log \frac{S}{K} + \left(r - \frac{\sigma^2}{2} \right) \tau \right\} \tag{4.40}$$

$$V_{\mathrm{P}}^{\mathrm{eur}}(S, \tau; K, r, \sigma) = -SF(-d_1) + Ke^{-r\tau}F(-d_2)$$

$$V_{\mathrm{C}}^{\mathrm{eur}}(S, \tau; K, r, \sigma) = SF(d_1) - Ke^{-r\tau}F(d_2)$$

(dividend-free case). F denotes the cumulated standard normal distribution function. For dividend-free options we only need an approximation formula for the American put $V_{\mathrm{P}}^{\mathrm{am}}$; the other cases are covered by the Black-Scholes formula.

This Section introduces into four analytic methods. The first two (Subsections 4.8.1, 4.8.2) are described in detail such that the implementation of the algorithms is an easy matter. Of the other approaches (Subsections 4.8.3, 4.8.4) only basic ideas are set forth.

4.8.1 Approximation Based on Interpolation

If a lower bound V^{low} and an upper bound V^{up} on the American put are available,

$$V^{\mathrm{low}} \leq V_{\mathrm{P}}^{\mathrm{am}} \leq V^{\mathrm{up}} \, ,$$

then the idea is to construct an α aiming at

$$V_{\mathrm{P}}^{\mathrm{am}} = \alpha V^{\mathrm{up}} + (1 - \alpha)V^{\mathrm{low}} \, .$$

This is the approach by [Joh83]. The parameter α, $0 \leq \alpha \leq 1$, defines the interpolation between V^{low} and V^{up}. Since $V_{\mathrm{P}}^{\mathrm{am}}$ depends on the market data S, τ, K, r, σ, the single parameter α and the above interpolation can not be expected to provide an exact value of $V_{\mathrm{P}}^{\mathrm{am}}$. (An exact value would mean that an exact formula for $V_{\mathrm{P}}^{\mathrm{am}}$ would exist.) Rather a formula for α is developed such that the interpolation formula $\alpha V^{\mathrm{up}} + (1 - \alpha)V^{\mathrm{low}}$ provides a good approximation for a wide range of market data. The closer the bounds V^{low} and V^{up} are, the better is the approximation.

An immediate candidate for the lower bound V^{low} is the value $V_{\mathrm{P}}^{\mathrm{eur}}$ provided by the Black-Scholes formula. Thus,

$$V_{\mathrm{P}}^{\mathrm{eur}}(S, \tau; K) \leq V_{\mathrm{P}}^{\mathrm{am}}(S, \tau; K) \leq V_{\mathrm{P}}^{\mathrm{am}}(S, \tau; Ke^{r\tau})$$

The latter of the inequalities is the monotonicity with respect to the strike, see Appendix D1. Following [Mar78], an American put with strike $Ke^{r\tau}$ rising exponentially with τ at the risk-free rate is not exercised, so

$$V_{\mathrm{P}}^{\mathrm{am}}(S, \tau; Ke^{r\tau}) = V_{\mathrm{P}}^{\mathrm{eur}}(S, \tau; Ke^{r\tau}) \, ,$$

which serves as upper bound. This allows to apply the Black-Scholes formula (4.40) to the European option and provides the upper bound to $V_P^{am}(S, t; K)$. The resulting approximation formula is

$$\overline{V} := \alpha V_P^{eur}(S, \tau; Ke^{r\tau}) + (1 - \alpha)V_P^{eur}(S, \tau; K) . \qquad (4.41)$$

The parameter α will depend on S, τ, K, r, σ, so does $\overline{V}$. Actually, the Black-Scholes formula (4.40) suggests that α and $\overline{V}$ only depend on the three parameters S/K, $r\tau$, and $\sigma^2\tau$. The approximation must be constructed such that the lower bound $(K - S)^+$ of the payoff is obeyed. As we will see, all depends on the free boundary S_f, which must be approximated as well.

[Joh83] carried out a regression analysis with two free parameters a_0, a_1 based on computed values of V_P^{am}. The result is

$$\alpha := \left(\frac{r\tau}{a_0 r\tau + a_1}\right)^\beta \quad , \quad \beta := \frac{\ln(S/S_f)}{\ln(K/S_f)} , \qquad (4.42)$$
$$a_0 = 3.9649 \quad , \quad a_1 = 0.032325 .$$

The ansatz for α is designed such that for $S = K$ and $\beta = 1$ upper and lower bound behavior and calculated option values can be matched with reasonable accuracy with only two parameters a_0, a_1. The S-dependent β is introduced to improve the approximation for $S \neq K$. Obviously, $S = S_f \Rightarrow \beta = 0 \Rightarrow \alpha = 1$, which captures the upper bound. And for the lower bound, $\alpha = 0$ is reached for $r\tau = 0$.

To estimate S_f, observe the extreme cases

$$S_f = K \quad \text{for} \quad \tau = 0$$
$$S_f = K\frac{2r}{\sigma^2 + 2r} \quad \text{for} \quad T \to \infty .$$

(For the latter case see Exercise 4.8 and Appendix A5.) This motivates to set the approximation $\overline{S}_f$ for S_f as

$$\overline{S}_f := K\left(\frac{2r}{\sigma^2 + 2r}\right)^\gamma , \qquad (4.43)$$

for a suitably modeled exponent γ. To match the extreme cases, γ should vanish for $\tau = 0$, and $\gamma \approx 1$ for large values of τ. [Joh83] suggests

$$\gamma := \frac{\sigma^2\tau}{b_0\sigma^2\tau + b_1} , \qquad (4.44)$$
$$b_0 = 1.04083 \quad , \quad b_1 = 0.00963 .$$

The constants b_0 and b_1 are again obtained by a regression analysis.

The analytic expressions of (4.43), (4.44) provide an approximation of S_f, and then by (4.42), (4.41) an approximation of V_P^{am} for $S > S_f$, based on the Black-Scholes formulas (4.40) for V_P^{eur}.

Algorithm 4.16   (interpolation)

For given S, τ, K, r, σ evaluate $\gamma, \overline{S}_f, \beta, \alpha$.

Evaluate the Black-Scholes formula for V_P^{eur}

for the arguments in (4.41).

Then $\overline{V}$ from (4.41) is an approximation to V_P^{am} .

Numerical experiments show that the approximaton quality of $\overline{S}_f$ is poor. But for S not too close to $\overline{S}_f$ the approximation quality of $\overline{V}$ is quite good. As reported in [Joh83], the error is small for $r\tau \leq 0.125$, which is satisfied for average values of the risk-free rate r and time to maturity τ. For larger values of $r\tau$, when the gap between lower and upper bound widens, the approximation works less well. An extension to options on dividend-paying assets is given in [Blo86].

4.8.2 Quadratic Approximation

Next we describe an analytic method due to [MaM86]. Recall that in the continuation region both V_P^{am} and V_P^{eur} obey the Black-Scholes equation. Since this equation is linear, also the difference

$$p(S,\tau) := V_P^{\text{am}}(S,\tau) - V_P^{\text{eur}}(S,\tau) \tag{4.45}$$

satisfies the Black-Scholes equation. The relation $V^{\text{am}} \geq V^{\text{eur}}$ suggests to interpret the difference p as *early-exercise premium*. Since both V_P^{am} and V_P^{eur} have the same payoff, the terminal condition for $\tau = 0$ is zero, $p(S,0) = 0$. The closeness of $p(S,\tau)$ to zero should scale roughly by

$$H(\tau) := 1 - e^{-r\tau}. \tag{4.46}$$

This motivates introducing a scaled version f of p,

$$p(S,\tau) =: H(\tau)f(S, H(\tau)) \tag{4.47}$$

For the analysis we repeat the Black-Scholes equation, here for $p(S,\tau)$, where subscripts denote partial differentiation, and $q := \frac{2r}{\sigma^2}$:

$$-\frac{q}{r}p_\tau + S^2 p_{SS} + qSp_S - qp = 0 \tag{4.48}$$

Substituting

$$p_S = Hf_S \ , \ p_{SS} = Hf_{SS} \ , \ p_\tau = H_\tau f + Hf_H H_\tau$$

and using

$$\frac{1}{r}H_\tau = 1 - H$$

yields after a short calculation (the reader may check) for $H \neq 0$ the modified version of the Black-Scholes equation

$$S^2 f_{SS} + qSf_S - \frac{q}{H}f\left[1 + H(1 - H)\frac{f_H}{f}\right] = 0 . \qquad (4.49)$$

Note that this is the "full" equation, nothing is simplified yet. No partial derivative with respect to t shows up, but instead the f_H-term.

At this point, following [MaM86], we introduce a simplifying approximation. The factor $H(H - 1)$ for the H varying in the range $0 \leq H < 1$ is a quadratic term with maximum value of $1/4$, and close to zero for $\tau \approx 0$ and for large values of τ, compare (4.46). This suggests that the term

$$H(1 - H)\frac{f_H}{f} \qquad (4.50)$$

may be small compared to 1, and to neglect it in (4.49). (This motivates the name "quadratic approximation.") The resulting equation

$$S^2 f_{SS} + qSf_S - \frac{q}{H}f = 0 \qquad (4.51)$$

is an ordinary differential equation with analytical solution, parametrized by H. An analysis similar as in Exercise 4.8 leads to the solution

$$f(S) = \alpha S^\lambda , \text{ where } \lambda := -\frac{1}{2}\left\{(q - 1) + \sqrt{(q - 1)^2 + \frac{4q}{H}}\right\} . \qquad (4.52)$$

Combining (4.45), (4.47) and (4.52) we deduce for $S > S_f$ the approximation $\overline{V}$

$$V_P^{am}(S, \tau) \approx \overline{V}(S, \tau) := V_P^{eur}(S, \tau) + \alpha H(\tau)S^\lambda \qquad (4.53)$$

The parameter α must be such that $\overline{V}$ reaches the payoff at S_f,

$$V_P^{eur}(S_f, \tau) + \alpha H S_f^\lambda = K - S_f . \qquad (4.54)$$

To fix the two unknowns S_f and α let us warm up the high-contact condition. This requires the partial derivative of $\overline{V}$ with respect to S. The main part is

$$\frac{\partial V_P^{eur}(S, \tau)}{\partial S} = F(d_1) - 1$$

where F is the cumulated normal distribution function, and d_1 (and below d_2) are the expressions defined by (4.40). d_1 and d_2 depend on all relevant market parameters; we emphasize the dependence on S by writing $d_1(S)$. This gives the high-contact condition

$$\frac{\partial \overline{V}(S_f, \tau)}{\partial S} = F(d_1(S_f)) - 1 + \alpha\lambda H S_f^{\lambda-1} = -1 ,$$

and immediately α in terms of S_f:

$$\alpha = -\frac{F(d_1(S_f))}{\lambda H S_f^{\lambda-1}} \ . \tag{4.55}$$

Substituting into (4.53) yields an equation for S_f,

$$V_P^{eur}(S_f, \tau) - F(d_1(S_f))\frac{1}{\lambda}S_f = K - S_f \ ,$$

which in view of the put-call parity (A4.11a) reads

$$S_f F(d_1) - Ke^{-r\tau}F(d_2) - S_f + Ke^{-r\tau} - F(d_1)\frac{S_f}{\lambda} - K + S_f = 0 \ .$$

This can be summarized to

$$S_f F(d_1)\left[1 - \frac{1}{\lambda}\right] + Ke^{-r\tau}\left[1 - F(d_2)\right] - K = 0 \ . \tag{4.56}$$

Note again that d_1 and d_2 vary with S_f. Hence (4.56) is an implicit equation for S_f and must be solved iteratively. This can be done, for example, by Newton's method ($\longrightarrow$ Appendix C1). In this way a sequence of approximations $S_1, S_2, \dots$ to S_f is constructed. As initial seed $S_0 = K$ may be used, or the more ambitious construction in [BaW87], which exploits information on the limiting case $\tau \to \infty$ ($\longrightarrow$ Exercise 4.8). We summarize

Algorithm 4.17   (quadratic approximation)

For given S, τ, K, r, σ evaluate $q = \dfrac{2r}{\sigma^2}$, $H = 1 - e^{-r\tau}$

and λ from (4.52).

Solve (4.56) iteratively for S_f.

(This involves a sub-algorithm, from which $F(d_1(S_f))$

should be saved.)

Evaluate $V_P^{eur}(S, \tau)$ using the Black-Scholes formula (4.40).

$$\overline{V} := V_P^{eur}(S, \tau) - \frac{1}{\lambda}S_f F(d_1(S_f))\left(\frac{S}{S_f}\right)^{\lambda} \tag{4.57}$$

is the approximation for $S > S_f$,

and $\overline{V} = K - S$ for $S \leq S_f$.

Note that $\lambda < 0$, and λ depends on τ via $H(\tau)$. The time-consuming part of the quadratic-approximation method consists of the numerical root finding procedure. ($\longrightarrow$ Exercise 4.14, Exercise 4.15)

4.8.3 Analytic Method of Lines

In solving PDEs numerically, the *method of lines* is a well-known approach. It is based on a semi-discretization, where the domain (here the (S, τ) half strip) is replaced by a set of parallel lines, each defined by a constant value of τ. To this end, the interval $0 \le \tau \le T$ is discretized into $\nu_{\max}$ sub-intervals by $\tau_\nu := \nu \Delta \tau$, $\Delta \tau := T/\nu_{\max}$, $\nu = 1, \ldots, \nu_{\max} - 1$. We write the Black-Scholes equation as in Section 4.5.3,

$$-\frac{\partial V(S, \tau)}{\partial \tau} + \mathcal{L}_{\mathrm{BS}}(V(S, \tau)) = 0 \; , \tag{4.58}$$

where the negative sign compensates for the transition from t to τ, and replace the partial derivative $\partial V/\partial \tau$ by the difference quotient

$$\frac{V(S, \tau) - V(S, \tau - \Delta \tau)}{\Delta \tau} \; .$$

This gives a semi-discretized version of (4.58), namely the ordinary differential equation

$$w(S, \tau - \Delta \tau) - w(S, \tau) + \Delta \tau \, \mathcal{L}_{\mathrm{BS}}(w(S, \tau)) = 0 \; ,$$

which holds for $S > S_{\mathrm{f}}$. Here we use the notation w rather than V to indicate that a discretization error is involved. This semi-discretized version is applied for each of the parallel lines, $\tau = \tau_\nu$, $\nu = 1, \ldots, \nu_{\max} - 1$. (The cover figure of this book motivates the procedure.) For each line $\tau = \tau_\nu$, the function $w(S, \tau_{\nu-1})$ is known from the previous line, starting from the known payoff for $\tau = 0$. The equation to be solved for each line τ_ν is

$$\frac{1}{2} \Delta \tau \, \sigma^2 S^2 \frac{\partial^2 w}{\partial S^2} + \Delta \tau \, r S \frac{\partial w}{\partial S} - (1 + \Delta \tau \, r) w = -w(\cdot, \tau_{\nu-1}) \tag{4.59}$$

This is an ordinary differential equation, with boundary conditions for $S_{\mathrm{f}}(\tau_\nu)$ and $S \to \infty$. The solution is obtained analytically, similar as in Exercise 4.8. Hence there is no truncation error in S-direction. The right-hand function $-w(S, \tau_{\nu-1})$ is known, and is an inhomogeneous term of the ODE.

The resulting *analytic method of lines* is carried out in [CaF95]. The above describes the basic idea. An additional complication arises from the early exercise curve, which separates each of the parallel lines into two parts. Since for the previous line $\tau_{\nu-1}$ the separation point lies more "on the right" (recall that for a put the curve $S_{\mathrm{f}}(\tau)$ is monotonically decreasing for growing τ), the inhomogeneous term $w(\cdot, \tau_{\nu-1})$ consists of two parts as well, but separated differently. Accordingly, the analytic solution of (4.59) consists of three parts, defined for

$$0 < S < S_{\mathrm{f}}(\tau_\nu) \; , \; S_{\mathrm{f}}(\tau_\nu) \le S < S_{\mathrm{f}}(\tau_{\nu-1}) \; , \; S_{\mathrm{f}}(\tau_{\nu-1}) \le S \; .$$

On the left-hand interval, w equals the payoff. For the middle interval the inhomogeneous term is given by the payoff. The unknown separation points $S_f(\tau_\nu)$ are again fixed by the high-contact conditions. The resulting analytic method of lines is quite involved, for details see [CaF95].

4.8.4 Methods Evaluating Probabilities

Another approach is based on compound options, see [GeJ84]. Hypothetical options are considered that can only be exercised at discrete dates. Let us denote the values of such options $\overline{V}_1, \overline{V}_2, \overline{V}_3, \ldots$, where

$$\overline{V}_1 : \text{exercise possible only at time } T$$

$$\overline{V}_2 : \text{exercise possible only at } T/2 \text{ and } T$$

$$\overline{V}_3 : \text{exercise possible only at } T/3,\ 2T/3 \text{ and } T$$

and so on. In this basic version of the method, the exercise points characterizing $\overline{V}_k$ are equally spaced. In this way, an infinite sequence of options on options is constructed, the limit of which is the American put value. Obviously $\overline{V}_1$ is the value of a European-style option. — For $\overline{V}_2$ the domain splits into two parts, above and below $T/2$. For $t > T/2$ no early exercise is possible, hence the value is given by the European put value $V_{\mathrm{P}}^{\mathrm{eur}}(S, t; K)$, where the payoff at T is $(K - S)^+$. At $T/2$ an approximation $\overline{S}_f(T/2)$ of the free boundary is given by the crossing of the payoff. That is, $\overline{S}_f(T/2)$ is defined as zero S of the implicit equation

$$V_{\mathrm{P}}^{\mathrm{eur}}(S, T/2; K) = K - S \ .$$

Now the decision at $T/2$ is essentially clear: Exercise when $S_{T/2} < \overline{S}_f(T/2)$, and hold when $S_{T/2} > \overline{S}_f(T/2)$. Since exercising would be meaningless for $S_T > K$, a better argument is to take the option in case $S_{T/2} > \overline{S}_f(T/2)$. This is an option on an option, called *compound option*. As motivated by the univariate integral in Exercise 3.9, which integrates the payoff, the integral over an option is a *bivariate* integral. The corresponding formula for $\overline{V}_2$ involves bivariate normal distribution functions.

Similarly, for $\overline{V}_3$ trivariate integrals are involved, and in general multivariate integrals for $\overline{V}_k$. For $\overline{V}_2, \overline{V}_3, \ldots$ analytic formulas are available [GeJ84]. Using extrapolation to the limit, the three-point formula

$$\overline{V} := \frac{1}{2}(9\overline{V}_3 - 8\overline{V}_2 + \overline{V}_1)$$

gives a good approximation to the value of the option. Since the evaluation of $\overline{V}_k$ requires the evaluation of the k-dimensional integrals of cumulative normal distributions, the method is too expensive for larger k. To improve the efficiency, [BuJ92] suggest to use a two-point formula $\overline{V} := 2\overline{V}_2^* - V_1^*$ where the exercise points for $\overline{V}_1^*$ and $\overline{V}_2^*$ are chosen optimally.

Notes and Comments

on Section 4.1:

General references on numerical PDEs include [Sm78], [Vi81], [CL90], [Th95], [Mo96]. For references on modeling of dividends consult [WDH96], [Kwok98]. A special solution of (4.2) is

$$y(x,\tau) = \frac{1}{2\sqrt{\pi\tau}} \exp\left(-\frac{x^2}{4\tau}\right) .$$

For small values of τ, the transformation (4.3) may take bad values in the argument of the exponential function because q_δ can be too large. An overflow will be the result. Then the transformation

$$\tau := \tfrac{1}{2}\sigma^2(T-t)$$
$$x := \log\left(\tfrac{S}{K}\right) + \left(r - \delta - \tfrac{\sigma^2}{2}\right)(T-t)$$
$$y(x,\tau) := e^{-rt}V(S,t)$$

can be used as alternative [BaP96]. Again (4.2) results, but initial conditions and boundary conditions must be adapted appropriately. As will be seen in Section 6.4, the quantities q and q_δ are basically the Péclet number. It turns out that large values of the Péclet number are a general source of difficulties. For other transformations see [ZhWC04].

on Section 4.2:

We follow the notation $w_{i\nu}$ for the approximation at the node (x_i, τ_ν), to stress the surface character of the solution y over a two-dimensional domain. In the literature a frequent notation is w_i^ν, which emphasizes the different character of the space variable (here x) and the time variable (here τ). Our vectors $w^{(\nu)}$ with componentes $w_i^{(\nu)}$ come close to this convention.

Summarizing the Black-Scholes equation to

$$\frac{\partial V}{\partial t} + \mathcal{L}_{\mathrm{BS}} = 0 \tag{4.60}$$

where $\mathcal{L}_{\mathrm{BS}}$ represents the other terms of the equation, see Section 4.5.3, motivates an interpretation of the finite-difference schemes in the light of numerical ODEs. There the forward approach is known as *explicit Euler method* and the backward approach as *implicit Euler method*.

on Section 4.3:

Crank and Nicolson suggested their approach in 1947. Theorem 4.4 discusses three main principles of numerical analysis, namely order (of convergence), stability, and efficiency. A Crank-Nicolson variant has been developed that is consistent with the *volatility smile*, which reflects the dependence of the volatility on the strike [AB97].

In view of the representation (4.12) the Crank-Nicolson approach corresponds to the ODE *trapezoidal rule*. Following these lines suggests to apply other ODE approaches, some of which lead to methods that relate more than two time levels. In particular, backward difference formula (BDF) are of interest, which evaluate $\mathcal{L}$ at only one time level. The relevant second-order discretization is listed in the end of Section 4.2.1. Using this formula (BDF2) for the time discretization, a three-term recursion involving $w^{(\nu+1)}$, $w^{(\nu)}$, $w^{(\nu-1)}$ replaces the two-term recursion (4.15b) ($\longrightarrow$ Exercise 4.10). But multistep methods such as BDF may not behave well close to the exercise boundary, where we encounter a lack of smoothness. This suggests to consider other alternatives with better stability properties then Crank-Nicolson. Crank-Nicolson is A-stable, several other methods are L-stable, which better damp out high-frequency oscillation, see [Cash84], [IkT04]. For numerical ODEs we refer to [La91], [HNW93]. From the ODE analysis circumstances are known where the implicit Euler behaves superior to the trapezoidal rule. The latter method may show a *slowly damped* oscillating error. Accordingly, in several PDE situations the fully implicit method of Section 4.2.5 behaves better than Crank-Nicolson [Ran84], [ZVF00]. The explicit scheme corresponds to the trinomial-tree method mentioned in Section 1.4.

on Section 4.4:

If European options are evaluated via the analytic formula (A4.10), the boundary conditions in (4.19) are of no practical interest. When boundary conditions are not clear, it often helps to set $V_{SS} = 0$ (or $y_{xx} = 0$).

on Section 4.5:

For a proof of the Black-Scholes inequality, see [LL96], p.111. The obstacle problem in this chapter is described following [WDH96]. Also the smooth pasting argument of Exercise 4.9 is based on that work. For other arguments concerning smooth pasting see [Moe76], and [Kwok98]. There you find a discussion of $S_f(t)$, and of the behavior of this curve for $t \to T$. There are several different possibilites to implement the boundary conditions at $x_{\min}$, $x_{\max}$, see [TR00], p. 122. For the front-fixing transformation, see [Cra84], [NiST02], [ZhWC04].

The general definition of a linear complementarity problem is

$$\mathsf{AB} = 0, \quad \mathsf{A} \geq 0, \quad \mathsf{B} \geq 0,$$

where A and B are abbreviations of more complex expressions. A general reference on free boundaries and on linear complementarity is [EO82].

The Figure 4.16 shows a detail of approximations to an early-exercise curve. The finite-difference calculated points are connected by straight lines (dashed). The figure also shows a local approximation valid close to maturity (for reference see [MR97]): For $t < T$ and $t \to T$, the asymptotic behavior of S_f is

$$S_f(t) \sim K \left(1 - \sigma \sqrt{(t-T) \log(T-t)}\right)$$

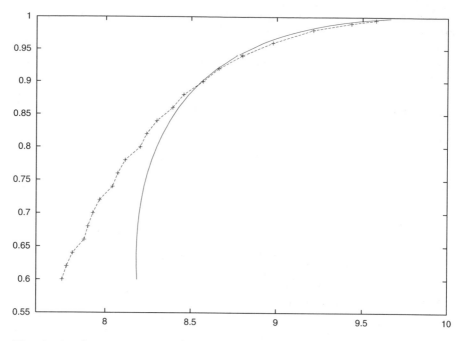

Fig. 4.16. Approximations of an early-exercise curve of an American put ($T = 1$, $\sigma = 0.3$, $K = 10$); dashed: finite-difference approximation, solid: asymptotic behavior for $t \approx T$

for an American put without dividends.

For a proof of the high-contact condition or *smooth-fit principle* see [Moe76], p.114. For a discussion of the smoothness of the free boundary S_f see [MR97] and the references therein.

on Section 4.6:

By choosing the θ in (4.29) one fixes at which position along the time axis the second-order spacial derivatives are focused. With

$$\theta = \frac{1}{2} - \frac{1}{12}\frac{\Delta x^2}{\Delta \tau}$$

a scheme results that is fourth-order accurate in x-direction. The application on American options requires careful compensation of the discontinuities [Mayo00].

Based on the experience of this author, an optimal choice of the relaxation parameter ω_R in Algorithm 4.13 can not be given. The simple strategy $\omega_R = 1$ appears recommendable.

on Section 4.7:

Since the accuracy of the results is not easily guaranteed, it does seem advisable to hesitate before exposing wealth to a chance of loss or damage. After having implemented a finite-difference algorithm it is a must to compare the results with the numbers obtained by means of other algorithms. The lacking smoothness of solutions near $S \approx K$, $t \approx T$ due to the nonsmooth payoff can be largely improved by solving for the difference function (4.45) rather than for $V(S,t)$. The lacking smoothness along the early-exercise curve can be avoided by using the front-fixing approach. Both ideas can be combined and give a successful method [ZhWC04].

The question how accurate different methods are has become a major concern in recent research; see for instance [CoLV02]. Clearly one compares a finite-difference European option with the analytic formula (A4.10). The latter is to be preferred, except the surface is the ultimate object. The correctness of codes can be checked by testing the validity of symmetry relations (A5.3).

We have introduced finite differences mainly in view of calculating American options. For exotic options PDEs occur, the solutions of which depend on three or more independent variables [WDH96], [Bar97], [TR00]; see also Chapter 6. For bounds on the error caused by truncating the infinite x- or S-interval, see [KaN00].

on Section 4.8:

For the case $H = 0$ an extra analysis is required [MaM86]. The method has been extended to the more general situation of commodity options, where the cost of carry is involved [BaW87]. Integral representations are based on an inhomogeneous differential equation. Recall from Section 4.5.3 that for a put the equation

$$\frac{\partial V}{\partial t} + \mathcal{L}_{\mathrm{BS}}(V) = \delta S - rK$$

holds for $S < S_{\mathrm{f}}(t)$; the homogeneous Black-Scholes equation holds for $S > S_{\mathrm{f}}(t)$. Both versions can be written in one equation, and an integral representation of the solution is presented in [Jam92], see also [Kwok98] and the references therein, and [Hei05] for a condition number.

It is possible to give an analytic expression of the premium p in terms of the free boundary S_{f} [Kwok98]. The formula for p consists of an integral with an integrand resembling the Black-Scholes formula (A4.10c). The dependence on S_{f} is via coefficients d_1^* and d_2^*, which are generalized from (4.40). Since the value of the option for S_{f} is given by the payoff, an integral equation is available which determines the free boundary implicitly [Kwok98]. By means of a suitable recursion, a pointwise approximation of S_{f} is possible, but costly. But here the border line to numerical approaches is definitely crossed.

For numerical comparisons of various methods see [AiC97], [BrD97]. For comparable accuracy, simple binomial approaches appear to have an edge over analytic methods.

on other methods:

Here we give a few hints on methods neither belonging to this chapter on finite differences, nor to Chapters 5 or 6. General hints can be found in [RT97], in particular with the references of [BrD97]. Closely related to linear complementarity problems are minimization methods. An efficient realization by means of methods of linear optimization is suggested in [DH99]. The uniform grid can only be the first step toward more flexible approaches, such as the finite elements to be introduced in Chapter 5. For spectral methods see [ZhWC04]. For penalty methods see [NiST02], [FV02]. Another possibility to enhance the power of finite differences is the *multigrid* approach; for general expositions see [Ha85], [TOS01]; for application to finance see [Oo03].

Exercises

Exercise 4.1   Continuous Dividend Flow

Assume that a stock pays a dividend D once per year. Calculate a corresponding continuous dividend rate δ under the assumptions

$$\dot{S} = (\mu - \delta)S \;, \;\; \mu = 0, \quad S(1) = S(0) - D > 0.$$

Generalize the result to general growth rates μ and arbitrary day t_D of dividend payment.

Exercise 4.2   Stability of the Fully Implicit Method

The backward-difference method is defined via the solution of the equation (4.11). Prove the stability.

Hint: Use the results of Section 4.2.4 and $w^{(\nu)} = A^{-1}w^{(\nu-1)}$.

Exercise 4.3   Crank-Nicolson Order $O(\Delta^2)$

Let the function $y(x, \tau)$ solve the equation

$$y_\tau = y_{xx}$$

and be sufficiently smooth. With the difference quotient

$$\delta_x^2 w_{i\nu} := \frac{w_{i+1,\nu} - 2w_{i\nu} + w_{i-1,\nu}}{\Delta x^2}$$

the local truncation error ϵ of the Crank-Nicolson method is defined

$$\epsilon := \frac{y_{\nu+1,i} - y_{i\nu}}{\Delta\tau} - \frac{1}{2}\left(\delta_x^2 y_{i\nu} + \delta_x^2 y_{i,\nu+1}\right).$$

Show

$$\epsilon = O(\Delta\tau^2) + O(\Delta x^2).$$

Exercise 4.4   Boundary Conditions of a European Call

Prove (4.19).

Hints: Either transform the Black-Scholes equation (4.1) with

$$S := \bar{S} \exp(\delta(T - t))$$

into a dividend-free version to obtain the dividend version of (4.18), or apply the dividend version of the put-call parity. The rest follows with transformation (4.3).

Exercise 4.5   Boundary Conditions of American Options

Show that the boundary conditions of American options satisfy

$$\lim_{x \to \pm \infty} y(x, \tau) = \lim_{x \to \pm \infty} g(x, \tau),$$

where g is defined in Problem 4.7.

Exercise 4.6   Gauß-Seidel as Special Case of SOR

Let the $n \times n$ matrix $A = ((a_{ij}))$ additively be partitioned into $A = D - L - U$, with D diagonal matrix, L strict lower triangular matrix, U strict upper triangular matrix, $x \in \mathbb{R}^n$, $b \in \mathbb{R}^n$. The *Gauß-Seidel method* is defined by

$$(D - L)x^{(k)} = Ux^{(k-1)} + b$$

for $k = 1, 2, \ldots$. Show that with

$$r_i^{(k)} := b_i - \sum_{j=1}^{i-1} a_{ij} x_j^{(k)} - \sum_{j=i}^{n} a_{ij} x_j^{(k-1)}$$

and for $\omega_\mathrm{R} = 1$ the relation

$$x_i^{(k)} = x_i^{(k-1)} + \omega_\mathrm{R} \frac{r_i^{(k)}}{a_{ii}}$$

holds. For general $1 < \omega_\mathrm{R} < 2$ this defines the SOR (successive overrelaxation) method.

Exercise 4.7

Implement Algorithms 4.13 and 4.14.

Test example: Example 1.6 and others.

Exercise 4.8   Perpetual Put Option

For $T \to \infty$ it is sufficient to analyze the ODE

$$\frac{\sigma^2}{2} S^2 \frac{d^2V}{dS^2} + (r - \delta)S \frac{dV}{dS} - rV = 0.$$

Consider an American put with high contact to the payoff $V = (K - S)^+$ at $S = S_f$. Show:

a) Upon substituting the boundary condition for $S \to \infty$ one obtains

$$V(S) = c\left(\frac{S}{K}\right)^{\lambda_2},$$

where $\lambda_2 = \frac{1}{2}(1 - q_\delta - \sqrt{(q_\delta - 1)^2 + 4q})$, $q = \frac{2r}{\sigma^2}$, $q_\delta = \frac{2(r-\delta)}{\sigma^2}$ and c is a positive constant.

Hint: Apply the transformation $S = Ke^x$. (The other root λ_1 drops out.)

b) V is convex.

For $S < S_f$ the option is exercised; then its intrinsic value is $S - K$. For $S > S_f$ the option is not exercised and has a value $V(S) > S - K$. The holder of the option decides when to exercise. This means, the holder makes a decision on the high contact S_f such that the value of the option becomes maximal [Mer73].

c) Show: $V'(S_f) = -1$, if S_f maximizes the value of the option.

Hint: Determine the constant c such that $V(S)$ is continuous in the contact point.

Exercise 4.9   Smooth Pasting of the American Put

Suppose a portfolio consists of an American put and the corresponding underlying. Hence the value of the portfolio is $\Pi := V_P^{am} + S$, where S satisfies the SDE (1.33). S_f is the value for which we have high contact, compare (4.22).

a) Show that

$$d\Pi = \begin{cases} 0 & \text{for } S < S_f \\ \left(\dfrac{\partial V_P^{am}}{\partial S} + 1\right)\sigma S\, dW + O(dt) & \text{for } S > S_f. \end{cases}$$

b) Use this to argue

$$\frac{\partial V_P^{am}}{\partial S}(S_f(t), t) = -1.$$

Hint: Use $dS > 0 \Rightarrow dW > 0$ for small dt. Assume $\frac{\partial V}{\partial S} > -1$ and construct an arbitrage strategy for $dS > 0$.

Exercise 4.10   Semi-Discretization

For a semi-discretization of the Black-Scholes equation (1.2) consider the semi-discretized domain

$$0 \le t \le T \quad, \quad S = S_i := i\Delta S \quad, \quad \Delta S := \frac{S_{max}}{m} \quad, \quad i = 0, 1, \dots, m$$

for some value S_{max}. On this set of parallel lines define for $1 \le i \le m - 1$ functions $w_i(t)$ as approximation to $V(S_i, t)$.

a) Using the standard second-order difference schemes of Section 4.2.1, derive the system

$$\dot{w} + Bw = 0 , \qquad (4.61)$$

which up to boundary conditions approximates (1.2). Here w is the vector $(w_1, \ldots, w_{m-1})^{\text{tr}}$. Show that B is a tridiagonal matrix, and calculate its coefficients.

b) Use the BDF2 formula of Section 4.2.1 to show that

$$w^{(\nu)} = 4w^{(\nu-1)} - 3w^{(\nu-2)} + 2\Delta t B w^{(\nu-2)}$$

is a valid scheme to integrate (4.61). Computer persons are encouraged to implement this scheme.

Exercise 4.11   Front-Fixing for American Options

Apply the transformation

$$\zeta := \frac{S}{S_f(t)} \quad , \quad y(\zeta, t) := V(S, t)$$

to the Black-Scholes equation (4.1).

a) Show

$$\frac{\partial y}{\partial t} + \frac{\sigma^2}{2} \zeta^2 \frac{\partial^2 y}{\partial \zeta^2} + \left[(r - \delta) - \frac{1}{S_f} \frac{dS_f}{dt} \right] \zeta \frac{\partial y}{\partial \zeta} - ry = 0$$

b) Set up the domain for (ζ, t) and formulate the boundary conditions for an American call. (Assume $\delta > 0$.)

c) (Project) Set up a finite-difference scheme to solve the derived boundary-value problem. The curve $S_f(t)$ is implicitly defined by the above PDE, with final value $S_f(T) = \max(K, \frac{r}{\delta}K)$.

Exercise 4.12   Brennan-Schwartz Algorithm

Let A be a tridiagonal matrix as in (C1.5) and b and g vectors. The system of equations $Aw = b$ is to be solved such that the side condition $w \geq g$ is obeyed componentwise. Assume $w_i = g_i$ for $1 \leq i \leq i_f$ and $w_i > g_i$ for $i_f < i \leq n$, where i_f is unknown.

a) Formulate an algorithm similar as in (C1.6) that solves $Aw = b$ in the backward/forward approach. In the final forward loop, for each i the calculated candidate w_i is tested for $w_i \geq g_i$: In case $w_i < g_i$ the calculated value w_i is corrected to $w_i = g_i$.

b) Apply the algorithm specifically to the case of a put with A, b, g from Section 4.6.1. For the case of a call adapt the forward/backward algorithm (C1.6). Incorporate this approach into Algorithm 4.13 by replacing the SOR-loop.

Exercise 4.13   Extrapolation of Higher Order

Similar as in Section 4.7 assume an error model

$$\eta^* = \eta(\Delta) - \gamma_1 \Delta^2 - \gamma_2 \Delta^3$$

and three calculated values

$$\eta_1 := \eta(\Delta) \quad , \quad \eta_2 := \eta\left(\frac{\Delta}{2}\right) \quad , \quad \eta_3 := \eta\left(\frac{\Delta}{4}\right) \; .$$

Show that

$$\eta^* = \frac{1}{21}(\eta_1 - 12\eta_2 + 32\eta_3) \; .$$

Exercise 4.14

a) Derive (4.49).

b) Derive (4.56).

Exercise 4.15   Analytic Method for the American Put

(Project) Implement both the Algorithm 4.16 and Algorithm 4.17. Think of how to combine them into a single hybrid algorithm.

Chapter 5 Finite-Element Methods

The finite-difference approach with equidistant grids is easy to understand and straightforward to implement. The resulting uniform rectangular grids are comfortable, but in many applications not flexible enough. Steep gradients of the solution require a finer grid such that the difference quotients provide good approximations of the differentials. On the other hand, a flat gradient may be well modeled on a coarse grid. Such a flexibility of the grid is hard to obtain with finite-difference methods.

An alternative type of methods for solving PDEs that does provide the desired flexibility is the class of finite-element methods. A "finite element" designates a mathematical topic such as a subinterval and defined thereupon a piece of function. There are alternative names as *variational methods*, or *weighted residuals*, or *Galerkin methods*. These names hint at underlying principles that serve to derive suitable equations. As these different names suggest, there are several different approaches leading to finite elements. The methods are closely related.

The flexibility of finite-element methods becomes important in higher-dimensional spaces. For the one-dimensional situation of standard options, the possible improvement of a finite-element method over the standard methods of the previous chapter ist not that significant. With the focus on standard options, Chapter 5 may be skipped on first reading. But for several exotic options ($\longrightarrow$ Chapter 6), finite elements are widely applicable and recommendable.

Faced with the huge field of finite-element methods, in this chapter we confine ourselves to a brief overview on several approaches and ideas (in Section 5.1). Then in Section 5.2, we describe the approximation with the simplest finite elements, namely piecewise straight-line segments. These approaches will be applied to the calculation of standard options in Section 5.3. Finally, in Section 5.4, we will introduce into error estimates. Methods more subtle than just the Taylor expansion of the truncation error are required to show that quadratic convergence is possible with unstructured grids and nonsmooth solutions. Many of the ideas extend to multidimensional scenarios.

Fig. 5.1. Discretization of a continuum

5.1 Weighted Residuals

Many of the principles on which finite-element methods are based, can be interpreted as weighted residuals. What does this mean? There lies a duality in a discretization. This may be illustrated by means of Figure 5.1, which shows a partition of an x-axis. This discretization is either represented by

(a) discrete grid points x_i, or by
(b) a set of subintervals.

Each of the two ways to see a discretization leads to a different approach of constructing an approximation w. Let us illustrate this with the one-dimensional situation of Figure 5.2. An approximation w based on finite differences is founded on the grid points and primarily consists of discrete points (Figure 5.2a). Finite elements are founded on subdomains (intervals in Figure 5.2b) with piecewise defined functions, which are defined by suitable criteria and constitute a global approximation w. In a narrower sense, a finite element is a pair consisting of one piece of subdomain and the corresponding function defined thereupon, mostly a polynomial. Figure 5.2 reflects the respective basic approaches; in a second step the isolated points of a finite-difference calculation can be extended to continuous piecewise functions by means of interpolation ($\longrightarrow$ Appendix C1). — A two-dimensional domain can be partitioned into triangles, for example, where w is again represented with piecewise polynomials. Another aspect of the flexibility of finite elements is that triangles easily approximate irregular domains.

As will be shown next, the approaches of finite-element methods use integrals. If done properly, integrals require less smoothness. This often matches applications better and adds to the flexibility of finite-element methods. The integrals may be derived in a natural way from minimum principles, or are constructed artificially. Finite elements based on polynomials make the calculation of the integrals easy.

5.1.1 The Principle of Weighted Residuals

To explain the principle of weighted residuals we discuss the formally simple case of the differential equation

$$Lu = f. \tag{5.1}$$

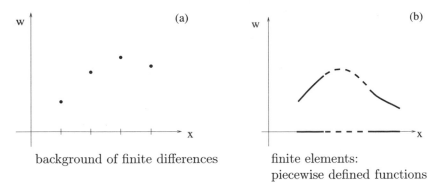

(a)

(b)

background of finite differences

finite elements:
piecewise defined functions

Fig. 5.2. Two kinds of approximations (one-dimensional situation)

Here L symbolizes a linear differential operator. Important examples are

$$Lu := -u'' \quad \text{for} \quad u(x), \quad \text{or} \tag{5.2a}$$

$$Lu := -u_{xx} - u_{yy} \quad \text{for} \quad u(x,y). \tag{5.2b}$$

Solutions u of the differential equation are studied on a domain $\mathcal{D} \subseteq \mathbb{R}^n$. The piecewise approach starts with a partition of the domain into a finite number of subdomains $\mathcal{D}_k$,

$$\mathcal{D} = \bigcup_k \mathcal{D}_k . \tag{5.3}$$

All boundaries should be included, and approximations to u are calculated on the closure $\bar{\mathcal{D}}$. The partition is assumed disjoint up to the boundaries of $\mathcal{D}_k$, so $\mathcal{D}_j^\circ \cap \mathcal{D}_k^\circ = \emptyset$ for $j \neq k$. In the one-dimensional case ($n = 1$), for example, the $\mathcal{D}_k$ are subintervals of a whole interval $\mathcal{D}$. In the two-dimensional case, (5.3) may describe a partition into triangles.

The ansatz for approximations w to a solution u is a basis representation,

$$w := \sum_{i=1}^N c_i \varphi_i . \tag{5.4}$$

In the case of one independent variable x the $c_i \in \mathbb{R}$ are constant coefficients, and the φ_i are functions of x. The φ_i are called **basis functions**, or *trial functions* or *shape functions*. Typically the $\varphi_1, ..., \varphi_N$ are prescribed, whereas the free parameters $c_1, ..., c_N$ are to be determined such that $w \approx u$.

One strategy to determine the c_i is based on the residual function

$$R := Lw - f. \tag{5.5}$$

We look for a w such that R becomes "small." Since the φ_i are considered prescribed, in view of (5.4) N conditions or equations must be established to define and calculate the unknown $c_1, ..., c_N$. To this end we weight the residual by introducing N weighting functions *(test functions)* $\psi_1, ..., \psi_N$ and require

$$\int_{\mathcal{D}} R\psi_j \, dx = 0 \quad \text{for } j = 1, ..., N \tag{5.6}$$

This amounts to the requirement that the residual be orthogonal to the set of weighting functions ψ_j. The "dx" in (5.6) symbolizes the integration that matches $\mathcal{D} \subseteq \mathbb{R}^n$; frequently it will be dropped. The system of equations (5.6) for the model problem (5.1) consists of the N equations

$$\int_{\mathcal{D}} Lw\psi_j = \int_{\mathcal{D}} f\psi_j \quad (j = 1, ..., N) \tag{5.7}$$

for the N unknowns $c_1, ..., c_N$, which are part of w. Often the equations in (5.7) are written using a formulation with inner products,

$$(Lw, \psi_j) = (f, \psi_j),$$

defined as the corresponding integrals in (5.7). For linear L the ansatz (5.4) implies

$$\int Lw\psi_j = \int \left(\sum_i c_i L\varphi_i \right) \psi_j = \sum_i c_i \underbrace{\int L\varphi_i \psi_j}_{=:a_{ij}}.$$

The integrals a_{ij} constitute a matrix A. The $r_j := \int f\psi_j$ set up a vector r and the coefficients c_j a vector $c = (c_1, ..., c_N)^{tr}$. This allows to rewrite the system of equations in vector notation as

$$Ac = r. \tag{5.8}$$

This outlines the general principle, but leaves open the questions how to handle boundary conditions and how to select the basis functions φ_i and the weighting functions ψ_j. The freedom to choose trial functions φ_i and test functions ψ_j allows to construct several different methods. For the time being suppose that these functions have sufficient potential to be differentiated or integrated. We will enter a discussion of relevant function spaces in Section 5.4.

5.1.2 Examples of Weighting Functions

Let us begin with listing important examples of how to select weighting functions ψ:

1.) **Galerkin method**, also Bubnov-Galerkin method:
    Choose $\psi_j := \varphi_j$. Then $a_{ij} = \int L\varphi_i \varphi_j$

2.) **collocation:**

Choose $\psi_j := \delta(x - x_j)$. Here δ denotes Dirac's delta function, which in $\mathbb{R}^1$ satisfies $\int f\delta(x - x_j)dx = f(x_j)$. As a consequence,

$$\int Lw\psi_j = Lw(x_j),$$

$$\int f\psi_j = f(x_j).$$

That is, a system of equations $Lw(x_j) = f(x_j)$ results, which amounts to evaluating the differential equation at selected points x_j.

3.) **least squares:**

Choose

$$\psi_j := \frac{\partial R}{\partial c_j}$$

This choice of test functions deserves its name *least-squares*, because to minimize $\int (R(c_1, ..., c_N))^2$ the necessary criterion is the vanishing of the gradient, so

$$\int_{\mathcal{D}} R\frac{\partial R}{\partial c_j} = 0 \quad \text{for all } j.$$

4.) **subdomain:**

Choose

$$\psi_k = \begin{cases} 1 & \text{in } \mathcal{D}_k \\ 0 & \text{for } x \notin \mathcal{D}_k \end{cases}$$

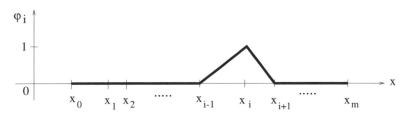

Fig. 5.3. "Hat function": simple choice of finite elements

5.1.3 Examples of Basis Functions

For the choice of suitable basis functions φ_i our concern will be to meet two aims: The resulting methods must be accurate, and their implementation should become efficient. We defer the aspect of accuracy to Section 5.4, and concentrate on the latter requirement, which can be focused on the sparsity of matrices. In particular, if the matrix A of the linear equations is sparse, then the system can be solved efficiently even when it is large. In order to achieve sparsity we require that $\varphi_i \equiv 0$ on most of the subdomains $\mathcal{D}_k$. Figure 5.3

illustrates an example for the one-dimensional case $n = 1$. This *hat function* of Figure 5.3 is the simplest example related to finite elements. It is piecewise linear, and each function φ_i has a support consisting of only two subintervals, $\varphi_i(x) \neq 0$ for $x \in$ support. A consequence is

$$\int_D \varphi_i \varphi_j = 0 \quad \text{for } |i - j| > 1, \tag{5.9}$$

as well as an analogous relation for $\int \varphi_i' \varphi_j'$. We will discuss hat functions in the following Section 5.2. More advanced basis functions are constructed using piecewise polynomials of higher degree. In this way, basis functions can be obtained with $\mathcal{C}^1$- or $\mathcal{C}^2$-smoothness ($\longrightarrow$ Exercise 5.1). Recall from interpolation ($\longrightarrow$ Appendix C1) that polynomials of degree three can lead to $\mathcal{C}^2$-smooth splines.

Remark on $Lu = -u''$, $\quad u, \varphi, \psi \in \{u : u(0) = u(1) = 0\}$:

Integration by parts implies formally

$$\int_0^1 \varphi'' \psi = -\int_0^1 \varphi' \psi' = \int_0^1 \varphi \psi'',$$

because the boundary conditions $u(0) = u(1) = 0$ let the nonintegral terms vanish. These three versions of the integral can be distinguished by the smoothness requirements on φ and ψ, and by the question whether the integrals exist. One will choose the integral version that corresponds to the underlying method, and to the smoothness of the solution. For example, for Galerkin's approach the elements a_{ij} of A consist of the integrals

$$-\int_0^1 \varphi_i' \varphi_j'.$$

We will return to the topic of function spaces in Section 5.4 (with Appendix C3).

5.2 Galerkin Approach with Hat Functions

As mentioned before, the uniformness of an equidistant grid leads to simple schemes of finite-difference methods. But equidistant x_i may not be comfortable when applied to options and to the PDE $y_\tau = y_{xx}$.

Example: $K = 50$, $m = 20$, $x_{\min} = -5 = -x_{\max}$. The transformation $S = Ke^x$ for equidistant x_i leads to the discrete values of S:

$$S_1 = 0.5554 \ , \quad S_2 = 0.915 \ , \quad \dots,$$
$$S_9 = 30.32 \ , \quad S_{10} = 50. \ , \quad S_{11} = 82.436 \ , \quad \dots,$$
$$S_{19} = 4500.85$$

In the area of interest $S \approx K$ (that is, $x \approx 0$), the S_i are unreasonably "odd" numbers. And in view of the accuracy, the distance between S_9 and S_{10} is too large compared to $S_2 - S_1$. For the application to options it would be more convenient to work with equidistant S_i and hence nonuniform x_i, if equidistance is desirable at all. Any required flexibility is provided by finite-elements methods.

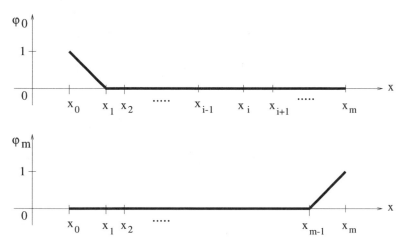

Fig. 5.4. Special "hat functions" φ_0 and φ_m

5.2.1 Hat Functions

We now explain the prototype of a finite-element method. This simple approach makes use of the hat functions, which we define formally (compare Figures 5.3 and 5.4).

Definition 5.1   (hat functions)
For $1 \le i \le m - 1$ set

$$\varphi_i(x) := \begin{cases} \dfrac{x - x_{i-1}}{x_i - x_{i-1}} & \text{for } x_{i-1} \le x < x_i \\[2mm] \dfrac{x_{i+1} - x}{x_{i+1} - x_i} & \text{for } x_i \le x < x_{i+1} \\[2mm] 0 & \text{elsewhere} \end{cases}$$

and for the boundary functions

$$\varphi_0(x) := \begin{cases} \dfrac{x_1 - x}{x_1 - x_0} & \text{for } x_0 \le x < x_1 \\[2mm] 0 & \text{elsewhere} \end{cases}$$

$$\varphi_m(x) := \begin{cases} \dfrac{x - x_{m-1}}{x_m - x_{m-1}} & \text{for } x_{m-1} \le x \le x_m \\[2mm] 0 & \text{elsewhere.} \end{cases}$$

These $m + 1$ hat functions satisfy the following properties.

Properties 5.2   (hat functions)

(a) The $\varphi_0, ..., \varphi_m$ form a basis of the space of polygons

$$\{g \in C^0[x_0, x_m] : g \text{ straight line on } \mathcal{D}_k := [x_k, x_{k+1}]$$
$$\text{for all } k = 0, ..., m - 1\}.$$

That is to say, for each polygon v on $\mathcal{D}_0, ..., \mathcal{D}_{m-1}$ there are unique coefficients $c_0, ..., c_m$ with

$$v = \sum_{i=0}^{m} c_i \varphi_i.$$

(b) On $\mathcal{D}_k$ only φ_k and $\varphi_{k+1} \neq 0$ are nonzero. Hence

$$\varphi_i \varphi_k = 0 \quad \text{for } |i - k| > 1.$$

(c) A simple approximation of the integral $\int_{x_0}^{x_m} f\varphi_j \, dx$ can be calculated as follows:
Substitute f by the interpolating polygon

$$f_\mathrm{p} := \sum_{i=0}^{m} f_i \varphi_i \quad , \text{ where } \quad f_i := f(x_i),$$

and obtain for each j the approximating integral

$$I_j := \int_{x_0}^{x_m} f_\mathrm{p}\varphi_j \, dx = \int_{x_0}^{x_m} \sum_{i=0}^{m} f_i\varphi_i\varphi_j \, dx = \sum_{i=0}^{m} f_i \underbrace{\int_{x_0}^{x_m} \varphi_i\varphi_j \, dx}_{=:b_{ji}}$$

The b_{ij} constitute a symmetric matrix B and the f_i a vector $\bar{f}$. If we arrange all integrals I_j $(0 \le j \le m)$ into a vector, then all integrals can be written in a compact way in vector notation as

$$B\bar{f}.$$

(d) The "large" $(m + 1)^2$–matrix $B := (b_{ij})$ can be set up $\mathcal{D}_k$-elementwise by (2×2)-matrices. The (2×2)-matrices are those integrals that integrate only over a single subdomain $\mathcal{D}_k$. For each $\mathcal{D}_k$ exactly the four integrals

$$\int_{x_k}^{x_{k+1}} \begin{pmatrix} \varphi_k^2 & \varphi_k\varphi_{k+1} \\ \varphi_{k+1}\varphi_k & \varphi_{k+1}^2 \end{pmatrix} dx$$

are nonzero. (The integral over a matrix is understood elementwise.)
These are the integrals on $\mathcal{D}_k$, where the integrand is a product of the
factors

$$\frac{x_{k+1} - x}{x_{k+1} - x_k} \quad \text{and} \quad \frac{x - x_k}{x_{k+1} - x_k}.$$

The four numbers

$$\frac{1}{(x_{k+1} - x_k)^2} \int_{x_k}^{x_{k+1}} \begin{pmatrix} (x_{k+1} - x)^2 & (x_{k+1} - x)(x - x_k) \\ (x - x_k)(x_{k+1} - x) & (x - x_k)^2 \end{pmatrix} dx$$

result. With $h_k := x_{k+1} - x_k$ integration yields the *element-mass matrix*
($\longrightarrow$ Exercise 5.2)

$$\frac{1}{6} h_k \begin{pmatrix} 2 & 1 \\ 1 & 2 \end{pmatrix}$$

(e) Analogously, integrating $\varphi_i' \varphi_j'$ yields

$$\int_{x_k}^{x_{k+1}} \begin{pmatrix} \varphi_k'^2 & \varphi_k' \varphi_{k+1}' \\ \varphi_{k+1}' \varphi_k' & \varphi_{k+1}'^2 \end{pmatrix} dx$$

$$= \frac{1}{h_k^2} \int_{x_k}^{x_{k+1}} \begin{pmatrix} (-1)^2 & (-1)1 \\ 1(-1) & 1^2 \end{pmatrix} dx = \frac{1}{h_k} \begin{pmatrix} 1 & -1 \\ -1 & 1 \end{pmatrix}.$$

These matrices are called *element-stiffness matrices*. They are used to set
up the matrix A.

5.2.2 Assembling

The next step is to assemble the matrices A and B. It might to be tempting
to organize this task as follows: Run a double loop on all i, j and check for
each (i, j) on which $\mathcal{D}_k$ the integral

$$\int_{\mathcal{D}_k} \varphi_i \varphi_j = 0$$

is nonzero. It turns out that such a procedure is cumbersome as compared
to the alternative of running a single loop on all k and calculate all relevant
integrals on $\mathcal{D}_k$.

To this end, we split the integrals

$$\int_{x_0}^{x_m} = \sum_{k=0}^{m-1} \int_{\mathcal{D}_k}$$

to construct the $(m+1) \times (m+1)$-matrices $A = (a_{ij})$ and $B = (b_{ij})$ *additively*
out of the (2×2)-element matrices. As noted earlier, on the subinterval

$$\mathcal{D}_k = \{x \ : \ x_k \leq x \leq x_{k+1}\}$$

only those integrals of $\varphi_i'\varphi_j'$ and $\varphi_i\varphi_j$ are nonzero, for which

$$i, j \in \mathcal{I}_k := \{k, k+1\} \tag{5.10}$$

holds. $\mathcal{I}_k$ is the set of indices of those basis functions that are nonzero on $\mathcal{D}_k$. The *assembling algorithm* performs a loop over the subinterval index $k = 0, 1, \ldots, m-1$ and distributes the (2×2)-element matrices additively to the positions (i, j) from (5.10). Before the assembling is started, the matrices A and B must be initialized with zeros. For $k = 0, ..., m-1$ one obtains for A the $(m+1)^2$-matrix

$$\begin{pmatrix} \frac{1}{h_0} & -\frac{1}{h_0} & & & \\ -\frac{1}{h_0} & \frac{1}{h_0} + \frac{1}{h_1} & -\frac{1}{h_1} & & \\ & -\frac{1}{h_1} & \frac{1}{h_1} + \frac{1}{h_2} & -\frac{1}{h_2} & \\ & & -\frac{1}{h_2} & \ddots & \ddots \\ & & & & \ddots \end{pmatrix} \tag{5.11a}$$

The matrix B is assembled in an analogous way. In the one-dimensional situation the matrices are tridiagonal. For an equidistant grid with $h = h_k$ the matrix A specializes to

$$A = \frac{1}{h} \begin{pmatrix} 1 & -1 & & & & & 0 \\ -1 & 2 & -1 & & & & \\ & -1 & 2 & \ddots & & & \\ & & \ddots & \ddots & \ddots & & \\ & & & & \ddots & 2 & -1 \\ 0 & & & & & -1 & 1 \end{pmatrix} \tag{5.11b}$$

and B to

$$B = \frac{h}{6} \begin{pmatrix} 2 & 1 & & & & 0 \\ 1 & 4 & 1 & & & \\ & 1 & 4 & \ddots & & \\ & & \ddots & \ddots & \ddots & \\ & & & \ddots & 4 & 1 \\ 0 & & & & 1 & 2 \end{pmatrix} \tag{5.11c}$$

5.2.3 A Simple Application

In order to demonstrate the procedure, let us consider the simple model boundary-value problem

$$Lu := -u'' = f(x) \quad \text{with} \quad u(x_0) = u(x_m) = 0. \tag{5.12}$$

We perform a Galerkin approach and substitute $w := \sum_{i=0}^{m} c_i\varphi_i$ into the differential equation. In view of (5.7) this leads to

$$\sum_{i=0}^{m} c_i \int_{x_0}^{x_m} L\varphi_i\varphi_j \, dx = \int_{x_0}^{x_m} f\varphi_j \, dx.$$

Next we apply integration by parts on the left-hand side, and invoke Property 5.2(c) on the right-hand side. The resulting system of equations is

$$\sum_{i=0}^{m} c_i \underbrace{\int_{x_0}^{x_m} \varphi_i'\varphi_j' \, dx}_{a_{ij}} = \sum_{i=0}^{m} f_i \underbrace{\int_{x_0}^{x_m} \varphi_i\varphi_j \, dx}_{b_{ij}}, \quad j = 0, 1, ..., m. \qquad (5.13)$$

This system is preliminary because the homogenous boundary conditions $u(x_0) = u(x_m) = 0$ are not yet taken into account.

At this state, the preliminary system of equations (5.13) can be written as

$$Ac = B\bar{f}. \qquad (5.14)$$

It is easy to see that the matrix A from (5.11b) is singular, because $A(1, 1, ..., 1)^{tr} = 0$. This singularity reflects the fact that the system (5.14) does not have a unique solution. This is consistent with the differential equation $-u'' = f(x)$: If $u(x)$ is solution, then also $u(x)+\alpha$ for arbitrary α. Unique solvability is attained by satisfying the boundary conditions; a solution u of $-u'' = f$ must be fixed by at least one essential boundary condition. For our example (5.12) we know in view of $u(x_0) = u(x_m) = 0$ the coefficients $c_0 = c_m = 0$. This information can be inserted into the system of equations in such a way that the matrix A changes to a nonsingular matrix without losing symmetry. For $c_0 = 0$ we replace the first equation of the system (5.14) by $(1, 0, \ldots, 0)^{tr} \, c = 0$. Adding the first equation to the second produces a zero in the first column of A. Analogously we realize $c_m = 0$ in the last row and column of A. Now the c_0 and c_m are decoupled, and the inner part of size $(m-1) \times (m-1)$ of A remains. Finally, for the special case of an equidistant grid, the system of equations is

$$\begin{pmatrix} 2 & -1 & & & 0 \\ -1 & 2 & \ddots & & \\ & \ddots & \ddots & \ddots & \\ & & \ddots & 2 & -1 \\ 0 & & & -1 & 2 \end{pmatrix} \begin{pmatrix} c_1 \\ c_2 \\ \vdots \\ c_{m-2} \\ c_{m-1} \end{pmatrix} =$$

$$\frac{h^2}{6} \begin{pmatrix} 1 & 4 & 1 & & & 0 \\ & 1 & 4 & 1 & & \\ & & \ddots & \ddots & \ddots & \\ & & & 1 & 4 & 1 \\ 0 & & & & 1 & 4 & 1 \end{pmatrix} \begin{pmatrix} \bar{f}_0 \\ \bar{f}_1 \\ \vdots \\ \bar{f}_{m-1} \\ \bar{f}_m \end{pmatrix} \qquad (5.15)$$

In (5.15) we have used an equidistant grid for sake of a lucid exposition. Our main focus is the nonequidistant version, which is also implemented easily. In case nonhomogeneous boundary conditions are prescribed, appropriate values of c_0 or c_m are predefined. The importance of finite-element methods in structural engineering has lead to call the global matrix A the *stiffness matrix*, and B is called the *mass matrix*.

5.3 Application to Standard Options

Analogously as the simple obstacle problem (Section 4.5.3) also the problem of calculating American options can be formulated as variational problem. Compare Problem 4.7, and recall the $g(x,\tau)$ defined in Section 4.6. The class of comparison functions is defined as

$$\mathcal{K} = \{v \in \mathcal{C}^0 : \tfrac{\partial v}{\partial x} \text{ piecewise } \mathcal{C}^0 ,$$
$$v(x,\tau) \geq g(x,\tau) \text{ for all } x,\tau , \ v(x,0) = g(x,0) , \tag{5.16}$$
$$v(x_{\max},\tau) = g(x_{\max},\tau), \ v(x_{\min},\tau) = g(x_{\min},\tau)\}.$$

Let y denote the exact solution of Problem 4.7. As solution of the partial differential equation, y is $\mathcal{C}^2$-smooth, and so $y \in \mathcal{K}$. For the American option and its partial differential inequality, y is $\mathcal{C}^2$-smooth on the continuation region, and again $y \in \mathcal{K}$. From

$$v \geq g, \quad \frac{\partial y}{\partial \tau} - \frac{\partial^2 y}{\partial x^2} \geq 0$$

we deduce

$$\int_{x_{\min}}^{x_{\max}} \left(\frac{\partial y}{\partial \tau} - \frac{\partial^2 y}{\partial x^2} \right) (v - g) \, dx \geq 0.$$

Invoking the complementarity

$$\int_{x_{\min}}^{x_{\max}} \left(\frac{\partial y}{\partial \tau} - \frac{\partial^2 y}{\partial x^2} \right) (y - g) \, dx = 0$$

and subtraction gives

$$\int_{x_{\min}}^{x_{\max}} \left(\frac{\partial y}{\partial \tau} - \frac{\partial^2 y}{\partial x^2} \right) (v - y) \, dx \geq 0.$$

Integration by parts leads to the inequality

$$\int_{x_{\min}}^{x_{\max}} \left(\frac{\partial y}{\partial \tau}(v - y) + \frac{\partial y}{\partial x}\left(\frac{\partial v}{\partial x} - \frac{\partial y}{\partial x} \right) \right) dx - \frac{\partial y}{\partial x}(v - y) \bigg|_{x_{\min}}^{x_{\max}} \geq 0.$$

The nonintegral term vanishes, because at the boundary for $x_{\min}$, $x_{\max}$, in view of $v = g$, $y = g$ the equality $v = y$ holds. The final result is

$$\int_{x_{\min}}^{x_{\max}} \left(\frac{\partial y}{\partial \tau} \cdot (v - y) + \frac{\partial y}{\partial x} \left(\frac{\partial v}{\partial x} - \frac{\partial y}{\partial x} \right) \right) dx \geq 0 \quad \text{for all } v \in \mathcal{K}. \quad (5.17)$$

The exact y is characterized by the fact that for all comparison functions $v \in \mathcal{K}$ the inequality (5.17) holds. For the special choice $v = y$ the integral takes its minimal value zero. A more general question is, whether a minimum (infimum) of the integral in (5.17) can be obtained with a $\hat{y}$ from $\mathcal{K}$ that is not $\mathcal{C}^2$-smooth. (Recall that the American option is widely $\mathcal{C}^2$-smooth, except across the early-exercise curve.) We look for a $\hat{y} \in \mathcal{K}$ such that the integral in (5.17) is minimal for all $v \in \mathcal{K}$. This formulation of our problem is called *weak version*, because it does *not* use $\hat{y} \in \mathcal{C}^2$. Solutions $\hat{y}$ of this minimization problem, which are globally continuous but only piecewise $\in \mathcal{C}^1$ are called *weak solutions*. The original partial differential equation requires $y \in \mathcal{C}^2$ and hence more smoothness. Such $\mathcal{C}^2$-solutions are called *strong solutions* or *classical solutions* ($\longrightarrow$ Section 5.4).

Now we approach the inequality (5.17) with finite-element methods. As a first step to approximately solve the minimum problem, assume approximations for $\hat{y}$ and v in the similar forms

$$\sum_i w_i(\tau) \varphi_i(x) \quad \text{for } \hat{y},$$

$$\sum_i v_i(\tau) \varphi_i(x) \quad \text{for } v.$$

The reduced smoothness of these expressions match the requirements of $\mathcal{K}$. The above setting assumes the independent variables τ and x to be *separated*. As a consequence of this simple approach, the same x-grid is applied for all τ, which results in a rectangular grid in the (x, τ)-plane. The time dependence is incorporated in the coefficient functions w_i and v_i. Plugging into (5.17) gives

$$\int \left\{ \left(\sum_i \frac{dw_i}{d\tau} \varphi_i \right) \left(\sum_j (v_j - w_j) \varphi_j \right) + \right.$$

$$\left. \left(\sum_i w_i \varphi_i' \right) \left(\sum_j (v_j - w_j) \varphi_j' \right) \right\} dx$$

$$= \sum_i \sum_j \frac{dw_i}{d\tau} (v_j - w_j) \int \varphi_i \varphi_j dx + \sum_i \sum_j w_i (v_j - w_j) \int \varphi_i' \varphi_j' dx \geq 0.$$

Translated into vector notation this is equivalent to

$$\left(\frac{dw}{d\tau} \right)^{tr} B(v - w) + w^{tr} A(v - w) \geq 0$$

or

$$(v - w)^{tr} \left(B\frac{dw}{d\tau} + Aw \right) \geq 0.$$

The matrices A and B are defined via the assembling described above; for equidistant steps the special versions in (5.11b), (5.11c) arise.

As a second step, the time is discretized. To this end let us define the vectors

$$w^{(\nu)} := w(\tau_\nu), \quad v^{(\nu)} := v(\tau_\nu).$$

Consequently, for all ν the inequalities

$$\left(v^{(\nu+1)} - w^{(\nu+1)} \right)^{tr} \left(B\frac{1}{\Delta\tau}(w^{(\nu+1)} - w^{(\nu)}) + \theta A w^{(\nu+1)} + (1-\theta)Aw^{(\nu)} \right) \geq 0 \tag{5.18}$$

must hold. As in Chapter 4 this is a Crank-Nicolson-type method for $\theta = 1/2$.

Side Conditions

$y(x, \tau) \geq g(x, \tau)$ amounts to

$$\sum w_i(\tau)\varphi_i(x) \geq g(x, \tau).$$

For hat functions φ_i (with $\varphi_i(x_i) = 1$ and $\varphi_i(x_j) = 0$ for $j \neq i$) and $x = x_j$ this implies $w_j(\tau) \geq g(x_j, \tau)$. With $\tau = \tau_\nu$ we have

$$w^{(\nu)} \geq g^{(\nu)}; \quad \text{analogously } v^{(\nu)} \geq g^{(\nu)}.$$

Rearranging (5.18) leads to

$$\left(v^{(\nu+1)} - w^{(\nu+1)} \right)^{tr} \left((B + \Delta\tau\theta A) w^{(\nu+1)} + (\Delta\tau(1-\theta)A - B) w^{(\nu)} \right) \geq 0.$$

With the abbreviations

$$r := (B - \Delta\tau(1-\theta)A)w^{(\nu)}$$
$$C := B + \Delta\tau\theta A$$

the inequality can be rewritten as

$$(v^{(\nu+1)} - w^{(\nu+1)})^{tr} \left(Cw^{(\nu+1)} - r \right) \geq 0.$$

For each time level ν we must find a solution that satisfies in addition the side condition

$$w^{(\nu+1)} \geq g^{(\nu+1)} \quad \text{for all } v^{(\nu+1)} \geq g^{(\nu+1)}.$$

In summary, the algorithm is

Algorithm 5.3   (finite elements for standard options)

$\theta := 1/2$. Calculate $w^{(0)}$.

For $\nu = 1, ..., \nu_{\max}$:

Calculate $r = (B - \Delta\tau(1 - \theta)A)w^{(\nu-1)}$ and $g = g^{(\nu)}$

Construct a w such that for all $v \geq g$

$\quad (v - w)^{tr}(Cw - r) \geq 0, \quad w \geq g.$

Set $w^{(\nu)} := w$

Let us emphasize again the main step, which is the kernel of this algorithm and the main labor:

(FE)
$$\begin{array}{l} \text{Construct } w \text{ such that for all } v \geq g \\ (v - w)^{tr}(Cw - r) \geq 0, \quad w \geq g \end{array} \qquad (5.19)$$

This task (FE) can be reformulated into a task we already solved in Section 4.6. To this end recall the finite-difference equation (4.32), replacing A by C and b by r, and label it with (FD):

(FD)
$$\begin{array}{l} Cw - r \geq 0, \quad w \geq g \\ (Cw - r)^{tr}(w - g) = 0 \end{array} \qquad (5.20)$$

Theorem 5.4   (equivalence)

The solution of the problem (FE) is equivalent to the solution of problem (FD).

Proof:

a) **(FD)** $\implies$ **(FE)**:

Let w solve (FD), so $w \geq g$, and

$$(v - w)^{tr}(Cw - r) = (v - g)^{tr} \underbrace{(Cw - r)}_{\geq 0} - \underbrace{(w - g)^{tr}(Cw - r)}_{=0}$$

hence $(v - w)^{tr}(Cw - r) \geq 0$ for all $v \geq g$

b) **(FE)** $\implies$ **(FD)**:

Let w solve (FE), so $w \geq g$, and

$$v^{tr}(Cw - r) \geq w^{tr}(Cw - r) \quad \text{for all } v \in \mathcal{K}$$

Suppose the kth component of $Cw - r$ is negative, and make v_k arbitrarily large. Then the left-hand side becomes arbitrarily small, which is a contradiction to $w \geq g$. So $Cw - r \geq 0$. Now

$$w \geq g \implies (w - g)^{tr}(Cw - r) \geq 0$$

Set in (FE) $v = g$, then $(w - g)^{tr}(Cw - r) \leq 0$.
Therefore $(w - g)^{tr}(Cw - r) = 0$.

Consequence: The solution of the finite-element problem (FE) can be calculated with the projected SOR with which we have solved problem (FD) in Section 4.6.

Implementation

Following the equivalence of Theorem 5.4 and the exposition of the projected SOR in Section 4.6, the kernel of the finite-element Algorithm 5.3 can be written as follows

(FE′)

> Solve $Cw = r$ such that
> componentwise $w \geq g$.

The vector v is not calculated. The boundary conditions on w are set up in the same way as discussed in Section 4.4 and summarized in Algorithm 4.13. Consequently, the finite-element algorithm parallels Algorithm 4.13 closely in the special case of an equidistant x-grid; there is no need to repeat this algorithm ($\longrightarrow$ Exercise 5.3). In the general nonequidistant case, the off-diagonal and the diagonal elements of the tridiagonal matrix C vary with i, and the formulation of the SOR-loop gets more involved. The details of the implementation are technical and omitted. The Algorithm 4.14 is the same in the finite-element case.

The computational results match those of Chapter 4 and need not be repeated. The costs of the presented simple version of a finite-element approach are slightly lower than that of the finite-difference approach.

5.4 Error Estimates

The similarity of the finite-element equation (5.15) with the finite-difference equations suggests that the errors might be of the same order. In fact, numerical experiments confirm that the finite-element approach with the linear basis functions from Definition 5.1 produces errors decaying quadratically with the mesh size. Applying the finite-element Algorithm 5.3 and entering the calculated data into a diagram as Figure 4.14, confirms the quadratic order experimentally. The proof of this order of the error is more difficult for finite-element methods because weak solutions assume less smoothness. This section explains some basic ideas of how to derive error estimates. We begin with reconsidering some of the related topics that have been introduced in previous sections.

5.4.1 Classical and Weak Solutions

Our exposition will be based on the model problem (5.12). That is, the simple second-order differential equation

$$-u'' = f(x) \quad \text{for } \alpha < x < \beta \tag{5.21a}$$

with homogeneous Dirichlet-boundary conditions

$$u(\alpha) = u(\beta) = 0 \tag{5.21b}$$

will serve as illustration. The differential equation is of the form $Lu = f$, compare (5.2). The domain $\mathcal{D} \subseteq \mathbb{R}^n$ on which functions u are defined specializes for $n = 1$ to the open and bounded interval $\mathcal{D} = \{x \in \mathbb{R}^1 : \alpha < x < \beta\}$. For continuous f, solutions of the differential equation (5.21a) satisfy $u \in \mathcal{C}^2(\mathcal{D})$. In order to have operative boundary conditions, solutions u must be continuous on $\mathcal{D}$ including its boundary, which is denoted $\partial\mathcal{D}$. Therefore we require $u \in \mathcal{C}^0(\bar{\mathcal{D}})$ where $\bar{\mathcal{D}} := \mathcal{D} \cup \partial\mathcal{D}$. In summary, classical solutions of second-order differential equations require

$$u \in \mathcal{C}^2(\mathcal{D}) \cap \mathcal{C}^0(\bar{\mathcal{D}}). \tag{5.22}$$

The function space $\mathcal{C}^2(\mathcal{D}) \cap \mathcal{C}^0(\bar{\mathcal{D}})$ must be reduced further to comply with the boundary conditions.

For weak solutions the function space is larger ($\longrightarrow$ Appendix C3). For functions u and v let us define the inner product

$$(u, v) := \int_{\mathcal{D}} uv \, dx. \tag{5.23}$$

Classical solutions u of $Lu = f$ satisfy

$$(Lu, v) = (f, v) \quad \text{for all } v. \tag{5.24}$$

Specifically for the model problem (5.21) integration by parts leads to

$$(Lu, v) = -\int_\alpha^\beta u'' v \, dx = -u'v \Big|_\alpha^\beta + \int_\alpha^\beta u'v' \, dx.$$

The nonintegral term on the right-hand side of the equation vanishes in case also v satisfies the homogeneous boundary conditions (5.21b). The remaining integral is a **bilinear form**, which we abbreviate

$$b(u, v) := \int_\alpha^\beta u'v' \, dx. \qquad (5.25)$$

Bilinear forms as $b(u, v)$ from (5.25) are linear in each of the two arguments u and v. For example, $b(u_1+u_2, v) = b(u_1, v)+b(u_2, v)$ holds. For several classes of more general differential equations analogous bilinear forms are obtained. Formally, (5.24) can be rewritten as

$$b(u, v) = (f, v), \qquad (5.26)$$

where we assume that v satisfies the homogeneous boundary conditions (5.21b).

The equation (5.26) has been derived out of the differential equation, for the solutions of which we have assumed smoothness in the sense of (5.22). Many "solutions" of practical importance do not satisfy (5.22) and, accordingly, are not classical. In several applications, u or derivatives of u have discontinuities. For instance consider the obstacle problem of Section 4.5.3: The second derivative u'' of the solution fails to be continuous at α and β. Therefore $u \notin C^2(-1, 1)$ no matter how smooth the data function is, compare Figure 4.10. As mentioned earlier, integral relations require less smoothness.

In the derivation of (5.26) the integral version resulted as a consequence of the primary differential equation. This is contrary to wide areas of applied mathematics, where an integral relation is based on first principles, and the differential equation is derived in a second step. For example, in the calculus of variations a minimization problem may be described by an integral performance measure, and the differential equation is a necessary criterion [St86]. This situation suggests considering the integral relation as an equation of its own right rather than as offspring of a differential equation. This leads to the question, *what is the maximal function space* such that (5.26) with (5.23), (5.25) is meaningful? That means to ask, for which functions u and v do the integrals exist? For a more detailed background we refer to Appendix C3. For the introductory exposition of this section it may suffice to sketch the maximum function space briefly. The suitable function space is denoted $\mathcal{H}^1$, the version equipped with the boundary conditions is denoted $\mathcal{H}_0^1$. This *Sobolev space* consists of those functions that are continuous on $\mathcal{D}$ and that are *piecewise differentiable* and satisfy the boundary conditions (5.21b). This function space corresponds to the class of functions $\mathcal{K}$ in (5.16). By means of the Sobolev space $\mathcal{H}_0^1$ a weak solution of $Lu = f$ is defined, where L is a second-order differential operator and b the corresponding bilinear form.

Definition 5.5   (weak solution)
  $u \in \mathcal{H}_0^1$ is called weak solution [of $Lu = f$], if $b(u, v) = (f, v)$ holds for all $v \in \mathcal{H}_0^1$.

This definition implicitly expresses the task: find a $u \in \mathcal{H}_0^1$ *such that* $b(u, v) = (f, v)$ *for all* $v \in \mathcal{H}_0^1$. This problem is called *variational problem*. The model problem (5.21) serves as example for $Lu = f$; the corresponding bilinear form $b(u, v)$ is defined in (5.25) and (f, v) in (5.23). For the integrals (5.23) to exist, we in addition require f to be square integrable ($f \in \mathcal{L}^2$, compare Appendix C3). Then (f, v) exists because of the Schwarzian inequality (C3.7). In a similar way, weak solutions are introduced for more general problems; the formulation of Definition 5.5 applies.

5.4.2 Approximation on Finite-Dimensional Subspaces

For a practical computation of a weak solution the infinite-dimensional space $\mathcal{H}_0^1$ is replaced by a finite-dimensional subspace. Such finite-dimensional subspaces are spanned by basis functions φ_i. The simplest examples are the hat functions of Section 5.2. Reminding of the important role splines play as basis functions, the finite-dimensional subspaces are denoted $\mathcal{S}$. The hat functions $\varphi_0, ..., \varphi_m$ span the space of polygons, compare Property 5.2(a). Recall that each polygon v can be represented as linear combination

$$v = \sum_{i=0}^{m} c_i \varphi_i.$$

The coefficients c_i are uniquely determined by the values of v at the nodes, $c_i = v(x_i)$. The hat functions are called linear elements because they consist of piecewise straight lines. Apart from linear elements, for example, also quadratic or cubic elements are used, which are piecewise polynomials of second or third degree [Zi77], [Ci91], [Sc91]. The attainable accuracy is different for basis functions consisting of higher-degree polynomials. The spaces $\mathcal{S}$ are called *finite-element spaces*.

Since by definition the functions of the Sobolev space $\mathcal{H}_0^1$ fulfill the homogeneous boundary conditions, each subspace does so as well. The subscript $_0$ indicates the realization of the homogeneous boundary conditions (5.21b)[1]. A finite-dimensional subspace of $\mathcal{H}_0^1$ is defined by

$$\mathcal{S}_0 := \{ v = \sum_i c_i \varphi_i \ : \ \varphi_i \in \mathcal{H}_0^1 \}. \tag{5.27}$$

The properties of $\mathcal{S}_0$ are determined by the basis functions φ_i. As mentioned earlier, basis functions with small supports give rise to sparse matrices. The partition (5.3) is implicitly included in the definition $\mathcal{S}_0$ because this information is contained in the definition of the φ_i. For our purposes the hat functions suffice. The larger m is, the better $\mathcal{S}_0$ approximates the space $\mathcal{H}_0^1$, since a finer discretization (smaller $\mathcal{D}_k$) allows to approximate the functions from $\mathcal{H}_0^1$ better by polygons. We denote the largest diameter of the $\mathcal{D}_k$ by

---

[1] In this subsection the meaning of the index $_0$ is twofold: It is the index of the "first" hat function, and serves as symbol of the homogeneous boundary conditions (5.21b).

h, and ask for convergence. That is, we study the behavior of the error for $h \to 0$ (basically $m \to \infty$).

In analogy to the variational problem expressed in connection with Definition 5.5, a *discrete* weak solution w is defined by replacing the space $\mathcal{H}_0^1$ by a finite-dimensional subspace $\mathcal{S}_0$:

Problem 5.6   (discrete weak solution)
    Find a $w \in \mathcal{S}_0$ such that $b(w, v) = (f, v)$ for all $v \in \mathcal{S}_0$.

The quality of the approximation depends on the discretization fineness h of $\mathcal{S}_0$. This is emphasized by writing w_h. The transition from the continuous variational problem following Definition 5.5 to the discrete Problem 5.6 is sometimes called the *principle of Rayleigh-Ritz*.

5.4.3 Céa's Lemma

Having defined a weak solution u and a discrete approximation w, we turn to the error $u - w$. To measure the distance between functions in $\mathcal{H}_0^1$ we use the norm $\| \ \|_1$ ($\longrightarrow$ Appendix C3). That is, our first aim is to construct a bound on $\|u - w\|_1$. Let us suppose that the bilinear form is continuous and $\mathcal{H}^1$-*elliptic*:

Assumptions 5.7   (continuous $\mathcal{H}^1$-elliptic bilinear form)
    (a) There is a $\gamma_1 > 0$ such that
        $|b(u, v)| \leq \gamma_1 \|u\|_1 \|v\|_1$ for all $u, v \in \mathcal{H}^1$
    (b) There is a $\gamma_2 > 0$ such that
        $b(v, v) \geq \gamma_2 \|v\|_1^2$ for all $v \in \mathcal{H}^1$

The assumption (a) is the continuity, and the property in (b) is called $\mathcal{H}^1$-ellipticity. Under the Assumptions 5.7, the problem to find a weak solution following Definition 5.5, possesses exactly one solution $u \in \mathcal{H}_0^1$; the same holds true for Problem 5.6. This is guaranteed by the Theorem of Lax-Milgram, see [Ci91], [BrS02]. In view of $\mathcal{S}_0 \subseteq \mathcal{H}_0^1$,

$$b(u, v) = (f, v) \quad \text{for all } v \in \mathcal{S}_0.$$

Subtracting $b(w, v) = (f, v)$ and invoking the bilinearity implies

$$b(w - u, v) = 0 \quad \text{for all } v \in \mathcal{S}_0. \tag{5.28}$$

The property of (5.28) is called *error projection property*. The Assumptions 5.7 and the error projection are the basic ingredients to obtain a bound on the error $\|u - w\|_1$:

Lemma 5.8   (Céa)
    Suppose the Assumptions 5.7 are satisfied. Then

$$\|u - w\|_1 \leq \frac{\gamma_1}{\gamma_2} \inf_{v \in \mathcal{S}_0} \|u - v\|_1. \tag{5.29}$$

Proof: $v \in \mathcal{S}_0$ implies $\tilde{v} := w - v \in \mathcal{S}_0$. Applying (5.28) for $\tilde{v}$ yields

$$b(w - u, w - v) = 0 \quad \text{for all } v \in \mathcal{S}_0.$$

Therefore

$$\begin{aligned} b(w - u, w - u) &= b(w - u, w - u) - b(w - u, w - v) \\ &= b(w - u, v - u). \end{aligned}$$

Applying the assumptions shows

$$\begin{aligned} \gamma_2 \|w - u\|_1^2 \leq |b(w - u, w - u)| &= |b(w - u, v - u)| \\ &\leq \gamma_1 \|w - u\|_1 \|v - u\|_1, \end{aligned}$$

from which

$$\|w - u\|_1 \leq \frac{\gamma_1}{\gamma_2} \|v - u\|_1$$

follows. Since this holds for all $v \in \mathcal{S}_0$, the assertion of the lemma is proven.

Let us check whether the Assumptions 5.7 are fulfilled by the model problem (5.21). For (a) this follows from the Schwarzian inequality (C3.7) with the norms

$$\|u\|_1 = \left(\int_\alpha^\beta (u^2 + u'^2) dx \right)^{1/2}, \quad \|u\|_0 = \left(\int_\alpha^\beta u^2 \, dx \right)^{1/2},$$

because

$$\left(\int_\alpha^\beta u'v' \, dx \right)^2 \leq \left(\int_\alpha^\beta u'^2 \, dx \right) \left(\int_\alpha^\beta v'^2 \, dx \right) \leq \|u\|_1^2 \, \|v\|_1^2.$$

The Assumption 5.7(b) can be derived from the inequality of the Poincaré-type

$$\int_\alpha^\beta v^2 dx \leq (\beta - \alpha)^2 \int_\alpha^\beta v'^2 dx,$$

which in turn is proven with the Schwarzian inequality. Adding $\int v'^2 dx$ on both sides leads to the constant γ_2 of (b). So Céa's lemma applies to the model problem.

The next question is, how small the infimum in (5.29) may be. This is equivalent to the question, how close the subspace $\mathcal{S}_0$ can approximate the space $\mathcal{H}_0^1$. We will show that for hat functions and $\mathcal{S}_0$ from (5.27) the infimum is of the order $O(h)$. Again h denotes the maximum mesh size, and the notation w_h reminds us that the discrete solution depends on the grid. Following Céa's lemma, we need an upper bound for the infimum of $\|u - v\|_1$. Such a bound is found easily by a specific choice of v, which is taken as an arbitrary interpolating polygon u_I. Then by (5.29)

$$\|u - w_h\|_1 \le \frac{\gamma_1}{\gamma_2} \inf_{v \in \mathcal{S}_0} \|u - v\|_1 \le \frac{\gamma_1}{\gamma_2} \|u - u_I\|_1. \tag{5.30}$$

It remains to bound the error of interpolating polygons. This bound is provided by the following lemma, which is formulated for $\mathcal{C}^2$-smooth functions u:

Lemma 5.9   (error of an interpolating polygon)

For $u \in \mathcal{C}^2$ let u_I be an arbitrary interpolating polygon and h the maximal distance between two consecutive nodes. Then

(a) $\max_x |u(x) - u_I(x)| \le \frac{h^2}{8} \max |u''(x)|$

(b) $\max_x |u'(x) - u_I'(x)| \le h \max |u''(x)|$

We leave the proof to the reader ($\longrightarrow$ Exercise 5.4). The assumption $u \in \mathcal{C}^2$ in Lemma 5.9 can be weakened to $u'' \in \mathcal{L}^2$ [SF73]. Lemma 5.9 asserts

$$\|u - u_I\|_1 = O(h),$$

which together with (5.30) implies the claimed error statement

$$\|u - w_h\|_1 = O(h). \tag{5.31}$$

Recall that this assertion is based on a continuous and $\mathcal{H}^1$-elliptic bilinear form and on hat functions φ_i. The $O(h)$-order in (5.31) is dominated by the unfavorable $O(h)$-order of the first-order derivative in Lemma 5.9(b). This low order is at variance with the actually observed $O(h^2)$-order attained by the approximation w_h itself (not its derivative). So the error statement (5.31) is not yet the final result. In fact, the square order can be proven with a tricky idea due to Nitsche. The final result is

$$\|u - w_h\|_0 \le Ch^2 \|u\|_2 \tag{5.32}$$

for a constant C.

The derivations of this section have been focused on the model problem (5.21) with a second-order differential equation and one independent variable x ($n = 1$), and have been based on linear elements. Most of the assertions can be generalized to higher-order differential equations, to higher-dimensional domains ($n > 1$), and to nonlinear elements. For example, in case the elements in $\mathcal{S}$ are polynomials of degree k, and the differential equation is of order $2l$, $\mathcal{S} \subseteq \mathcal{H}^l$, and the corresponding bilinear form on $\mathcal{H}^l$ satisfies the Assumptions 5.7 with norm $\| \ \|_l$, then the inequality

$$\|u - w_h\|_l \le Ch^{k+1-l} \|u\|_{k+1}$$

holds. This general statement includes for $k = 1$, $l = 1$ the special case of equation (5.32) discussed above. For the analysis of the general case, we refer to [Ci91], [Ha92]. This includes boundary conditions more general than the homogeneous Dirichlet conditions of (5.21b).

Notes and Comments

on Section 5.1:
As an alternative to the piecewise defined finite elements one may use polynomials φ_j that are defined globally on $\mathcal{D}$, and that are pairwise orthogonal. Then the orthogonality is the reason for the vanishing of many integrals. Such type of methods are called *spectral methods*. Since the φ_j are globally smooth on $\mathcal{D}$, spectral methods can produce high accuracies. Rayleigh–Ritz–approaches choose the φ_j as eigenfunctions of L. For symmetric L this leads to diagonal matrices A.

on Section 5.2:
In the early stages of their development, finite-element methods have been applied intensively in structural engineering. In this field, stiffness matrix and mass matrix have a physical meaning leading to these names [Zi77].

The construction of the global matrices by assembling the local element matrices is easy for the one-dimensional application ($x \in \mathbb{R}^1$), because the numbering of the subintervals (with k) and the numbering of the nodes (with i or j) interlace in a unique way. In the two-dimensional case, with for instance triangles $\mathcal{D}_k$, the assignment of the element is more complicated, and the index set $\mathcal{I}_k$ does not have such a simple structure as in (5.10). For two-dimensional hat functions (3×3)-element matrices must be distributed. To this end, for each $\mathcal{D}_k$ the index set of all adjoining nodes must be stored. For example, for triangles $\mathcal{D}_k$ each index set is of the form

$$\mathcal{I}_k = \{i_1, i_2, i_3\},$$

where i_1, i_2, i_3 are the numbers of the three nodes in the corners of $\mathcal{D}_k$. This generalizes (5.10), see for instance [Sc91].

on Section 5.3:
The approximation $\sum w_i(\tau)\varphi_i(x)$ for $\hat{y}$ is a one-dimensional finite-element approach. The geometry of the grid and the accuracy resemble the finite-difference approach. A two-dimensional approach as in

$$\sum w_i \varphi_i(x, \tau)$$

with two-dimensional hat functions and constant w_i is more involved and more flexible. The exposition of Section 5.3 widely follows [WDH96].

on Section 5.4:
The finite-dimensional function space $\mathcal{S}_0$ in (5.27) is assumed to be *sub*space of $\mathcal{H}_0^1$. Elements with this property are called *conforming elements*. A more accurate notation for $\mathcal{S}_0$ of (5.27) is $\mathcal{S}_0^1$. In the general case, conforming elements are characterized by $\mathcal{S}^l \subseteq \mathcal{H}^l$. In the respresentation of v in equation

(5.27) the range of indices i is left out intentionally to avoid discussing the technical issue of how to organize different types of boundary conditions.

There are also smooth basis functions φ, for example, cubic Hermite polynomials. For sufficiently smooth solutions, such basis functions produce higher accuray than hat functions do. For the accuracy of finite-element methods consult, for example, [SF73], [Ci91], [Ha92], [BaS01], [BrS02].

on other methods:

Finite-element methods are frequently used in the area of computational finance, in particular in multidimensional situations. For different types of derivatives special methods have been developed. For applications, computational results and accuracies see also [Top00], [Top05]. The accuracy aspect is also treated in [FuST02]. Galerkin methods are used with wavelet functions in [MaPS02].

Exercises

Exercise 5.1   Cubic B-Spline

Suppose an equidistant partition of an interval be given with mesh-size $h = x_{k+1} - x_k$. Cubic B-splines have a support of four subintervals. In each subinterval the spline is a piece of polynomial of degree three. Apart from special boundary splines, the cubic B-splines φ_i are determined by the requirements

$$\varphi_i(x_i) = 1$$
$$\varphi_i(x) \equiv 0 \quad \text{for } x < x_{i-2}$$
$$\varphi_i(x) \equiv 0 \quad \text{for } x > x_{i+2}$$
$$\varphi \in C^2(-\infty, \infty).$$

To construct the φ_i proceed as follows:
a) Construct a spline $S(x)$ that satisfies the above requirements for the special nodes
$$\tilde{x}_k := -2 + k \quad \text{for } k = 0, 1, ..., 4.$$
b) Find a transformation $T_i(x)$, such that $\varphi_i = S(T_i(x))$ satisfies the requirements for the original nodes.
c) For which i, j does $\varphi_i \varphi_j = 0$ hold?

Exercise 5.2   Finite-Element Matrices

For the *hat functions* φ from Section 5.2 calculate for arbitrary subinterval $\mathcal{D}_k$ all nonzero integrals of the form

$$\int \varphi_i \varphi_j dx, \quad \int \varphi_i' \varphi_j dx, \quad \int \varphi_i' \varphi_j' dx$$

and represent them as local 2×2 matrices.

Exercise 5.3   Calculating Options with Finite Elements

Design an algorithm for the pricing of standard options by means of finite elements. To this end proceed as outlined in Section 5.3. Start with a simple version using an equidistant discretization step Δx. If this is working properly change the algorithm to a version with nonequidistant x-grid. Distribute the nodes x_i closer around $x = 0$. Always put a node at the strike.

Exercise 5.4

Prove Lemma 5.9, and for $u \in \mathcal{C}^2$ the assertion $\|u - w_h\|_1 = O(h)$.

Chapter 6 Pricing of Exotic Options

In Chapter 4 we discussed the pricing of vanilla options (standard options) by means of finite differences. The methods were based on the simple partial differential equation (4.2),

$$\frac{\partial y}{\partial \tau} = \frac{\partial^2 y}{\partial x^2},$$

which was obtained from the Black-Scholes equation (4.1) for $V(S,t)$ via the transformations (4.3). These transformations could be applied because $\frac{\partial V}{\partial t}$ in the Black-Scholes equation is a linear combination of terms of the type

$$c_j S^j \frac{\partial^j V}{\partial S^j}$$

with constants c_j, $j = 0, 1, 2$.

Exotic options lead to partial differential equations that are not of the simple structure of the basic Black-Scholes equation (4.1). In the general case, the transformations (4.3) are no longer useful and the PDEs must be solved directly. Thereby numerical instabilities or spurious solutions may occur, which do not play any role for the methods of Chapter 4. To cope with the "new" difficulties, Chapter 6 introduces ideas and tools not needed in Chapter 4. Exotic options often involve higher-dimensional problems. This significantly adds to the complexity. The aim of this chapter will not be to formulate algorithms, but to give an outlook and lead the reader to the edge of several aspects of recent research. Some of the many possible methods will be exemplified on Asian options.

Sections 6.1 and 6.2 give a brief overview on important types of exotic options. An exhaustive discussion of the wide field of exotic options is far beyond the scope of this book. Section 6.3 introduces approaches for path-dependent options, with the focus on Asian options. Then numerical aspects of convection-diffusion problems are discussed (in Section 6.4), and upwind schemes are analyzed (in Section 6.5). After these preparations the Section 6.6 arrives at the state of the art high-resolution methods.

6.1 Exotic Options

So far, this book has concentrated on standard options. These are the American or European call or put options with payoff functions (1.1C) or (1.1P) as discussed in Section 1.1, based on a single underlying asset. The corresponding model is summarized by Assumption 1.2. The options traded on official exchanges are mainly standard options; their prices and terms are quoted in relevant newspapers.

All nonstandard options are called exotic options. That is, at least one of the features of a standard option is violated. One of the main possible differences between standard and exotic options lies in the payoff; examples are given in this section. Another extension from standard to exotic is an increase in the dimension, from single-factor to multifactor options; this will be discussed in Section 6.2. The distinction between put und call, and between European and American remains valid for exotic options. Financial institutions have been imaginative in designing exotic options to meet the needs of clients. Many of the products have a highly complex structure. Exotic options are harder to price than standard options. Exotic options are traded outside the exchanges (OTC).

Next we list a selection of some important types of exotic options. For more explanation we refer to [Hull00], [Wi98].

Compound Option: Compound options are options on options. Depending on whether the options are put or call, there are four main types of compound options. For example, the option may be a call on a call.

Chooser Option: After a specified period of time the holder of a chooser option can choose whether the option is a call or a put. The value of a chooser option at this time is

$$\max\{V_C, V_P\}$$

Binary Option: Binary options have a discontinuous payoff, for example

$$V_T = Q \cdot \begin{cases} 1 & \text{if } S_T < K \\ 0 & \text{if } S_T \geq K \end{cases}$$

for a fixed amount Q.

Path-Dependent Options

Options where the payoff depends not only on S_T but also on the path of S_t for previous times $t < T$ are called *path dependent*. Important path-dependent options are the *barrier option*, the *lookback option*, and the *Asian option*.

Barrier Option: For a barrier option the payoff is contingent on the underlying asset's price S_t reaching a certain threshold value B, which is called barrier. Barrier options can be classified depending on whether S_t reaches B from above *(down)* or from below *(up)*. Another feature of a barrier option is whether it ceases to exist when B is reached *(knock out)* or conversely comes into existence *(knock in)*. Obviously, for a down option, $S_0 > B$ and for an up option $S_0 < B$. Depending on whether the barrier option is a put or a call, a number of different types are possible. For example, the payoff of a European *down-and-out* call is

$$V_T = \begin{cases} (S_T - K)^+ & \text{in case } S_t > B \text{ for all } t \\ 0 & \text{in case } S_t \leq B \text{ for some } t \end{cases}$$

The value of the option before the barrier has been triggered still satisfies the Black-Scholes equation. The details of the barrier feature come in through the specification of the boundary conditions, see [Wi98].

Lookback Option: The payoff of a lookback option depends on the maximum or minimum value the asset price S_t reaches during the life of the option. For example, the payoff of a lookback option is

$$\max_t S_t - S_T \,.$$

Average Option / Asian Option: The payoff from an Asian option depends on the average price of the underlying asset. This will be discussed in more detail in Section 6.3.

Pricing of Exotic Options

Several types of exotic options can be reduced to the Black-Scholes equation. In these cases the numerical methods of Chapter 4 are adequate. For a number of options of the European type the Black-Scholes evaluation formula (A4.10) can be applied. For related reductions of exotic options we refer to [Hull00], [WDH96], [Kwok98]. Approximations are possible with binomial methods or with Monte Carlo simulation. The Algorithm 3.6 applies, only the calculation of the payoff (step 2) must be adapted to the exotic option.

6.2 Options Depending on Several Assets

The options listed in Section 6.1 depend on one underlying asset. Options depending on several assets are called multifactor options. Two large groups of multifactor options are the *rainbow options* and the *baskets*. The subdivision into the groups is by their payoff. Assume n assets are underlying, with prices $S_1, \ldots, S_n$. Different from the notation in previous chapters, the index refers to the number of the asset.

Rainbow options compare the value of individual assets [Smi97]. Examples of payoffs are

$\max(S_1, \ldots, S_n)$	"n-color better-of option"
$\min(S_1, S_2)$	"two-color worse-of option"
$\max(S_2 - S_1, 0)$	"outperformance option"
$\max(\min(S_1 - K, \ldots, S_n - K), 0)$	"min call option"

A basket is an option with payoff depending on a portfolio of assets. An example is the payoff of a basket call,

$$\left(\sum_{i=1}^{n} c_i S_i - K \right)^{+} ,$$

where the weights c_i are given by the portfolio. It is recommendable to sketch the above payoffs for $n = 2$.

For the pricing of multifactor options the instruments introduced in the previous chapters apply. This holds for the four large classes of methods discussed before, namely the PDE methods, the tree methods, the evaluation of integrals, and the Monte Carlo methods. Each class subdivides into further methods. For the choice of an appropriate method, the dimension n is essential. For large values of n, in particular PDE methods suffer from the *curse of dimension*. At present state it is not possible to decide at what level the threshold of n might be, above which PDE methods are too expensive. At least for $n = 2$ and $n = 3$, PDE approaches are competitive. In this multidimensional situation, finite-elements are recommendable.

For the **PDE methods** the relevant PDEs *and* boundary conditions are required. The equations are of the Black-Scholes type. To extend the one-factor model, an appropriate generalization of geometric Brownian motion is needed. We begin with the two-factor model, with the prices of the two assets S_1 and S_2. The assumption of a constant-coefficient GBM is then expressed as

$$dS_1 = \mu_1 S_1 \, dt + \sigma_1 S_1 \, dW^{(1)}$$
$$dS_2 = \mu_2 S_2 \, dt + \sigma_2 S_2 \, dW^{(2)} \qquad (6.1a)$$
$$\mathsf{E}(dW^{(1)} dW^{(2)}) = \rho \, dt ,$$

where ρ is the correlation between the two assets. Note that the third equation in (6.1a) is equivalent to $\mathsf{Cov}(dW^{(1)}, dW^{(2)}) = \rho \, dt$, because $\mathsf{E}(dW^{(1)}) = \mathsf{E}(dW^{(2)}) = 0$. Compared to more general systems as in (1.41), the version (6.1a) with correlated Wiener processes has the advantage that each asset price has its own growth factor μ and volatility σ, which can be estimated from data. The correlation ρ is given by the correlation of the returns $\frac{dS}{S}$, since

$$\mathsf{Cov}\left(\frac{dS_1}{S_1}, \frac{dS_2}{S_2} \right) = \mathsf{E}(\sigma_1 dW^{(1)} \sigma_2 dW^{(2)}) = \rho \sigma_1 \sigma_2 \, dt . \qquad (6.1b)$$

The two-dimensional version of the Itô-Lemma ($\longrightarrow$ Appendix B2) and a no-arbitrage argument yield the Black-Scholes-type PDE

$$\frac{\partial V}{\partial t} + \frac{1}{2} \sigma_1^2 S_1^2 \frac{\partial^2 V}{\partial S_1^2} + r S_1 \frac{\partial V}{\partial S_1} - rV$$
$$+ \frac{1}{2} \sigma_2^2 S_2^2 \frac{\partial^2 V}{\partial S_2^2} + r S_2 \frac{\partial V}{\partial S_2} + \rho \sigma_1 \sigma_2 S_1 S_2 \frac{\partial^2 V}{\partial S_1 \partial S_2} = 0. \qquad (6.1c)$$

This is a PDE on (S_1, S_2, t). Usually, the time variable is not counted when the dimension is discussed. In this sense, the PDE (6.1c) is two-dimensional, whereas the classic Black-Scholes PDE is considered as one-dimensional.

The general n-factor model is analogous. The appropriate model is a straightforward generalization of (6.1a),

$$dS_i = (\mu_i - \delta_i)S_i\, dt + \sigma_i S_i\, dW^{(i)}\,, \quad i = 1, \ldots, n$$

$$\mathsf{E}(dW^{(i)} dW^{(j)}) = \rho_{ij}\, dt\,, \quad i, j = 1, \ldots, n$$

(6.2a)

where ρ_{ij} is the correlation between asset i and asset j, and δ_i denotes the dividend flow paid by the ith asset. For a simulation of such a stochastic vector process see Section 2.3.3. The Black-Scholes-type PDE of the model (6.2a) is

$$\frac{\partial V}{\partial t} + \frac{1}{2}\sum_{i,j=1}^{n} \rho_{ij}\sigma_i\sigma_j S_i S_j \frac{\partial^2 V}{\partial S_i \partial S_j} + \sum_{i=1}^{n}(r - \delta_i)S_i \frac{\partial V}{\partial S_i} - rV = 0\,. \quad (6.2b)$$

Boundary conditions depend on the specific type of option. For example in the two-dimensional situation, one boundary can be defined by the plane $S_1 = 0$ and the other by the plane $S_2 = 0$. It may be appropriate to apply the Black-Scholes vanilla formula (A4.10) along these planes, or to define one-dimensional sub-PDEs only for the purpose to calculate the values of $V(S_1, 0, t)$ and $V(0, S_2, t)$ along the boundary planes.

For **tree methods**, the binomial method can be generalized canonically. But already for $n = 2$ the recombining standard tree with M time levels requires $\frac{1}{3}M^3 + O(M^2)$ nodes, and for $n = 3$ the number of nodes is of the order $O(M^4)$. Tree methods also suffer from the curse of dimension. But obviously not all of the nodes of the canonical binomial approach are needed. The ultimate aim is to approximate the lognormal distribution, and this can be done with fewer nodes. Nodes in $\mathbb{R}^n$ should be constructed in such a way that the number of nodes grows comparably slower than the quality of the approximation of the distribution function. An example of a two-dimensional approach is presented in [Lyuu02]. Also the trinomial approach can be generalized to higher dimensions, or other geometrical structures as icosahedral volumes can be applied. For different tree approaches, see [McW01].

An advantage of tree methods and of **Monte Carlo methods** is that no boundary conditions are needed. The essential advantage of MC methods is that they are not affected by high dimensions. An example of a five-dimensional American-style option is calculated in [BrG04], [LanS01]. It is most inspiring to perform Monte Carlo experiments on exotic options. For European-style options, this amounts to a straightforward application of Section 3.5 ($\longrightarrow$ Exercise 6.1).

6.3 Asian Options

An Asian option[1] can be of European or American style. The price of the
Asian option depends on the average and hence on the history of S_t. We
choose this type of option to discuss some strategies of how to handle path-
dependent options. Let us first define different types of Asian options via
their payoff.

6.3.1 The Payoff

There are several ways how an average of past values of S_t can be formed.
If the price S_t is observed at discrete time instances t_i, say equidistantly
with time interval $h := T/n$, one obtains a times series $S_{t_1}, S_{t_2}, \ldots, S_{t_n}$. An
obvious choice of average is the arithmetic mean

$$\frac{1}{n} \sum_{i=1}^{n} S_{t_i} = \frac{1}{T} h \sum_{i=1}^{n} S_{t_i} \ .$$

If we imagine the observation as continuously sampled in the time period
$0 \le t \le T$, the above mean corresponds to the integral

$$\widehat{S} := \frac{1}{T} \int_0^T S_t dt \tag{6.3}$$

The arithmetic average is used mostly. Sometimes the geometric average is
applied, which can be expressed as

$$\left(\prod_{i=1}^{n} S_{t_i} \right)^{1/n} = \exp \left(\frac{1}{n} \log \prod_{i=1}^{n} S_{t_i} \right) = \exp \left(\frac{1}{n} \sum_{i=1}^{n} \log S_{t_i} \right) .$$

Hence the continuously sampled geometric average of the price S_t is the
integral

$$\widehat{S}_{\mathrm{g}} := \exp \left(\frac{1}{T} \int_0^T \log S_t dt \right) \ .$$

The averages $\widehat{S}$ and $\widehat{S}_{\mathrm{g}}$ are formulated for the time period $0 \le t \le T$, which
corresponds to a European option. To allow for early exercise at time $t < T$,
$\widehat{S}$ and $\widehat{S}_{\mathrm{g}}$ are modified appropriately, for instance to

$$\widehat{S} := \frac{1}{t} \int_0^t S_\theta d\theta.$$

With an average value $\widehat{S}$ like the arithmetic average of (6.3) the payoff of
Asian options can be written conveniently:

---

[1] Again, the name has no geographical relevance.

Definition 6.1   (Asian option)

With an average $\widehat{S}$ of the price evolution S_t the payoff functions of Asian options are defined as

$$
\begin{array}{ll}
(\widehat{S} - K)^+ & \textit{average price call} \\
(K - \widehat{S})^+ & \textit{average price put} \\
(S_T - \widehat{S})^+ & \textit{average strike call} \\
(\widehat{S} - S_T)^+ & \textit{average strike put}
\end{array}
$$

The price options are also called *rate options*, or *fixed strike options*; the strike options are also called *floating strike options*. Compared to the vanilla payoffs of (1.1P), (1.1C), for an Asian price option the average $\widehat{S}$ replaces S whereas for the Asian strike option $\widehat{S}$ replaces K. The payoffs of Definition 6.1 form surfaces on the quadrant $S > 0, \widehat{S} > 0$. The reader may visualize these payoff surfaces.

6.3.2 Modeling in the Black-Scholes Framework

The above averages can be expressed by means of the integral

$$
A_t := \int_0^t f(S_\theta, \theta) d\theta, \tag{6.4}
$$

where the function $f(S, t)$ corresponds to the type of chosen average. In particular $f(S, t) = S$ corresponds to the continuous arithmetic average (6.3), up to scaling by the length of interval. For Asian options the price V is a function of S, A and t, which we write $V(S, A, t)$. To derive a partial differential equation for V using a generalization of Itô's Lemma we require a differential equation for A. But this is given by (6.4), it lacks a stochastic dW_t-term,[2]

$$
dA = a_A(t)dt + b_A dW_t,
$$
$$
\text{with} \quad a_A(t) := f(S_t, t) \quad , \quad b_A := 0.
$$

For S_t the standard GBM of (1.33) is assumed. By the multidimensional version (B2.1) of Itô's Lemma adapted to $Y_t := V(S_t, A_t, t)$, the two terms in (1.43) or (1.44) that involve b_A as factors to $\frac{\partial V}{\partial A}, \frac{\partial^2 V}{\partial A^2}$ vanish. Accordingly,

$$
dV_t = \left(\frac{\partial V}{\partial t} + \mu S \frac{\partial V}{\partial S} + \frac{1}{2} \sigma^2 S^2 \frac{\partial^2 V}{\partial S^2} + f(S, t) \frac{\partial V}{\partial A} \right) dt + \sigma S \frac{\partial V}{\partial S} dW_t .
$$

The derivation of the Black-Scholes-type PDE goes analogously as outlined in Appendix A4 for standard options and results in

---

[2] The ordinary integral A_t is random but has zero quadratic variation [Shr04].

$$\frac{\partial V}{\partial t} + \frac{1}{2}\sigma^2 S^2 \frac{\partial^2 V}{\partial S^2} + rS\frac{\partial V}{\partial S} + f(S,t)\frac{\partial V}{\partial A} - rV = 0. \tag{6.5}$$

Compared to the original vanilla version (1.2), only one term in (6.5) is new, namely

$$f(S,t)\frac{\partial V}{\partial A}.$$

As we will see below, the lack of a second-order derivative with respect to A may cause numerical difficulties. The transformations (4.3) cannot be applied advantageously to (6.5). As an alternative to the definition of A_t in (6.4), one can scale by t. This leads to a different "new term" ($\longrightarrow$ Exercise 6.2).

6.3.3 Reduction to a One-Dimensional Equation

Solutions to (6.5) are defined on the domain

$$S > 0 \,,\; A > 0 \,,\; 0 \leq t \leq T$$

of the three-dimensional space. The extra dimension leads to significantly higher costs when (6.5) is solved numerically. This is the general situation. But in some cases it is possible to reduce the dimension. Let us discuss an example, concentrating on the case $f(S,t) = S$ of the arithmetic average.

We consider a European arithmetic average strike call with payoff

$$\left(S_T - \frac{1}{T}A_T\right)^+ = S_T\left(1 - \frac{1}{TS_T}\int_0^T S_\theta d\theta\right)^+.$$

An auxiliary variable R_t is defined by

$$R_t := \frac{1}{S_t}\int_0^t S_\theta d\theta \,,$$

and the payoff is rewritten

$$S_T\left(1 - \frac{1}{T}R_T\right)^+ = S_T * \text{function}(R_T, T).$$

This motivates trying a separation of the solution in the form

$$V(S, A, t) = S \cdot H(R, t) \tag{6.6}$$

for some function $H(R,t)$. In this role, R is an independent variable. But note that the integral R_t satisfies an SDE. From

$$R_{t+dt} = R_t + dR_t$$
$$dS_t = \mu S_t dt + \sigma S_t dW_t$$

the SDE

$$dR_t = (1 + (\sigma^2 - \mu)R_t)dt - \sigma R_t dW_t \qquad (6.7)$$

follows.

Substituting the separation ansatz (6.6) into the PDE (6.5) leads to a PDE for H,

$$\frac{\partial H}{\partial t} + \frac{1}{2}\sigma^2 R^2 \frac{\partial^2 H}{\partial R^2} + (1 - rR)\frac{\partial H}{\partial R} = 0 \qquad (6.8)$$

($\longrightarrow$ Exercise 6.2). To solve this PDE, boundary conditions are required.

A *right-hand boundary* condition for $R \to \infty$ follows from the payoff

$$H(R_T, T) = (1 - \tfrac{1}{T}R_T)^+ \ ,$$

which implies $H(R_T, T) = 0$ for $R_T \to \infty$. The integral R_t is bounded, hence $S \to 0$ for $R \to \infty$. For $S \to 0$ a European call option is not exercised, which suggests

$$H(R, t) = 0 \quad \text{for} \quad R \to \infty. \qquad (6.9)$$

At the *left-hand boundary* $R = 0$ we encounter more difficulties. Even if $R_0 = 0$ holds, the equation (6.7) shows that $dR_0 = dt$ and R_t will not stay at 0. So there is no reason to expect $R_T = 0$, and the value of the payoff cannot be predicted. Another kind of boundary condition is required.

To this end, we start from the PDE (6.8), which for $R \to 0$ is equivalent to

$$\frac{\partial H}{\partial t} + \frac{1}{2}\sigma^2 R^2 \frac{\partial^2 H}{\partial R^2} + \frac{\partial H}{\partial R} = 0 \ .$$

Assuming that H is bounded, one can prove that the term

$$R^2 \frac{\partial^2 H}{\partial R^2}$$

vanishes for $R \to 0$. The resulting boundary condition is

$$\frac{\partial H}{\partial t} + \frac{\partial H}{\partial R} = 0 \quad \text{for} \quad R \to 0. \qquad (6.10)$$

The vanishing of the second-order derivative term is shown by contradiction: Assuming a nonzero value of $R^2 \frac{\partial^2 H}{\partial R^2}$ leads to

$$\frac{\partial^2 H}{\partial R^2} = O\left(\frac{1}{R^2}\right),$$

which can be integrated twice to

$$H = O(\log R) + c_1 R + c_2.$$

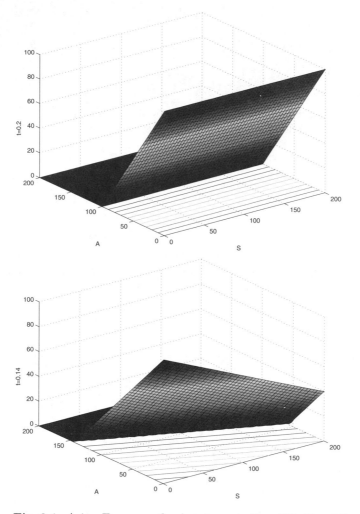

Fig. 6.1. Asian European fixed strike put, $K = 100$, $T = 0.2$, $r = 0.05$, $\sigma = 0.25$, payoff ($t = 0.2$) and three solution surfaces for $t = 0.14$, $t = 0.06$, and $t = 0$. (Figure continued on facing page)

This contradicts the boundedness of H for $R \to 0$.

For a numerical realization of the boundary condition (6.10) in the finite-difference framework of Chapter 4, we may use the second-order formula

$$\left.\frac{\partial H}{\partial R}\right|_{0\nu} = \frac{-3H_{0\nu} + 4H_{1\nu} - H_{2\nu}}{2\Delta R} + O(\Delta R^2). \tag{6.11}$$

The indices have the same meaning as in Chapter 4. We summarize the boundary-value problem of PDEs to

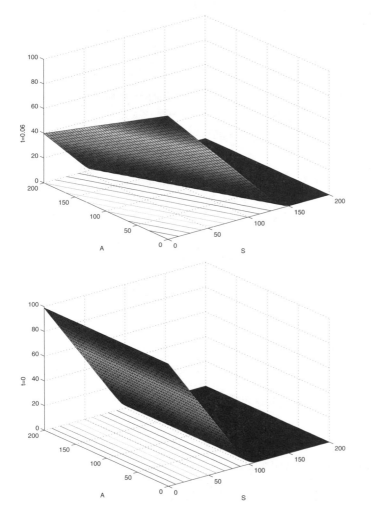

Fig. 6.1. continued

$$\frac{\partial H}{\partial t} + \frac{1}{2}\sigma^2 R^2 \frac{\partial^2 H}{\partial R^2} + (1 - rR)\frac{\partial H}{\partial R} = 0$$

$$H = 0 \quad \text{for} \quad R \to \infty$$

$$\frac{\partial H}{\partial t} + \frac{\partial H}{\partial R} = 0 \quad \text{for} \quad R = 0$$

$$H(R_T, T) = \left(1 - \frac{R_T}{T}\right)^+$$

(6.12)

Solving this problem numerically for $0 \leq t \leq T$, $R \geq 0$, gives $H(R,t)$, and via (6.6) the required values of V.

6.3.4 Discrete Monitoring

Instead of defining a continuous averaging as in (6.3), a realistic scenario is to assume that the average is monitored only at discrete time instances

$$t_1, t_2, \ldots, t_M \ .$$

These time instances are not to be confused with the grid times of the numerical discretization. The discretely sampled arithmetic average at t_k is given by

$$A_{t_k} := \frac{1}{k} \sum_{i=1}^{k} S_{t_i}, \ k = 1, \ldots, M \ . \tag{6.13}$$

A new average is updated from a previous one by

$$A_{t_k} = A_{t_{k-1}} + \frac{1}{k}(S_{t_k} - A_{t_{k-1}})$$

or

$$A_{t_{k-1}} = A_{t_k} + \frac{1}{k-1}(A_{t_k} - S_{t_k}).$$

The latter of these update formulas is relevant to us, because we integrate backwards in time. The discretely sampled A_t is constant between sampling times, and it jumps at t_k with the step

$$\frac{1}{k-1}(A_{t_k} - S_{t_k}).$$

For each k this jump can be written

$$A^-(S) = A^+(S) + \frac{1}{k-1}(A^+(S) - S), \ \text{where } S = S_{t_k}. \tag{6.14a}$$

A^- and A^+ denote the values of A immediately before and immediately after sampling at t_k. The no-arbitrage principle implies continuity of V at the sampling instances t_k in the sense of continuity of $V(S_t, A_t, t)$ for any realization of a random walk. In our setting, this continuity is written

$$V(S, A^+, t_k) = V(S, A^-, t_k) \ . \tag{6.14b}$$

But for a *fixed* (S, A) this equation defines a **jump** of V at t_k.

     The numerical application of the jump condition (6.14) is as follows: The A-axis is discretized into discrete values A_j, $j = 1, \ldots, J$. For each time period between two consecutive sampling instances, say for $t_{k+1} \to t_k$, the option's value is independent of A because in our discretized setting A_t is piecewise constant; accordingly $\frac{\partial V}{\partial A} = 0$. So J one-dimensional Black-Scholes equations are integrated separately and independently from t_{k+1} to t_k, one for each j. Each of the one-dimensional Black-Scholes problems has its own terminal condition. For each A_j, the "first" terminal condition is taken from the payoff surface for $t_M = T$. Proceeding backwards in time, at each sampling time t_k the J parallel one-dimensional Black-Scholes problems are halted because new terminal conditions must be derived from the jump condition (6.14). The new values for $V(S, A_j, t_k)$ that serve as terminal values (starting values for the backward integration) for the next time period $t_k \rightarrow t_{k-1}$, are defined by the jump condition, and are obtained by interpolation. Only at these sampling times the J standard one-dimensional Black-Scholes problems are coupled; the coupling is provided by the interpolation. In this way, a sequence of surfaces $V(S, A, t_k)$ is calculated for $t_M = T, \ldots, t_1 = 0$. Figure 6.1 shows[3] the payoff and three surfaces calculated for an Asian European fixed strike put. As this illustration indicates, there is a kind of rotation of this surface as t varies from T to 0.

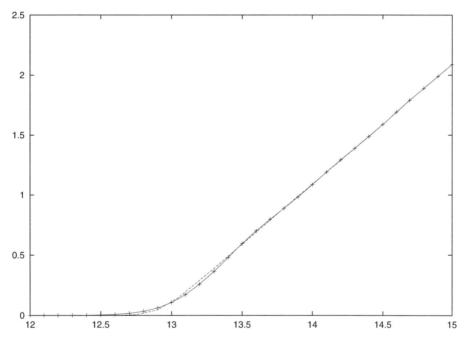

Fig. 6.2. European call, $K = 13$, $r = 0.15$, $\sigma = 0.01$, $T = 1$. Crank-Nicolson approximation $V(S, 0)$ with $\Delta t = 0.01$, $\Delta S = 0.1$ and centered difference scheme for $\frac{dV}{dS}$. Comparison with the exact Black-Scholes values (dashed).

---

[3] after interpolation; MATLAB graphics; courtesy of S. Göbel; similar [ZFV99]

6.4 Numerical Aspects

A direct numerical approach to the PDE (6.5) for functions $V(S, A, t)$ depending on three independent variables requires more effort than in the two-dimensional case. For example, a finite-difference approach uses a three-dimensional grid. And a separation ansatz as in Section 5.3 applies with two-dimensional basis functions. Although much of the required technology is widely analogous to the approaches discussed in Chapters 4 and 5, a thorough numerical treatment of higher-dimensional PDEs is beyond the scope of this book. Here we confine ourselves to PDEs with two independent variables, as in (6.8).

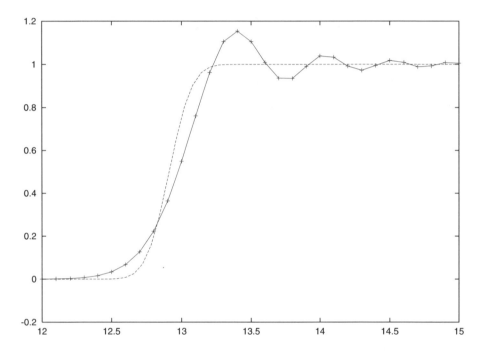

Fig. 6.3. Delta= $\frac{\partial V}{\partial S}$, otherwise the same data as in Figure 6.2

6.4.1 Convection-Diffusion Problems

Before entering a discussion on how to solve numerically a PDE like (6.8) without using transformations like (4.3), we perform an experiment with our well-known "classical" Black-Scholes equation (1.2). In contrast to the procedure of Chapter 4 we directly apply finite-difference quotients to (1.2). Here we use the second-order differences of Section 4.2.1 for a European call, and compare the numerical approximation with the exact solution (A4.10). Figure 6.2 shows the result for $V(S, 0)$. The lower part of the figure shows an oscillating error, which seems to be small. But differentiating magnifies oscillations. This is clearly visible in Figure 6.3, where the important hedge variable delta= $\frac{\partial V}{\partial S}$ is depicted. The wiggles are even worse for the second-order derivative gamma. These oscillations are financially unrealistic and are not tolerable, and we have to find its causes. The oscillations are *spurious* in that they are produced by the numerical scheme and are not solutions of the differential eqation. The spurious oscillations do not exist for the transformed version $y_\tau = y_{xx}$, which is illustrated by Figure 6.4.

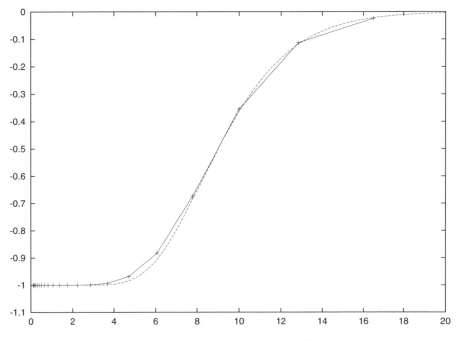

Fig. 6.4. European put, $K = 10$, $r = 0.06$, $\sigma = 0.30$, $T = 1$. Approximation delta= $\frac{\partial V}{\partial S}(S, 0)$ based on $y_\tau = y_{xx}$ with $m = 40$. Comparison with the exact Black-Scholes values (dashed).

In order to understand possible reasons why spurious oscillations may occur, we recall elementary fluid dynamics, where so-called convection-diffusion equations play an important role. For such equations, the second-order term is responsible for diffusion and the first-order term for convection. The ratio of convection to diffusion —scaled by a characteristic length— is the *Péclet number*, a dimensionless parameter characterizing the convection-diffusion problem. It turns out that the Péclet number is relevant for the understanding of underlying phenomena. Let us see what the Péclet number is for PDEs discussed so far in the text.

As a first example we take the original Black-Scholes equation (1.2), with diffusion term: $\qquad \dfrac{1}{2}\sigma^2 S^2 \dfrac{\partial^2 V}{\partial S^2}$

convection term: $\qquad rS\dfrac{\partial V}{\partial S}$

length scale: $\qquad \Delta S$

Péclet number: $\qquad \Delta S r S / \dfrac{1}{2}\sigma^2 S^2 = \dfrac{2r}{\sigma^2}\dfrac{\Delta S}{S}$

Since this dimensionless parameter involves the mesh size ΔS it is also called *mesh Péclet number*. (In case of a continuous dividend flow δ, replace r by $r - \delta$.) Experimental evidence indicates that the higher the Péclet number, the higher the danger that the numerical solution exhibits oscillations.

The PDE $y_\tau = y_{xx}$ has no convection term, hence its Péclet number is zero. Asian options described by the PDE (6.5) have a cumbersome situation: With respect to A there is no diffusion term (i.e., no second-order derivative), hence its Péclet number is ∞! For the original Black-Scholes equation the Péclet number basically is r/σ^2. It may become large when a small volatility σ is not compensated by a small riskless interest rate r. For the reduced PDE (6.8), the Péclet number is

$$\frac{\Delta R(1 - rR)}{\frac{1}{2}\sigma^2 R^2},$$

here a small σ can not be compensated by a small r.

These investigations of the Péclet numbers do not yet explain *why* spurious oscillations occur, but should open our eyes to the relation between convection and diffusion in the different PDEs. Let us discuss causes of the oscillations by means of a **model problem**. The model problem is the pure initial-value problem for a scalar function u defined on $t \geq 0$, $x \in \mathbb{R}$,

$$\frac{\partial u}{\partial t} + a\frac{\partial u}{\partial x} = b\frac{\partial^2 u}{\partial x^2} \quad , \quad u(x,0) = u_0(x). \tag{6.15}$$

For sake of simplicity, assume a and b are constants, with $b > 0$. This sign of b does not contradict the signs in (6.8) since there we have a terminal condition for $t = T$, whereas (6.15) prescribes an initial condition for $t = 0$. The equation (6.15) is meant to be integrated in forward time with discretization step size $\Delta t > 0$. So the equation (6.15) is a model problem representing a large class of convection-diffusion problems, to which the Black-Scholes equation (1.2) and the equation (6.8) belong. For the Black-Scholes equation, the simple transformation $S = Ke^x$, $t = T - \tau$, which works even for variable coefficients r, σ, produces (6.15) except for a further term $-ru$ on the right-hand side (compare Exercise 1.2). And for constant r, σ the transformed equation $y_\tau = y_{xx}$ is a member of the class (6.15), although it lacks convection. Discussing the stability properties of the model problem (6.15) will help us understanding how discretizations of (1.2) or (6.8) behave. For the analysis assume an equidistant grid on the x-range, with grid size $\Delta x > 0$ and nodes $x_j = j\Delta x$ for integers j.

6.4.2 Von Neumann Stability Analysis

First we apply to (6.15) the standard second-order centered space difference scheme in x-direction together with a forward time step, leading to

$$\frac{w_{j,\nu+1} - w_{j\nu}}{\Delta t} + a\frac{w_{j+1,\nu} - w_{j-1,\nu}}{2\Delta x} = b\delta_x^2 w_{j\nu} \tag{6.16}$$

with $\delta_x^2 w_{j\nu}$ defined as in (4.13). This scheme is called *Forward Time Centered Space* (FTCS). Instead of performing an eigenvalue-based stability analysis as in Chapter 4, we now apply the **von Neumann stability analysis**. This method expresses the approximations $w_{j\nu}$ of the ν-th time level by a sum of *eigenmodes* or Fourier modes,

$$w_{j\nu} = \sum_k c_k^{(\nu)} e^{ikj\Delta x}, \tag{6.17}$$

where i denotes the imaginary unit and k are the *wave numbers*. Substituting this expression into the FTCS-difference scheme (6.16) leads to a correspon-ding sum for $w_{j,\nu+1}$ with coefficients $c_k^{(\nu+1)}$. The linearity of the scheme (6.16) allows to find a relation

$$c_k^{(\nu+1)} = G_k c_k^{(\nu)},$$

where G_k is the *growth factor* of the mode with wave number k. In case $|G_k| \leq 1$ holds, it is guaranteed that the modes e^{ikx} in (6.17) are not ampli-fied, which means the method is stable.

Applying the von Neumann stability analysis to (6.16) leads to

$$G_k = 1 - 2\lambda + \left(\tfrac{\gamma}{2} + \lambda\right)e^{-ik\Delta x} + \left(\lambda - \tfrac{\gamma}{2}\right)e^{ik\Delta x},$$

where we use the abbreviations

$$\gamma := \frac{a\Delta t}{\Delta x} \quad , \quad \lambda := \frac{b\Delta t}{\Delta x^2} \quad , \quad \beta := \frac{a\Delta x}{b} . \tag{6.18}$$

Here β is the mesh Péclet number, and $\gamma = \beta\lambda$ is the famous *Courant number*. Using $e^{i\alpha} = \cos\alpha + i\sin\alpha$ and

$$s := \sin\frac{k\Delta x}{2} \quad , \quad \cos k\Delta x = 1 - 2s^2 \quad , \quad \sin k\Delta x = 2s\sqrt{1 - s^2}$$

we arrive at

$$G_k = 1 - 2\lambda + 2\lambda\cos k\Delta x - i\beta\lambda\sin k\Delta x \tag{6.19}$$

and

$$|G_k|^2 = (1 - 4\lambda s^2)^2 + 4\beta^2\lambda^2 s^2(1 - s^2).$$

A straightforward discussion of this polynomial on $0 \le s^2 \le 1$ reveals that $|G_k| \le 1$ for

$$0 \le \lambda \le \tfrac{1}{2} \quad , \quad \lambda\beta^2 \le 2. \tag{6.20}$$

The inequality $0 \le \lambda \le \tfrac{1}{2}$ brings back the stability criterion of Section 4.2.4. The inequality $\lambda\beta^2 \le 2$ is an additional restriction to the parameters λ and β, which depend on the discretization steps Δt, Δx and on the convection parameter a and the diffusion parameter b as defined in (6.18). The restriction due to the convection becomes apparent when we, for example, choose $\lambda = \tfrac{1}{2}$ for a maximal time step Δt. Then $|\beta| \le 2$ is a bound imposed on the mesh Péclet number, which restricts Δx to $\Delta x \le 2b/|a|$. A violation of this bound might be an explanation why the difference schemes of (6.16) applied to the Black-Scholes equation (1.2) exhibit faulty oscillations.[4] The bounds on $|\beta|$ and Δx are not active for problems without convection ($a = 0$). Note that the bounds give a severe restriction on problems with small values of the diffusion constant b. For $b = 0$ (no diffusion) the scheme (6.16) can not be applied at all. Although the constant-coefficient model problem (6.15) is not the same as the Black-Scholes equation (1.2) or the equation (6.8), the above analysis reflects the core of the difficulties. We emphasize that small values of the volatility represent small diffusion. So other methods than the standard finite-difference approach (6.16) are needed.

6.5 Upwind Schemes and Other Methods

The instability analyzed for the model combination (6.15)/(6.16) occurs when the mesh Péclet number is high and because the symmetric and centered difference quotient is applied to the first-order derivative. Next we discuss the extreme case of an infinite Péclet number of the model problem, namely $b = 0$. The resulting PDE is the prototypical equation

$$\frac{\partial u}{\partial t} + a\frac{\partial u}{\partial x} = 0 \ . \tag{6.21}$$

6.5.1 Upwind Scheme

The standard FTCS approach for (6.21) does not lead to a stable scheme. The PDE (6.21) has solutions in the form of *traveling waves*,

$$u(x,t) = F(x - at),$$

where $F(\xi) = u_0(\xi)$ in case initial conditions $u(x,0) = u_0(x)$ are incorporated. For $a > 0$, the profile $F(\xi)$ drifts in positive x-direction: the "wind blows

---

[4] In fact, the situation is more subtle. We postpone an outline of how *dispersion* is responsible for the oscillations to the Section 6.5.2.

to the right." Seen from a grid point (j, ν), the neighboring node $(j - 1, \nu)$ lies *upwind* and $(j + 1, \nu)$ lies *downwind*. Here the j indicates the node x_j and ν the time instant t_ν. Information flows from upstream to downstream nodes. Accordingly, the first-order difference scheme

$$\frac{w_{j,\nu+1} - w_{j\nu}}{\Delta t} + a \frac{w_{j\nu} - w_{j-1,\nu}}{\Delta x} = 0 \qquad (6.22)$$

is called *upwind discretization* $(a > 0)$. The scheme (6.22) is also called Forward Time Backward Space (FTBS) scheme.

Applying the von Neumann stability analysis to the scheme (6.22) leads to growth factors given by

$$G_k := 1 - \gamma + \gamma e^{-ik\Delta x}. \qquad (6.23)$$

Here $\gamma = \frac{a\Delta t}{\Delta x}$ is the Courant number from (6.18). The stability requirement is that $c_k^{(\nu)}$ remains bounded for any k and $\nu \to \infty$. So $|G_k| \le 1$ should hold. It is easy to see that

$$\gamma \le 1 \Rightarrow |G_k| \le 1 .$$

(The reader may sketch the complex G-plane to realize the situation.) The condition $|\gamma| \le 1$ is called the **Courant-Friedrichs-Lewy (CFL) condition**. The above analysis shows that this condition is sufficient to ensure stability of the upwind-scheme (6.22) applied to the PDE (6.21) with prescribed initial conditions.

In case $a < 0$, the scheme in (6.22) is no longer an upwind scheme. The upwind scheme for $a < 0$ is

$$\frac{w_{j,\nu+1} - w_{j\nu}}{\Delta t} + a \frac{w_{j+1,\nu} - w_{j\nu}}{\Delta x} = 0 \qquad (6.24)$$

The von Neumann stability analysis leads to the restriction $|\gamma| \le 1$, or $\lambda|\beta| \le 1$ if expressed in terms of the mesh Péclet number, see (6.18). This again emphasizes the importance of small Péclet numbers.

We note in passing that the FTCS scheme for $u_t + au_x = 0$, which is unstable, can be cured by replacing $w_{j\nu}$ by the average of its two neighbors. The resulting scheme

$$w_{j,\nu+1} = \tfrac{1}{2}(w_{j+1,\nu} + w_{j-1,\nu}) - \tfrac{1}{2}\gamma(w_{j+1,\nu} - w_{j-1,\nu}) \qquad (6.25)$$

is called *Lax-Friedrichs scheme*. It is stable if and only if the CFL condition is satisfied. A simple calculation shows that the Lax-Friedrichs scheme (6.25) can be rewritten in the form

$$\frac{w_{j,\nu+1} - w_{j\nu}}{\Delta t} = -a \frac{w_{j+1,\nu} - w_{j-1,\nu}}{2\Delta x} + \frac{1}{2\Delta t}(w_{j+1,\nu} - 2w_{j\nu} + w_{j-1,\nu}) .$$
$$(6.26)$$

This is a FTCS scheme with the additional term

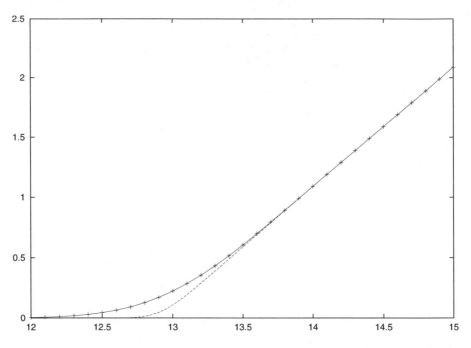

Fig. 6.5. European call, $K = 13$, $r = 0.15$, $\sigma = 0.01$, $T = 1$. Approximation $V(S,0)$, calculated with upwind scheme for $\frac{\partial V}{\partial S}$ and $\Delta t = 0.01$, $\Delta S = 0.1$. Comparison with the exact Black-Scholes values (dashed)

$$\frac{(\Delta x)^2}{2\Delta t}\, \delta_x^2 w_{j\nu}\,,$$

representing the PDE

$$u_t + au_x = \zeta u_{xx} \quad \text{with } \zeta = \Delta x^2/2\Delta t \,.$$

That is, the stabilization is accomplished by adding artificial diffusion. The scheme (6.26) is said to have *numerical dissipation*.

We return to the model problem (6.15) with $b > 0$. For the discretization of the $a\frac{\partial u}{\partial x}$ term we now apply the appropriate upwind scheme from (6.22) or (6.24), depending on the sign of the convection constant a. This noncentered first-order difference scheme can be written

$$\begin{aligned}
w_{j,\nu+1} = \quad &w_{j\nu} - \gamma \tfrac{1-\mathrm{sign}(a)}{2}\left(w_{j+1,\nu} - w_{j\nu}\right)\\
&-\gamma \tfrac{1+\mathrm{sign}(a)}{2}\left(w_{j\nu} - w_{j-1,\nu}\right) + \lambda(w_{j+1,\nu} - 2w_{j\nu} + w_{j-1,\nu})
\end{aligned}$$
(6.27)

with parameters γ, λ as defined in (6.18). For $a > 0$ the growth factors are

$$G_k = 1 - \lambda(2 + \beta)(1 - \cos k\Delta x) - i\lambda\beta \sin k\Delta x.$$

The analysis follows the lines of Section 6.4 and leads to the single stability criterion

$$\lambda \leq \frac{1}{2 + |\beta|} \, . \tag{6.28}$$

This inequality is valid for both signs of a ($\longrightarrow$ Exercise 6.3). For $\lambda \ll \beta$ the inequality (6.28) is less restrictive than (6.20). For example, a hypothetic value of $\lambda = \frac{1}{50}$ leads to the bound $|\beta| \leq 10$ for the FTCS scheme (6.16) and to the bound $|\beta| \leq 48$ for the upwind scheme (6.27).

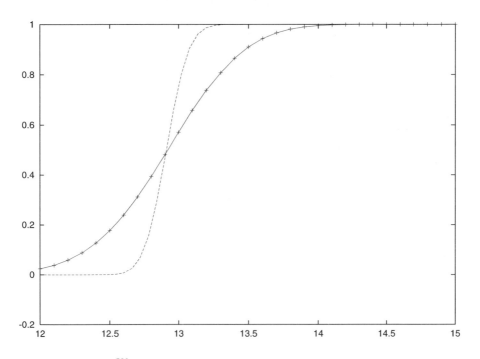

Fig. 6.6. Delta= $\frac{\partial V}{\partial S}(S, 0)$, same data as in Fig. 6.5

The Figures 6.5 and 6.6 show the Black-Scholes solution (dashed curve) and an approximation obtained by using the upwind scheme as in (6.27). No oscillations are visible, but the low order of the approximation can be seen from the moderate gradient, which does not reflect the steep gradient of the reality. The spurious wiggles have disappeared but the steep profile is heavily smeared. So the upwind scheme discussed above is a motivation to look for better methods (in Section 6.6).

6.5.2 Dispersion

The spurious wiggles are attributed to *dispersion*. Dispersion is the phenomenon of different modes traveling at different speeds. We explain dispersion for the simple PDE $u_t + au_x = 0$. Consider for $t = 0$ an initial profile u represented by a sum of Fourier modes, as in (6.17). Because of the linearity it is sufficient to study how the kth mode e^{ikx} is conveyed for $t > 0$. The differential equation $u_t + au_x = 0$ conveys the mode without change, because $e^{ik[x-at]}$ is a solution. For an observer who travels with speed a along the x-axis, the mode appears "frozen."

This does not hold for the numerical scheme. Here the amplitude and the phase of the kth mode may change. That is, the special initial profile

$$1 \cdot e^{ik[x-0]}$$

may change to

$$c(t) \cdot e^{ik[x-d(t)]} ,$$

where $c(t)$ is the amplitude and $d(t)$ the phase (up to the traveler distance at). Their values must be compared to those of the exact solution.

To be specific, apply Taylor's expansion to the upwind scheme for $u_t + au_x = 0$ $(a > 0)$,

$$\frac{w(x, t + \Delta t) - w(x, t)}{\Delta t} + a\frac{w(x, t) - w(x - \Delta x, t)}{\Delta x} = 0 ,$$

to derive the *equivalent differential equation*

$$w_t + aw_x = \zeta w_{xx} + \xi w_{xxx} + O(\Delta^2) ,$$

with

$$\zeta := \frac{a}{2}(\Delta x - a\Delta t) \ = \ \frac{a}{2}\Delta x(1 - \gamma) ,$$
$$\xi := \frac{a}{6}(-\Delta x^2 + 3a\Delta t\Delta x - 2a^2\Delta t^2) \ = \ \frac{a}{6}\Delta x^2(1 - \gamma)(2\gamma - 1) .$$

A solution can be obtained for the truncated PDE $w_t + aw_x = \zeta w_{xx} + \xi w_{xxx}$. Substituting $w = e^{i(\omega t + kx)}$ with undetermined frequency ω gives ω and

$$w = \exp\{-\zeta k^2 t\} \cdot \exp\{ik[x - t(\xi k^2 + a)]\} .$$

This defines the relevant amplitude $c(t)$ and phase shift $d(t)$, or rather

$$c_k(t) = \exp\{-\zeta k^2 t\}$$
$$d_k(t) = \xi k^2 t .$$

The $w = c_k(t)e^{ik[x-at-d_k(t)]}$ represents the solution of the applied upwind scheme. It is compared to the exact solution $u = e^{ik[x-at]}$ of the model problem, for which all modes propagate with the same speed a and without decay of the amplitude. The phase shift d_k in w due to a nonzero ξ becomes more relevant if the wave number k gets larger. That is, modes with different wave numbers drift across the finite-difference grid at different rates. Consequently, an initial signal represented by a sum of modes, changes its shape as it travels. The different propagation speeds of different modes e^{ikx} give rise to oscillations. This phenomenon is called dispersion. (Note that in our scenario of the simple model problem with upwind scheme, dispersion vanishes for $\gamma = 1$ and $\gamma = \frac{1}{2}$.)

A value of $|c(t)| < 1$ amounts to dissipation. If a high phase shift is compensated by heavy dissipation $(c \approx 0)$, then the dispersion is damped and may be hardly noticeable.

For several numerical schemes, related values of ζ and ξ have been investigated. For the influence of dispersion or dissipation see, for example, [St86], [Th95], [QSS00], [TR00]. Dispersion is to be expected for numerical schemes that operate on those versions of the Black-Scholes equation that have a convection term. This holds in particular for the θ-methods as described in Section 4.6.1, and for the upwind scheme. Numerical schemes for the convection-free version $y_\tau = y_{xx}$ do not suffer from dispersion.

6.6 High-Resolution Methods

The naive FTCS approach of the scheme (6.16) is only first-order in t-direction and suffers from severe stability restrictions. There are second-order approaches with better properties. A large class of schemes has been developed for so-called *conservation laws*, which in the one-dimensional situation are written

$$\frac{\partial u}{\partial t} + \frac{\partial}{\partial x}\, f(u) = 0\,. \tag{6.29}$$

The function $f(u)$ represents the *flux* in the equation (6.29), which originally was tailored to applications in fluid dynamics. We introduce the second-order method of Lax and Wendroff for the flux-conservative equation (6.29) because this method is too valuable to be discussed only for the simple special case $u_t + au_x = 0$. Then we present basic ideas of high-resolution methods.

6.6.1 The Lax-Wendroff Method

The Lax-Wendroff scheme is based on

$$u_{j,\nu+1} = u_{j\nu} + \Delta t \frac{\partial u_{j\nu}}{\partial t} + O(\Delta t^2) = u_{j\nu} - \Delta t \frac{\partial f(u_{j\nu})}{\partial x} + O(\Delta t^2)\,.$$

This expression makes use of (6.29) and replaces time derivatives by space derivatives. For suitably adapted indices this basic scheme is applied three times on a *staggered grid*. The staggered grid (see Figure 6.7) uses half steps of lengths $\frac{1}{2}\Delta x$ and $\frac{1}{2}\Delta t$ and intermediate mode numbers $j-\frac{1}{2}, j+\frac{1}{2}, \nu+\frac{1}{2}$. The main step is the second-order centered step (CTCS) with the center in the node $(j, \nu+\frac{1}{2})$ (square in Figure 6.7). This main step needs the flux function f evaluated at approximations w obtained for the two intermediate nodes $(j \pm \frac{1}{2}, \nu + \frac{1}{2})$, which are marked by crosses in Figure 6.7.

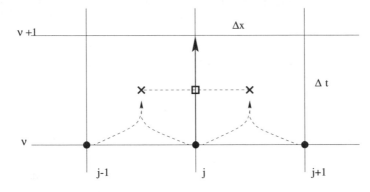

Fig. 6.7. Staggered grid for the Lax-Wendroff scheme.

Algorithm 6.2   (Lax-Wendroff)

$$
\begin{aligned}
w_{j+\frac{1}{2},\nu+\frac{1}{2}} &:= \tfrac{1}{2}(w_{j\nu} + w_{j+1,\nu}) - \tfrac{\Delta t}{2\Delta x}\left(f(w_{j+1,\nu}) - f(w_{j\nu})\right) \\
w_{j-\frac{1}{2},\nu+\frac{1}{2}} &:= \tfrac{1}{2}(w_{j-1,\nu} + w_{j\nu}) - \tfrac{\Delta t}{2\Delta x}\left(f(w_{j\nu}) - f(w_{j-1,\nu})\right) \\
w_{j,\nu+1} &:= w_{j\nu} - \tfrac{\Delta t}{\Delta x}\left(f(w_{j+\frac{1}{2},\nu+\frac{1}{2}}) - f(w_{j-\frac{1}{2},\nu+\frac{1}{2}})\right)
\end{aligned}
\tag{6.30}
$$

In this algorithm the half-step values $w_{j+\frac{1}{2},\nu+\frac{1}{2}}$ and $w_{j-\frac{1}{2},\nu+\frac{1}{2}}$ are provisional and discarded after $w_{j,\nu+1}$ is calculated. A stability analysis for the special case $f(u) = au$ in equation (6.29) (that is, of equation (6.21)) leads to the CFL condition as before. The Lax-Wendroff algorithm fits well discontinuities and steep fronts as the Black-Scholes delta-profile in Figures 6.3 and 6.6. But there are still spurious wiggles in the vicinity of steep gradients. The Lax-Wendroff scheme produces oscillatons near sharp fronts. We need to find a way to damp out the oscillations.

6.6.2 Total Variation Diminishing

Since $u_t + au_x$ convects an initial profile $F(x)$ with velocity a, a monotonicity of F will be preserved for all $t > 0$. So it makes sense to require also a numerical scheme to be *monotonicity preserving*. That is,

$$
\begin{aligned}
w_{j0} \leq w_{j+1,0} \ \text{ for all } j &\implies w_{j\nu} \leq w_{j+1,\nu} \ \text{ for all } j,\nu \geq 1 \\
w_{j0} \geq w_{j+1,0} \ \text{ for all } j &\implies w_{j\nu} \geq w_{j+1,\nu} \ \text{ for all } j,\nu \geq 1.
\end{aligned}
$$

A stronger requirement is that oscillations be diminished. To this end we define the *total variation* of the approximation vector $w^{(\nu)}$ at the ν-th time level as

$$\text{TV}(w^{(\nu)}) := \sum_j |w_{j+1,\nu} - w_{j\nu}|. \tag{6.31}$$

The aim is to construct a method that is *total variation diminishing* (TVD),

$$\text{TV}(w^{(\nu+1)}) \leq \text{TV}(w^{(\nu)}) \quad \text{for all } \nu.$$

Before we come to a criterion for TVD, note that the schemes discussed in this section are explicit and of the form

$$w_{j,\nu+1} = \sum_i c_i w_{j+i,\nu}. \tag{6.32}$$

The coefficients c_i decide whether such a scheme is monotonicity preserving or TVD.

Lemma 6.3 (monotonicity and TVD)
 (a) The scheme (6.32) is monotonicity preserving if and only if $c_i \geq 0$ for all c_i.
 (b) The scheme (6.32) is total variation diminishing (TVD) if and only if

$$c_i \geq 0 \quad \text{for all } c_i, \quad \text{and} \quad \sum_i c_i \leq 1.$$

The proof of (a) is left to the reader; for proving (b) the reader may find help in [We01], see also [Kr97]. As a consequence of Lemma 6.3 note that TVD implies monotonicity preservation. The Lax-Wendroff scheme satisfies $c_i \geq 0$ for all i only in the exceptional case $\gamma = 1$. For practical purposes, in view of nonconstant coefficients a, the Lax-Wendroff scheme is not TVD. For $f(u) = au$, the upwind scheme (6.22) and the Lax-Friedrichs scheme (6.25) are TVD for $|\gamma| \leq 1$ ($\longrightarrow$ Exercise 6.4).

6.6.3 Numerical Dissipation

For clarity we continue to discuss the matters for the linear scalar equation (6.21),

$$u_t + au_x = 0, \quad \text{for } a > 0.$$

For this equation it is easy to substitute the two provisional half-step values of the Lax-Wendroff algorithm into the equation for $w_{j,\nu+1}$. Then a straightforward calculation shows that the Lax-Wendroff scheme can be obtained by adding a diffusion term to the upwind scheme (6.22). The details are discussed next. For ease of notation, we define the difference

$$\delta_x^- w_{j\nu} := w_{j\nu} - w_{j-1,\nu}. \tag{6.33}$$

Then the upwind scheme is written

$$w_{j,\nu+1} = w_{j\nu} - \gamma\delta_x^- w_{j\nu}\,, \qquad \gamma = \frac{a\,\Delta t}{\Delta x}\,.$$

The reader may check that the Lax-Wendroff scheme is obtained by adding the term

$$-\delta_x^-\{\tfrac{1}{2}\gamma(1-\gamma)(w_{j+1,\nu} - w_{j\nu})\}\,. \qquad (6.34)$$

So the Lax-Wendroff scheme is rewritten

$$w_{j,\nu+1} = w_{j\nu} - \gamma\delta_x^- w_{j\nu} - \delta_x^-\{\tfrac{1}{2}\gamma(1-\gamma)(w_{j+1,\nu} - w_{j\nu})\}\,.$$

That is, the Lax-Wendroff scheme is the first-order upwind scheme plus the term (6.34), which is

$$-\tfrac{1}{2}\gamma(1-\gamma)(w_{j+1,\nu} - 2w_{j\nu} + w_{j-1,\nu})\,.$$

Hence the added term is —similar as for the Lax-Friedrichs scheme (6.26)— the discretized analogue of the artificial diffusion

$$-\tfrac{1}{2}a\Delta t(\Delta x - a\Delta t)u_{xx}\,.$$

Adding this artificial dissipation term (6.34) to the upwind scheme makes the scheme a second-order method.

The aim is to find a scheme that will give us neither the wiggles of the Lax-Wendroff scheme nor the smearing and low accuracy of the upwind scheme. On the other hand, we wish to benefit both from the second-order accuracy of the Lax-Wendroff scheme and from the smoothing capabilities of the upwind scheme. The idea is not to add the same amount of dissipation everywhere along the x-axis, but to add artificial dissipation in the right amount where it is needed. The resulting hybrid scheme will be of Lax-Wendroff type when the gradient is flat, and will be upwind-like at strong gradients of the solution. The decision on how much dissipation to add will be based on the solution.

In order to meet the goals, high-resolution methods control the artificial dissipation by introducing a *limiter* $\varphi_{j\nu}$ such that

$$w_{j,\nu+1} = w_{j\nu} - \gamma\delta_x^- w_{j\nu} - \delta_x^-\{\varphi_{j\nu}\tfrac{1}{2}\gamma(1-\gamma)(w_{j+1,\nu} - w_{j\nu})\}\,. \qquad (6.35)$$

Obviously this hybrid scheme specializes to the upwind scheme for $\varphi_{j\nu} = 0$ and is identical to the Lax-Wendroff scheme for $\varphi_{j\nu} = 1$. Accordingly, $\varphi_{j\nu} = 0$ may be chosen for strong gradients in the solution profile and $\varphi_{j\nu} = 1$ for smooth sections. To check the smoothness of the solution one defines the *smoothness parameter*

$$q_{j\nu} := \frac{w_{j\nu} - w_{j-1,\nu}}{w_{j+1,\nu} - w_{j\nu}}\,. \qquad (6.36)$$

The limiter $\varphi_{j\nu}$ will be a function of $q_{j\nu}$. We now drop the indices $j\nu$. For $q \approx 1$ the solution will be considered smooth, so we require the function $\varphi = \varphi(q)$ to satisfy $\varphi(1) = 1$ to reproduce the Lax-Wendroff scheme. Several strategies have been suggested to choose the limiter function $\varphi(q)$ such that the scheme (6.35) is total variation diminishing. For a thorough discussion of this matter we refer to [Sw84], [Kr97], [Th99]. One example of a limiter function is the van Leer limiter, which is defined by

$$\varphi(q) = \begin{cases} 0 & , \ q \leq 0 \\ \frac{2q}{1+q} & , \ q > 0 \end{cases} \tag{6.37}$$

The above principles of high-resolution methods have been successfully applied to financial engineering. The transfer of ideas from the simple problem (6.21) to the Black-Scholes world is quite involved. The methods are TVD for the Black-Scholes equation, which is in nonconservative form. Further the methods can be applied to nonuniform grids, and to implicit methods. The application of the Crank-Nicolson approach can be recommended. This amounts to solve nonlinear equations for each time step because the limiter introduces via (6.36), (6.37) a nonlinearity in $w^{(\nu+1)}$. Newton's method is applied to calculate the approximation $w^{(\nu+1)}$ in each time step ν [ZFV98].

Notes and Comments

on Section 6.1:

For barrier options we refer to [ZFV99], [SWH99], [PFVS00], [ZVF00]. For lookback options we mention [Kat95], [FVZ99], [Dai00].

on Section 6.2:

To see how the multidimensional volatilities of the model enter into a lumped volatility, consult [Shr04]. Other multidimensional PDEs arise when stochastic volatilities are modeled with SDEs, see [BaR94], [ZvFV98a], [Oo03], [HiMS04]. A list of exotic options with various payoffs is in Section 19.2 of [Deu01]. Also the n-dimensional PDEs can be transformed to simpler forms. This is shown for $n = 2$ and $n = 3$ in [Int05].

on Section 6.3:

Further higher-dimensional PDEs related to finance can be found in [TR00]. A reduction as in (6.8) from $V(S, A, t)$ to $H(R, t)$ is called *similarity reduction*. The derivation of the boundary-value problem (6.12) follows [WDH96]. For the discrete sampling discussed in Section 6.3.4 see [WDH96], [ZFV99]. The strategies introduced for Asian options work similarly for other path-dependent options.

on Section 6.4:

The von Neumann stability analysis is tailored to linear schemes and pure initial-value problems. It does not rigorously treat effects caused by boundary conditions. In this sense it provides a necessary stability condition for boundary-value problems. For a rigorous treatment of stability see [Th95], [Th99]. The stability analysis based on eigenvalues of iteration matrices as used in Chapter 4 is an alternative to the von Neumann analysis.

Spurious oscillations are special solutions of the difference equations and do not correspond to solutions of the differential equation. The spurious oscillations are not related to rounding errors. This may be studied analytically for the simple ODE model boundary-value problem $au' = bu''$, which is the steady state of (6.15), along with boundary conditions $u(0) = 0$, $u(1) = 1$. Here for mesh Péclet numbers $\frac{a\Delta x}{b} > 2$ the analytical solution of the discrete centered-space analog is oscillatory, whereas the solution $u(x)$ of the differential equation is monotone, see [Mo96]. The model problem is extensively studied in [PT83], [Mo96]. The mesh Péclet number is also called "algebraic Reynold's number of the mesh."

on Section 6.5:

It is recommendable to derive the equivalent differential equation in Section 6.5.2.

on Section 6.6:

The Lax-Wendroff scheme is an example of a *finite-volume method*. Another second-order scheme for (6.21) is the *leapfrog* scheme $\delta_t^2 w + a\delta_x^2 w = 0$, which involves three time levels. The discussion of monotonicity is based on investigations of Godunov, see [Kr97], [We01]. The Lax-Wendroff scheme for (6.21) and $\gamma \geq 0$ can also be written

$$w_j^{\nu+1} = w_j^\nu - \tfrac{1}{2}\gamma(w_{j+1}^\nu - w_{j-1}^\nu) + \tfrac{1}{2}\gamma^2(w_{j+1}^\nu - 2w_j^\nu + w_{j-1}^\nu) \ .$$

(This version adopts the frequent notation w_j^ν for our $w_{j\nu}$.) Here the diffusion term has a slightly different factor than (6.34). The numerical dissipation term is also called *artificial viscosity*. In [We01], p. 348, the Lax-Wendroff scheme is embedded in a family of schemes. A special choice of the family parameter yields a third-order scheme. The TVD criterion can be extended to implicit schemes and to schemes that involve more than two time levels. For the general analysis of numerical schemes for conservation laws (6.29) we refer to [Kr97].

on other methods:

Computational methods for exotic options are under rapid development. The universal binomial method can be adapted to exotic options [Kl01], [JiD04]. [TR00] gives an overview on a class of PDE solvers. For barrier options see [ZFV99], [ZVF00], [FuST02]. For two-factor barrier options and their finite-element solution, see [PFVS00]. PDEs for lookback options are given in [Bar97]. Using Monte Carlo for path-dependent options, considerable efficiency gains are possible with bridge techniques [RiW02], [RiW03]. We recommend to consult, for example, the issues of the *Journal of Computational Finance*.

Exercises

Exercise 6.1   Project: Monte Carlo Valuation of Exotic Options

Perform Monte Carlo valuations of barrier options, basket options, and Asian options, each European style.

Exercise 6.2   PDEs for Asian Options

a) Suppose that the value function $V(S, A, t)$ of an Asian option satisfies

$$dV = \left(\frac{\partial V}{\partial t} + S\frac{\partial V}{\partial A} + \mu S\frac{\partial V}{\partial S} + \frac{1}{2}\sigma^2 S^2 \frac{\partial^2 V}{\partial S^2} \right) dt + \sigma S\frac{\partial V}{\partial S}dW,$$

where S is the price of the asset and A its average. Construct a suitable riskless portfolio and derive the Black-Scholes equation

$$\frac{\partial V}{\partial t} + S\frac{\partial V}{\partial A} + \frac{1}{2}\sigma^2 S^2 \frac{\partial^2 V}{\partial S^2} + rS\frac{\partial V}{\partial S} - rV = 0 .$$

b) Use the transformation $V(S, A, t) = \tilde{V}(S, R, t) = SH(R, t)$, with $R = \frac{A}{S}$ and transform the Black-Scholes equation to

$$\frac{\partial H}{\partial t} + \frac{1}{2}\sigma^2 R^2 \frac{\partial^2 H}{\partial R^2} + (1 - rR)\frac{\partial H}{\partial R} = 0.$$

c) For

$$A_t := \frac{1}{t} \int_0^t S_\theta d\theta$$

show $dA = \frac{1}{t}(S - A)dt$ and derive the PDE

$$\frac{\partial V}{\partial t} + \frac{1}{2}\sigma^2 S^2 \frac{\partial^2 V}{\partial S^2} + rS\frac{\partial V}{\partial S} + \frac{1}{t}(S - A)\frac{\partial V}{\partial A} - rV = 0 .$$

Exercise 6.3   Upwind Scheme

Apply von Neumann's stability analysis to

$$\frac{\partial u}{\partial t} + a\frac{\partial u}{\partial x} = b\frac{\partial^2 u}{\partial x^2} , \quad b > 0$$

using the FTBS upwind scheme for the left-hand side and the centered second-order difference quotient for the right-hand side.

Exercise 6.4   TVD of a Model Problem

Analyze whether the upwind scheme (6.22), the Lax-Friedrichs scheme (6.25) and the Lax-Wendroff scheme (6.30) applied to the scalar partial differential equation

$$u_t + au_x, \quad a > 0, \ t \geq 0, \ x \in \mathbb{R}$$

satisfy the TVD property.
Hint: Apply Lemma 6.3.

Appendix A Financial Derivatives

A1 Investment and Risk

Basic markets in which money is invested trade in particular with

> equities (stocks),
> bonds, and
> commodites.

Front pages of *The Financial Times* or *The Wall Street Journal* open with charts informing about the trading in these key markets. Such charts symbolize and summarize myriads of buys and sales, and of individual gains and losses. The assets bought in the markets are collected and held in the portfolios of investors.

An easy way to buy or sell an asset is a spot contract, which is an agreement on the price assuming delivery on the same date. Typical examples are furnished by the trading of stocks on an exchange, where the spot price is paid the same day. On the spot markets, gain or loss, or risks are clearly visible. The spot contracts are contrasted with those contracts that agree today $(t = 0)$ to sell or buy an asset for a certain price at a certain *future time* $(t = T)$. Historically, the first objects traded in this way have been commodities, such as agricultural products, metals, or oil. For example, a farmer may wish to sell in advance the crop expected for the coming season. Later, such trading extended to stocks, currencies and other financial instruments. Today there is a virtually unlimited variety of contracts on objects and their future state, from credit risks to weather prediction.

The future price of the underlying asset is usually unknown, it may move up or down in an unexpected way. For example, scarcity of a product will result in higher prices. Or the prices of stocks may decline sharply. But the agreement must fix a price today, for an exchange of asset and payment that will happen in weeks or months. At maturity, the spot price usually differs from the agreed price of the contract. The difference between spot price and contract price may be significant. Hence contracts into the future are risky. Investors and portfolio managers hope their shares and markets perform well, and are concerned of risks that might weigh on their assets.

Different investments vary in their degree of uncertainty. The price of a stock may fall, and the company might default. The issuer of a bond may fail to meet the obligations in that he does not pay coupons or even fails to repay the principal amount. Some commodities like agricultural produce may spoil.

Financial risk of assets is defined as the degree of uncertainty of their return. *Market risks* are those risks that cannot be diversified away. Market risks are contrary to *default risks (credit risks)*.

No investment is really free of risks. But bonds can come close to the idealization of being riskless. If the seller of a bond has top ratings, then the return of a bond at maturity can be considered safe, and its value is known today with certainty. Such a bond is regarded as "riskless asset." The rate earned on a riskless asset is the *risk-free interest rate*. To avoid the complication of re-investing coupons, *zero-coupon bonds* are considered. The interest rate, denoted r, depends on the time to maturity T. The interest rate r is the continuously compounded interest which makes an initial investment S_0 grow to $S_0 e^{rT}$. We shall often assume that $r > 0$ is constant throughout that time period. A candidate for r is the LIBOR[1], which can be found in the financial press. In the mathematical finance literature, the term "bond" is used as synonym for a risk-free investment. Examples of bonds in real bond markets that come close to our idealized risk-free bond are provided by Treasury bills, which are short-term obligations of the US government, and by the long-term Treasury notes. See [Hull00] for further introduction, and consult for instance *The Wall Street Journal* for market diaries.

All other assets are risky, with equities being the most prominent examples. *Hedging* is possible to protect against financial loss. Many hedging instruments have been developed. Since these financial instruments depend on the particular asset that is to be hedged, they are called *derivatives*. Main types of derivatives are *futures, forwards, options,* and *swaps*[2]. They are explained below in some more detail. Tailoring and pricing derivates is the core of *financial engineering*. Hedging with derivatives is the way to bound financial risks and to protect investments.

The risks will play an important role in fixing the terms of the agreements, and in designing strategies for compensation.

A2 Financial Derivatives

Derivatives are instruments to assist and regulate agreements on transactions of the future. Derivatives can be traded on specialized exchanges.

Futures and **forwards** are agreements between two parties to buy or sell an asset at a certain time in the future for a certain delivery price. Both parties make a binding commitment, there is nothing to choose at a later time.

---

[1] London Interbank Offered Rate

[2] A comprehensive glossary of financial terms is provided by www.bloomberg.com/analysis

For forwards no premiums are required and no money changes hands until maturity. A basic difference between futures and forwards is that futures contracts are traded on exchanges and are more formalized, whereas forwards are traded in the over-the-counter market (OTC). Also the OTC market usually involves financial institutions. Large exchanges on which futures contracts are traded are the Chicago Board of Trade (CBOT), the Chicago Mercantile Exchange (CME), and the Eurex.

Options are *rights* to buy or sell underlying assets for an *exercise price (strike)*, which is fixed by the terms of the option contract. That is, the purchaser of the option is *not obligated* to buy or sell the asset. The decision will be based on the payoff, which is contingent on the underlying asset's behavior. The buying or selling of the underlying asset by exercising the option at a future date ($t = T$) must be distinguished from the purchase of the option (at $t = 0$, say), for which a premium ist paid. After the Chicago Board of Options Exchange (CBOE) opened in 1973, the volume of the trading with options has grown dramatically. Options are discussed in more detail in Section 1.1.

Swaps are contracts regulating an exchange of cash flows at different future times. A common type of swap is the *interest-rate swap*, in which two parties exchange interest payments periodically, typically fixed-rate payments for floating-rate payments. Counterparty A agrees to pay to counterparty B a fixed interest rate on some notional principal, and in return party B agrees to pay party A interest at a floating rate on the same notional principal. The principal itself is not exchanged. Each of the parties borrows the money at his market. The interest payment is received from the counterparty and paid to the lending bank. Since the interest payments are in the same currency, the counterparties only exchange the interest differences. The *swap rate* is the fixed-interest rate fixed such that the deal (initially) has no value to either party ("par swap"). For a *currency swap*, the two parties exchange cash flows in different currencies.

An important application of derivatives is **hedging**. Hedging means to eliminate or limit risks. For example, consider an investor who owns shares and wants protection against a possible decline of the price below a value K in the next three months. The investor could buy put options on this stock with strike K and a maturity that matches his three months time horizon. Since the investor can exercise his puts when the share price falls below K, it is guaranteed that the stock can be sold at least for the price K during the life time of the option. With this strategy the value of the stock is protected. The premium paid when purchasing the put option plays the role of an insurance premium. — Hedging is intrinsic for calls. The writer of a call must hedge his position to avoid being hit by rising asset prices. Generally speaking, options and other derivatives facilitate the transfer of financial risks.

What kind of principle is so powerful to serve as basis for a fair valuation of derivatives? The concept is **arbitrage**, or rather the assumption that arbitrage is not possible in an idealized market. Arbitrage means the existence of a portfolio, which requires no investment initially, and which with guarantee makes no loss but very likely a gain at maturity. Or shorter: arbitrage is a self-financing trading strategy with zero initial value and positive terminal value.

If an arbitrage profit becomes known, arbitrageurs will take advantage and try to lock in.[3] This makes the arbitrage profits shrink. In an idealized market, informations spread rapidly and arbitrage opportunites become apparent. So arbitrage cannot last for long. Hence, in efficient markets at most very small arbitrage opportunities are observed in practice. For the modeling of financial markets this leads to postulate the **no-arbitrage principle**: One assumes an idealized market such that arbitrage is ruled out. Arguments based on the no-arbitrage principle resemble indirect proofs in mathematics: Suppose a certain financial situation. If this assumed scenario enables constructing an arbitrage opportunity, then there is a conflict to the no-arbitrage principle. Consequently, the assumed scenario is impossible. See Appendix A3 for an example.

For valuing derivatives one compares the return of the risky financial investment with the return of an investment that is free of risk. For the comparison, one calculates the gain the same initial capital would yield when invested in bonds. To compare properly, one chooses a bond with time horizon T matching the terms of the derivative that is to be priced. Then, by the no-arbitrage principle, the risky investment should have the same price as the equivalent risk-free strategy. The construction and choice of derivatives to optimize portfolios and protect against extreme price movements is the essence of financial engineering.

The pricing of options is an ambitious task and requires sophisticated algorithms. Since this book is devoted to computational tools, mainly concentrating on options, the features of options are part of the text (Section 1.1 for standard options, and Section 6.1 for exotic options). This text will not enter further the discussion of forwards, futures, and swaps, with one exception: We choose the forward as an example (below) to illustrate the concept of arbitrage. For a detailed discussion of futures, forwards and swaps we refer to the literature, for instance to [Hull00], [BaR96], [MR97], [Wi98], [Shi99], [Lyuu02].

---

[3] This assumes that investors prefer more to less, the basis for a rational pricing theory [Mer73].

A3 Forwards and the No-Arbitrage Principle

As stated above, a forward is a contract between two parties to buy or sell an asset to be delivered at a certain time T in the future for a certain delivery price K. The time the parties agree on the forward contract (fixing T and K) is set to $t_0 = 0$. Since no premiums and no money change hands until maturity, the initial value of a forward is zero.

The party with the *long position* agrees to buy the underlying asset; the other party assumes the *short position* and agrees to sell the asset.

For the subsequent explanations S_t denotes the price of the asset in the time interval $0 \leq t \leq T$. To fix ideas, we assume just one interest rate r for borrowing or lending risk-free money over the time period $0 \leq t \leq T$. By the definition of the forward, at time of maturity T the party with the long position pays K to get the asset, which is then worth S_T.

Arbitrage Arguments

As will be shown next, the no-arbitrage principle enforces the forward price to be

$$K = S_0 e^{rT} . \tag{A3.1}$$

Thereby it is assumed that the asset does not produce any income (dividends) and does not cost anything until $t = T$.

Let us see how the no-arbitrage principle is invoked. We ask what the fair price K of a forward is at time $t = 0$, when the terms of a forward are settled. Then the spot price of the asset is S_0.

Assume first $K > S_0 e^{rT}$. Then an arbitrage strategy exists as follows: At $t = 0$ borrow S_0 at the interest rate r, buy the asset, and enter into a forward contract to sell the asset for the price K at $t = T$. When the time instant T has arrived, the arbitrageur completes the strategy by selling the asset $(+K)$ and by repaying the loan $(-S_0 e^{rT})$. The result is a riskless profit of $K - S_0 e^{rT} > 0$. This contradicts the no-arbitrage principle, so $K - S_0 e^{rT} \leq 0$ must hold.

Suppose next the complementary situation $K < S_0 e^{rT}$. In this case an investor who owns the asset[4] would sell it, invest the proceeds at interest rate r for the time period T, and enter a forward contract to buy the asset at $t = T$. In the end there would be a riskless profit of $S_0 e^{rT} - K > 0$. The conflict with the no-arbitrage principle implies $S_0 e^{rT} - K \leq 0$.

Combining the two inequalities $\leq$ and $\geq$ proves the equality. $[S_0 e^{r_1 T} \leq K \leq S_0 e^{r_2 T}$ in case of different rates $0 \leq r_1 \leq r_2$ for lending or borrowing]

One of the many applications of forwards is to hedge risks caused by foreign exchange.

---

[4] otherwise: *short sale*, selling a security the seller does not own.

Example   **(hedging against exchange rate moves)**

A U.S. corporation will receive one million euro in three months (on December 25), and wants to hedge against exchange rate moves. The corporation contacts a bank ("today" on September 25) to ask for the forward foreign exchange quotes. The three-month forward exchange rate is that $1.1428 will buy one euro, says the bank.[5] Why this? For completeness, on that day the spot rate is $1.1457. If the corporation and the bank enter into the corresponding forward contract on September 25, the corporation is obligated to sell one million euro to the bank for $1,142,800 on December 25. The bank then has a long forward contract on euro, and the corporation is in the short position.

Let us summarize the terms of the forward:
>   asset: one million euro
>   asset price S_t: the value of the asset in US $ $(S_0 = \$1,145,700)$
>   maturity $T = 1/4$ (three months)
>   delivery price K: $1,142,800 (forward price)

To understand the forward price in the above example, we need to generalize the basic forward price $S_0 e^{rT}$ to a situation where the asset produces income. In the foreign-exchange example, the asset earns the foreign interest rate, which we denote δ. To agree on a forward contract, $Ke^{-rT} = S_0 e^{-\delta T}$, so

$$K = S_0 e^{(r-\delta)T} . \tag{A3.2}$$

(See [Hull00].) On that date of the example the three-month interest rate in the U.S. was $r = 1\%$, and in the euro world $\delta = 2\%$. So

$$S_0 e^{(r-\delta)T} = 1145700 e^{-0.01\frac{1}{4}} = 1142800$$

which explains the three-month forward exchange rate of the example.

A4 The Black-Scholes Equation

The Classical Equation

This appendix applies Itô's lemma to derive the Black-Scholes equation. The first basic assumption is a geometric Brownian motion of the stock price. According to Model 1.13 the price S obeys the linear stochastic differential equation (1.33)

$$dS = \mu S\, dt + \sigma S\, dW \tag{A4.1}$$

with constant μ and σ. Further consider a portfolio consisting at time t of α_t shares of the asset with value S_t, and of β_t shares of the bond with value B_t. The bond is assumed riskless with

---

[5] September 25, 2003

$$dB = rBdt .   \quad (A4.2)$$

At time t the value of the portfolio is

$$\Pi_t := \alpha_t S_t + \beta_t B_t .   \quad (A4.3)$$

The portfolio is supposed to hedge a European option with value V_t, and payoff V_T at maturity T. So we assume *replication*,

$$\Pi_T = V_T = \text{ payoff } .   \quad (A4.4)$$

The European option cannot be traded before maturity; neither any investment is required in $0 < t < T$ for holding the option nor is there any payout stream. To compare the values of V_t and Π_t, and to apply no-arbitrage arguments, the portfolio should have an equivalent property. Suppose the portfolio is "closed" for $0 < t < T$ in the sense that no money is injected into or removed from the portfolio. This amounts to the *self-financing property*

$$d\Pi_t = \alpha_t dS_t + \beta_t dB_t .   \quad (A4.5)$$

That is, changes in the value of Π_t can only be due to changes in S or B.

Now the no-arbitrage principle is invoked. Replication (A4.4) and self-financing (A4.5) imply

$$\Pi_t = V_t \quad \text{for all } t \text{ in } 0 \le t \le T ,   \quad (A4.6)$$

because both investments have the same payout stream. So the replicating and self-financing portfolio is equivalent to the risky option. This has consequences for the quantites α_t and β_t of stocks and bonds, which are continuously readjusted.

Assuming a sufficiently smooth value function $\Pi_t = V(S,t)$, we infer from Itô's lemma (Section 1.8)

$$d\Pi = \left(\mu S \frac{\partial V}{\partial S} + \frac{\partial V}{\partial t} + \frac{1}{2}\sigma^2 S^2 \frac{\partial^2 V}{\partial S^2} \right) dt + \sigma S \frac{\partial V}{\partial S} dW .   \quad (A4.7)$$

On the other hand, substitute (A4.1) and (A4.2) into (A4.5) and obtain another version of $d\Pi$, namely

$$d\Pi = (\alpha\mu S + \beta r B)dt + \alpha\sigma S dW .   \quad (A4.8)$$

Because of uniqueness, the coefficients of both versions must match. Comparing the dW coefficients leads to the hedging strategy

$$\alpha_t = \frac{\partial V(S_t, t)}{\partial S} .   \quad (A4.9)$$

Matching the dt coefficients gives a relation for β, in which the stochastic $\alpha\mu S$ terms drop out. The βB term is replaced via (A4.3) and (A4.6), which amounts to

$$S\frac{\partial V}{\partial S} + \beta B = V \ .$$

This results in the renowned Black-Scholes equation (1.2),

$$\frac{\partial V}{\partial t} + \frac{1}{2}\sigma^2 S^2 \frac{\partial^2 V}{\partial S^2} + rS\frac{\partial V}{\partial S} - rV = 0.$$

The terminal condition is given by (A4.4).

Choosing in (A4.9) the *delta hedge* $\Delta(S, t) := \alpha = \frac{\partial V}{\partial S}$ provides a dynamic strategy to eliminate the risk that lies in stochastic fluctuations and in the unknown drift μ of the underlying asset. In this sense the modeling of V is risk neutral. The only remaining parameter reflecting stochastic behavior in the Black-Scholes equation is the volatility σ. Note that in the above derivation the standard understanding of constant coefficients μ, σ, r was actually not used. In fact the Black-Scholes equation holds also for time-varying deterministic functions $\mu(t), \sigma(t), r(t)$ ($\longrightarrow$ Exercise 1.19). For reference see, for example, [BaR96], [Du96], [Irle98], [HuK00], [Ste01].

The Solution and the Greeks

The **delta** $\Delta = \frac{\partial V}{\partial S}$ plays a crucial role for hedging portfolios. The corresponding number of units of the underlying asset makes the portfolio (A4.3) riskless. As will be shown next, there is a simple analytic formula for Δ in case of European options. (In reality, hedging must be done in discrete time.)

The Black-Scholes equation has a closed-form solution. For a European call with continuous dividend yield δ as in (4.1) (in Section 4.1) the formulas are

$$d_1 := \frac{\log\frac{S}{K} + \left(r - \delta + \frac{\sigma^2}{2}\right)(T - t)}{\sigma\sqrt{T - t}} \tag{A4.10a}$$

$$d_2 := d_1 - \sigma\sqrt{T - t} = \frac{\log\frac{S}{K} + \left(r - \delta - \frac{\sigma^2}{2}\right)(T - t)}{\sigma\sqrt{T - t}} \tag{A4.10b}$$

$$V_{\mathrm{C}}(S, t) = Se^{-\delta(T-t)}F(d_1) - Ke^{-r(T-t)}F(d_2), \tag{A4.10c}$$

where F denotes the standard normal cumulative distribution (compare Exercise 1.3 or Appendix D2). The value $V_{\mathrm{P}}(S, t)$ of a put is obtained by applying the put-call parity on (A4.10c), see Exercise 1.1. For a continuous dividend yield δ as in (4.1) the put-call parity of European options is

$$V_{\mathrm{P}} = V_{\mathrm{C}} - Se^{-\delta(T-t)} + Ke^{-r(T-t)} \tag{A4.11a}$$

from which

$$V_{\mathrm{P}} = -Se^{-\delta(T-t)}F(-d_1) + Ke^{-r(T-t)}F(-d_2) \tag{A4.11b}$$

follows.

For nonconstant but known coefficient functions $\sigma(t), r(t), \delta(t)$, the closed-form solution is modified by introducing integral mean values [Kwok98], [Øk98], [Wi98], [Zag02]. For example, replace the term $r(T-t)$ by the more general term $\int_t^T r(s)ds$, and replace

$$\sigma\sqrt{T-t} \quad \longrightarrow \quad \left(\int_t^T \sigma^2(s)ds\right)^{1/2}$$

Differentiating the Black-Scholes formula gives delta, $\Delta = \frac{\partial V}{\partial S}$, as

$$\begin{aligned}\Delta &= F(d_1) \qquad \text{for a European call,}\\ \Delta &= F(d_1) - 1 \text{ for a European put.}\end{aligned} \qquad (A4.12)$$

The delta of (A4.9) is the most prominent example of the "Greeks." Also other derivatives of V are denoted by Greek sounding names:

$$\text{gamma} = \frac{\partial^2 V}{\partial S^2}, \quad \text{theta} = \frac{\partial V}{\partial t}, \quad \text{vega} = \frac{\partial V}{\partial \sigma}, \quad \text{rho} = \frac{\partial V}{\partial r}$$

Analytic expressions can be obtained by differentiating (A4.10). The Greeks are important for a sensitivity analysis. A derivation of the Black-Scholes formula (A4.10) can be found in the literature; for references see [WDH96], [Shi99]. Essential parts of the derivation can also be collected from this book; see for instance Exercise 3.9 or Exercise 1.8.

Hedging a Portfolio in Case of a Jump Process

Next consider a jump-diffusion process as described in Section 1.9, summarized by equation (1.52). The portfolio is the same as above, see (A4.3), and the same assumptions such as replication and self-financing apply. Itô's lemma is applied in a piecewise fashion on the time intervals between jumps. Accordingly (A4.7) is modified by adding the jumps in V with jumps sizes

$$\Delta V := V(S_{\tau^+}, \tau) - V(S_{\tau^-}, \tau)$$

for all jump instances τ_j. Consequently the term $\Delta V dJ$ is added to (A4.7). On the other hand, (1.52) leads to add the term $\alpha(q-1)SdJ$ to (A4.8). Comparing coefficients of the dW terms in both expressions of Π again implies the hedging strategy (A4.9), namely $\alpha = \frac{\partial V}{\partial S}$. This allows to shorten both versions of Π by subtracting equal terms. Let us denote the resulting values of the reduced portfolios by $\tilde{\Pi}$. Then (A4.7) leads to

$$\tilde{\Pi} = \left(\frac{\partial V}{\partial t} + \frac{1}{2}\sigma^2 S^2 \frac{\partial^2 V}{\partial S^2}\right)dt + (V(qS,t) - V(S,t))dJ$$

and (A4.8) leads to

$$\tilde{\Pi} = \left(rV - rS\frac{\partial V}{\partial S}\right)dt + \frac{\partial V}{\partial S}(q-1)SdJ$$

(The reader may check.)

Different from the analysis leading to the classical Black-Scholes equation, $d\tilde{\Pi}$ is not deterministic and it does not make sense to equate both versions. The risk can not be perfectly hedged away to zero in the case of jump-diffusion processes. Following [Mer76], we apply the expectation operator over the random variable q to both versions of $\tilde{\Pi}$. Denote this expectation E, with

$$\mathsf{E}(X) = \int_{-\infty}^{\infty} x f_q(x)dx \tag{A4.13}$$

in case q_t has a density f_q that obeys $q > 0$. The expectations of both versions of $\mathsf{E}(\tilde{\Pi})$ can be equated. The result is

$$0 = \left(\frac{\partial V}{\partial t} + \frac{1}{2}\sigma^2 S^2 \frac{\partial^2 V}{\partial S^2} + rS\frac{\partial V}{\partial S} - rV\right)dt$$
$$+ \mathsf{E}\left([V(qS,t) - V(S,t) - (q-1)S\frac{\partial V}{\partial S}]\,dJ\right)\,.$$

Since all stochastic terms are assumed independent, the second part of the equation is

$$\mathsf{E}[...]\mathsf{E}(dJ).$$

Using from (1.50)

$$\mathsf{E}(dJ) = \lambda dt$$

and the abbreviation

$$c := \mathsf{E}(q-1)$$

this second part of the equation becomes

$$\{\mathsf{E}(V(qS,t)) - V(S,t) - cS\frac{\partial V}{\partial S}\}\,\lambda dt\,.$$

The integral $c = \mathsf{E}(q-1)$ does not depend on V. This number c can be calculated via (A4.13) as soon as a distribution for q is stipulated. For instance, one may assume a lognormal distribution, with relevant parameters fitted from marked data. (The parameters are not the same as those in (1.47).) With the precalculated number c, the resulting differential equation can be ordered into

$$\frac{\partial V}{\partial t} + \frac{1}{2}\sigma^2 S^2 \frac{\partial^2 V}{\partial S^2} + (r - \lambda c)S\frac{\partial V}{\partial S} - (\lambda + r)V + \lambda\mathsf{E}(V(qS,t)) = 0 \quad \text{(A4.14)}$$

Note that the last term is an integral taken over the unknown solution function $V(S,t)$. So the resulting equation is a partial integro-differential equation (PIDE). The standard Black-Scholes PDE is included for $\lambda = 0$. The integral can be discretized, for example, by the means of the composite trapezoidal rule ($\longrightarrow$ Appendix C1). A further discussion requires a model for the process q_t, see for example [Mer76], [Wi98], [Tsay02], [ConT04]. For computational approaches see [AnA00], [MaPS02].

A5 Early-Exercise Curve

This appendix briefly discusses properties of the early-exercise curve S_f of standard American put and call options, compare Section 4.5.1. The following holds for the

Put:
(1) S_f is continuously differentiable for $0 \leq t < T$.
(2) S_f is nondecreasing.
(3)   A lower bound is

$$S_f(t) > \frac{\lambda_2}{\lambda_2 - 1} K \text{ , where}$$

$$\lambda_2 = \frac{1}{2\sigma^2} \left\{ -\left(r - \delta - \frac{\sigma^2}{2}\right) - \sqrt{\left(r - \delta - \frac{\sigma^2}{2}\right)^2 + 2\sigma^2 r} \right\} \tag{A5.1}$$

(4)   An upper bound for $t < T$ is given by (4.23P),

$$S_f(t) < \lim_{\substack{t \to T \\ t < T}} S_f(t) = \min\left(K, \frac{r}{\delta} K\right) .$$

For proofs of (1) see [MR97], [Kwok98]. For the smoothness of the value function $V(S, t)$ on the continuation region, see [MR97]. Monotonicity of $V(S, t)$ with respect to time implies (2), as shown for instance in [Kwok98].

The monotonicity of S_f leads to conclude that a lower bound is obtained by $T \to \infty$. This limiting case is the perpetual option, compare Exercise 4.8. Specifically for $\delta = 0$, λ_2 simplifies, and the lower bound is $K \frac{q}{1+q}$, where $q := \frac{2r}{\sigma^2}$. For an illustration of a long horizon $T = 40$ see Figure A.1. Simple calculus shows that λ_2 is the same as the λ_2 in Exercise 4.8.

Here we give a **proof of property (4)**. For $t = T$ the value V_P^{am} equals the payoff, $V_P^{am}(S, T) = K - S$ for $S < K$. Substitute this into the Black-Scholes equation gives[6]

$$\frac{\partial V}{\partial t} + 0 - (r - \delta)S - rV = 0 \quad ,$$

or

$$\frac{\partial V(S, T)}{\partial t} = rK - \delta S .$$

---

[6] Recall the context: V means V_P^{am}.

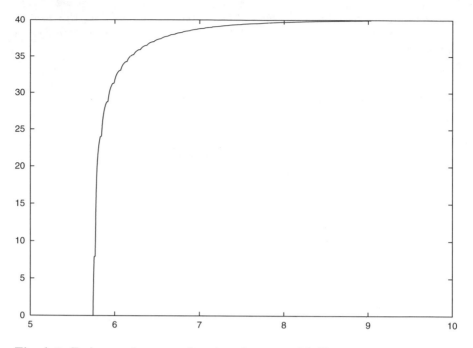

Fig. A.1. Early-exercise curve of an American put with $K = 10$, $r = 0.06$, $\sigma = 0.3$, $\delta = 0$, which leads to $\lambda_2 = -\frac{4}{3}$ and a lower bound of $\frac{4}{7}K$ (nonsmoothed output of a finite-difference calculation)

Observe that

$$\frac{\partial V(S,T)}{\partial t} \leq 0$$

because otherwise for t close to T a contradiction to $V \geq$ payoff results. Hence, for $t = T$ and $S < K$,

$$rK - \delta S \leq 0 \quad , \quad S \geq \frac{r}{\delta}K \ .$$

This makes sense only for $\delta > r$. We conclude

$$\frac{r}{\delta}K < S < K \quad \text{implies} \quad \frac{\partial V(S,T)}{\partial t} < 0 \ ,$$
$$\text{and } V > \text{payoff} \quad \text{for } t \approx T \ .$$

On the other hand, for

$$0 < S < \frac{r}{\delta}K \ , \quad \frac{\partial V(S,T)}{\partial t} = 0 \quad \text{and} \quad V = \text{payoff for } t \approx T \ .$$

For convenience denote

$$S_{\mathrm{f}}(T) := \lim_{\substack{t \to T \\ t < T}} S_{\mathrm{f}}(t) \ .$$

Where is $S_f(T)$ located on the maturity line $t = T$? Recall $V =$payoff for $S < S_f$ and $V >$payoff for $S > S_f$. Assume first $S_f(T) > \frac{r}{\delta}K$. Then for S in the area $\frac{r}{\delta}K < S < S_f(T)$ and $t \approx T$ we would simultaneously have $V >$payoff (since $S > \frac{r}{\delta}K$) and $V =$payoff (since $S < S_f(t)$), which is a contradiction.

Assume next $S_f(T) < \frac{r}{\delta}K$. Then for S inbetween, the contradiction is analogous. So the conclusion is

$$S_f(T) = \frac{r}{\delta}K \quad \text{for } \delta > r .$$

Finally we discuss the case $\delta \leq r$. By simple geometrical arguments, $S_f(T) > K$ cannot happen. Assume $S_f(T) < K$. Then for $S_f(T) < S < K$ and $t \approx T$

$$\underbrace{dV}_{\leq 0} = \underbrace{rK - \delta S}_{>0} \underbrace{dt}_{>0}$$

leads to a contradiction. So

$$S_f(T) = K \quad \text{for } \delta \leq r .$$

Both assertions are summarized to

$$\lim_{\substack{t \to T \\ t < T}} S_f(t) = \min\left(K, \frac{r}{\delta}K\right)$$

We conclude with listing the properties of an American

Call:
(1) S_f is continuously differentiable for $0 \leq t < T$.
(2) S_f is nonincreasing.
(3)   An upper bound is

$$S_f(t) < \frac{\lambda_1}{\lambda_1 - 1}K \text{ , where}$$

$$\lambda_1 = \frac{1}{2\sigma^2}\left\{ -\left(r - \delta - \frac{\sigma^2}{2}\right) + \sqrt{\left(r - \delta - \frac{\sigma^2}{2}\right)^2 + 2\sigma^2 r} \right\}$$

(A5.2)

(4) A lower bound for $t < T$ is given by (4.23C),

$$S_f(t) > \max\left(K, \frac{r}{\delta}K\right) .$$

Derivations are analogous as in the case of the American put. We note from properties (4) two extreme cases for $t \to T$:

$$\begin{aligned} \text{put}: & \quad r \to 0 \quad \Rightarrow \quad S_f \to 0 \\ \text{call}: & \quad \delta \to 0 \quad \Rightarrow \quad S_f \to \infty . \end{aligned}$$

The second assertion is another clue that for a call early exercise will never be optimal when no dividends are paid ($\delta = 0$).

By the way, the **symmetry** of the above properties is reflected by

$$S_{\mathrm{f,call}}(t; r, \delta) S_{\mathrm{f,put}}(t; \delta, r) = K^2$$
$$V_{\mathrm{C}}^{\mathrm{am}}(S, T - t; K, r, \delta) = V_{\mathrm{P}}^{\mathrm{am}}(K, T - t; S, \delta, r) \ . \tag{A5.3}$$

This put-call symmetry is derived in [Kwok98].

Appendix B Stochastic Tools

B1 Essentials of Stochastics

This appendix lists some basic instruments and notations of probability theory and statistics. For further foundations we refer to the literature, for example, [Fe50], [Fisz63], [Bi79], [Mik98], [JaP03], [Shr04].

Let Ω be a *sample space*. In our context Ω is mostly uncountable, for example, $\Omega = \mathbb{R}$. A subset of Ω is an *event* and an element $\omega \in \Omega$ is a sample point. The sample space Ω represents all possible scenarios. Classes of subsets of Ω must satisfy certain requirements to be useful for probability. One assumes that such a class $\mathcal{F}$ of events is a *σ-algebra* or a *σ-field*[1]. That is, $\Omega \in \mathcal{F}$, and $\mathcal{F}$ is closed under the formation of complements and countable unions. In our finance scenario, $\mathcal{F}$ represents the space of events that are observable in a market. If t denotes time, all informations available until t can be regarded as a σ-algebra $\mathcal{F}_t$. Then it is natural to assume a *filtration* —that is, $\mathcal{F}_t \subseteq \mathcal{F}_s$ for $t < s$.

The sets in $\mathcal{F}$ are also called *measurable sets*. A measure on these sets is the probability measure P, a real-valued function taking values in the interval $[0, 1]$ with the three axioms

$$P(A) \geq 0 \ \text{ for all events } A \in \mathcal{F} \ , \qquad P(\Omega) = 1 \ ,$$

$$P\left(\bigcup_{i=1}^{\infty} A_i\right) = \sum_{i=1}^{\infty} P(A_i) \ \text{ for any sequence of disjoint } A_i \in \mathcal{F} \ .$$

The tripel $(\Omega, \mathcal{F}, P)$ is called a *probability space*. An assertion is said to hold *almost everywhere* (P–a.e.) if it is wrong with probability 0.

A real-valued function X on Ω is called **random variable** if the sets

$$\{X \leq x\} := \{\omega \in \Omega \ : \ X(\omega) \leq x\} = X^{-1}((-\infty, x])$$

are measurable for all $x \in \mathbb{R}$. That is, $\{X \leq x\} \in \mathcal{F}$. This book does not explicitly indicate the dependence on the sample space Ω. We write X instead of $X(\omega)$, or X_t or $X(t)$ instead of $X_t(\omega)$ when the random variable depends on a parameter t.

---

[1] This notation with σ is not related with volatility.

For $x \in \mathbb{R}$ a **distribution function** $F(x)$ of X is defined by the probability P that $X \leq x$,

$$F(x) := \mathsf{P}(X \leq x). \tag{B1.1}$$

Distributions are nondecreasing, right-continuous, and satisfy the limits $\lim_{x \to -\infty} F(x) = 0$ and $\lim_{x \to +\infty} F(x) = 1$. Every absolutely continuous distribution F has a derivative almost everywhere, which is called **density function**. For all $x \in \mathbb{R}$ a density function f has the properties $f(x) \geq 0$ and

$$F(x) = \int\limits_{-\infty}^{x} f(t)dt . \tag{B1.2}$$

To stress the dependence on X, the distribution is also written F_X and the density f_X. If X has a density f then the kth *moment* is defined as

$$\mathsf{E}(x^k) := \int\limits_{-\infty}^{\infty} x^k f(x)dx = \int\limits_{-\infty}^{\infty} x^k dF(x) , \tag{B1.3}$$

provided the integrals exist. The most important moment of a distribution is the **expected value** or **mean**

$$\mu := \mathsf{E}(X) := \int_{-\infty}^{\infty} x f(x)dx . \tag{B1.4}$$

The **variance** is defined as the second central moment

$$\sigma^2 := \mathsf{Var}(X) := \mathsf{E}((X - \mu)^2) = \int_{-\infty}^{\infty} (x - \mu)^2 f(x)dx . \tag{B1.5}$$

A consequence is

$$\sigma^2 = \mathsf{E}(X^2) - \mu^2 .$$

The expectation depends on the underlying probability measure P, which is sometimes emphasized by writing $\mathsf{E_P}$. Here and in the sequel we assume that the integrals exist. The square root $\sigma = \sqrt{\mathsf{Var}(X)}$ is the *standard deviation* of X. For $\alpha, \beta \in \mathbb{R}$ and two random variables X, Y on the same probability space, expectation and variance satisfy

$$\mathsf{E}(\alpha X + \beta Y) = \alpha\mathsf{E}(X) + \beta\mathsf{E}(Y)$$
$$\mathsf{Var}(\alpha X + \beta) = \mathsf{Var}(\alpha X) = \alpha^2 \mathsf{Var}(X). \tag{B1.6}$$

The *covariance* of two random variables X and Y is

$$\mathsf{Cov}(X, Y) := \mathsf{E}\left((X - \mathsf{E}(X))(Y - \mathsf{E}(Y))\right) = \mathsf{E}(XY) - \mathsf{E}(X)\mathsf{E}(Y),$$

from which

$$\mathsf{Var}(X \pm Y) = \mathsf{Var}(X) + \mathsf{Var}(Y) \pm 2\mathsf{Cov}(X, Y) \tag{B1.7}$$

follows. More general, the covariance between the components of a *vector X* is the matrix

$$\mathsf{Cov}(X) = \mathsf{E}[(X - \mathsf{E}(X))(X - \mathsf{E}(X))^{tr}] = \mathsf{E}(XX^{tr}) - \mathsf{E}(X)\mathsf{E}(X)^{tr}, \quad \text{(B1.8)}$$

where the expectation E is applied to each component. The diagonal carries the variances of the components X_i. Back to the scalar world: Two random variables X and Y are called *independent* if

$$\mathsf{P}(X \le x, Y \le y) = \mathsf{P}(X \le x)\mathsf{P}(Y \le y).$$

For independent random variables X and Y the equations

$$\mathsf{E}(XY) = \mathsf{E}(X)\mathsf{E}(Y),$$
$$\mathsf{Var}(X + Y) = \mathsf{Var}(X) + \mathsf{Var}(Y)$$

are valid; analogous assertions hold for more than two independent random variables.

Normal distribution (Gaussian distribution): The density of the normal distribution is

$$f(x) = \frac{1}{\sigma\sqrt{2\pi}} \exp\left(-\frac{(x - \mu)^2}{2\sigma^2}\right). \quad \text{(B1.9)}$$

$X \sim \mathcal{N}(\mu, \sigma^2)$ means: X is normally distributed with expectation μ and variance σ^2. An implication is $Z = \frac{X-\mu}{\sigma} \sim \mathcal{N}(0, 1)$, which is the *standard normal distribution*, or $X = \sigma Z + \mu \sim \mathcal{N}(\mu, \sigma^2)$. The values of the corresponding distribution function $F(x)$ can be approximated by analytic expressions ($\longrightarrow$ Appendix D2) or numerically ($\longrightarrow$ Exercise 1.3). For multidimensional Gaussian, see Section 2.3.3.

Uniform distribution over an interval $a \le x \le b$:

$$f(x) = \frac{1}{b - a} \quad \text{for } a \le x \le b \; ; \; f = 0 \quad \text{elsewhere.} \quad \text{(B1.10)}$$

The uniform distribution has expected value $\frac{1}{2}(a+b)$ and variance $\frac{1}{12}(b-a)^2$. If the uniform distribution is considered over a higher-dimensional domain $\mathcal{D}$, then the value of the density is is the inverse of the volume of $\mathcal{D}$,

$$f = \frac{1}{\text{vol}(\mathcal{D})} \cdot 1_{\mathcal{D}}$$

For example, on a unit disc we have $f = 1/\pi$.

Estimates of mean and variance of a normally distributed random variable X from a sample of M realizations $x_1, ..., x_M$ are given by

$$\hat{\mu} := \frac{1}{M} \sum_{k=1}^{M} x_k$$

$$\hat{s}^2 := \frac{1}{M-1} \sum_{k=1}^{M} (x_k - \hat{\mu})^2 \qquad (\text{B1.11})$$

These expressions of the sample mean $\hat{\mu}$ and the sample variance $\hat{s}^2$ satisfy $\mathsf{E}(\hat{\mu}) = \mu$ and $\mathsf{E}(\hat{s}^2) = \sigma^2$. That is, $\hat{\mu}$ and $\hat{s}^2$ are unbiased estimates. For the computation see Exercise 1.4, or [PTVF92].

Central Limit Theorem: Suppose $X_1, X_2, ...$ are independent and identically distributed (i.i.d.) random variables, and $\mu := \mathsf{E}(X_i)$, $S_n := \sum_{i=1}^{n} X_i$, $\sigma^2 = \mathsf{E}(X_i - \mu)^2$. Then for each a

$$\lim_{n \to \infty} \mathsf{P}\left(\frac{S_n - n\mu}{\sigma\sqrt{n}} \le a\right) = \frac{1}{\sqrt{2\pi}} \int_{-\infty}^{a} e^{-z^2/2} dz. \qquad (\text{B1.12})$$

The **weak law of large numbers** states that for all $\epsilon > 0$

$$\lim_{n \to \infty} \mathsf{P}\left(\left|\frac{S_n}{n} - \mu\right| > \epsilon\right) = 0,$$

and the strong law says $\mathsf{P}(\lim_n \frac{S_n}{n} = \mu) = 1$.

For a **discrete probability space** the sample space Ω is countable. The expectation and the variance of a discrete random variable X with realizations x_i are given by

$$\mu = \mathsf{E}(X) = \sum_{\omega \in \Omega} X(\omega) \mathsf{P}(\omega) = \sum_{i} x_i \mathsf{P}(X = x_i)$$

$$\sigma^2 = \sum_{i} (x_i - \mu)^2 \mathsf{P}(X = x_i) \qquad (\text{B1.13})$$

Occasionally, the underlying probability measure P is mentioned in the notation. For example, a Bernoulli experiment[2] with $\Omega = \{\omega_1, \omega_2\}$ and $\mathsf{P}(\omega_1) = p$ has expectation

$$\mathsf{E}_\mathsf{P}(X) = p X(\omega_1) + (1-p) X(\omega_2).$$

The probability that for n Bernoulli trials the event ω_1 occurs exactly k times, is

$$\mathsf{P}(X = k) = b_{n,p}(k) := \binom{n}{k} p^k (1-p)^{n-k} \quad \text{for } 0 \le k \le n. \qquad (\text{B1.14})$$

---

[2] repeated independent trials, where only two possible outcomes are possible for each trial, such as tossing a coin

The *binomial coefficient* defined as

$$\binom{n}{k} = \frac{n!}{(n-k)!k!}$$

states in how many ways k elements can be chosen out of a population of size n. For the **binomial distribution** $b_{n,p}(k)$ the mean is $\mu = np$, and the variance $\sigma^2 = np(1-p)$. The probability that event ω_1 occurs at least M times is

$$P(X \geq M) = B_{n,p}(M) := \sum_{k=M}^{n} \binom{n}{k} p^k (1-p)^{n-k} . \tag{B1.15}$$

This follows from the axioms of the probability measure.

For the **Poisson distribution** the probability that an event occurs exactly k times within a specified (time) interval is given by

$$P(X = k) = \frac{a^k}{k!} e^{-a} \quad \text{for } k = 0, 1, 2, \ldots \tag{B1.16}$$

and a constant $a > 0$. Its mean and variance are both a.

Convergence in the mean: A sequence X_n is said to converge in the (square) mean to X, if $E(X_n^2) < \infty$, $E(X^2) < \infty$ and if

$$\lim_{n \to \infty} E((X - X_n)^2) = 0.$$

A notation for convergence in the mean is

$$\text{l.i.m.}_{n \to \infty} X_n = X.$$

B2 Advanced Topics

General Itô Formula

Let $dX_t = a(.)dt + b(.)dW_t$, where X_t is n-dimensional, $a(.)$ too, and $b(.)$ $(n \times m)$ matrix and W_t m-dimensional, with uncorrelated components, see (1.42). Let g be twice continuously differentiable, defined for (X, t) with values in $\mathbb{R}$. Then $g(X, t)$ is an Itô process with

$$dg = \left[\frac{\partial g}{\partial t} + g_x^{tr} a + \frac{1}{2} \text{ trace } (b^{tr} g_{xx} b) \right] dt + g_x^{tr} b \, dW_t . \tag{B2.1}$$

g_x is the gradient vector of the first-order partial derivatives with respect to x, and g_{xx} is the matrix of the second-order derivatives, all evaluated at (X, t). The matrix $b^{tr} g_{xx} b$ is $m \times m$. (Recall that the trace of a matrix is the sum of the diagonal elements.)

(B2.1) is derived via Taylor expansion. The linear terms $g_x^{tr}dX$ are straightforward. The quadratic terms are

$$\frac{1}{2}dX^{tr}g_{xx}dX \quad ,$$

from which the order dt terms remain

$$\frac{1}{2}(bdW)^{tr}g_{xx}bdW = \frac{1}{2}dW^{tr}b^{tr}g_{xx}bdW =: \frac{1}{2}dW^{tr}AdW \ .$$

These remaining terms are

$$\frac{1}{2} \text{ trace } (A)dt \ .$$

A matrix manipulation shows that the elements of $b^{tr}g_{xx}b$ are

$$\sum_{i=1}^{n}\sum_{j=1}^{n} g_{x_ix_j}b_{il}b_{jk} \quad \text{for } l,k = 1,\ldots,m$$

This is different from $bb^{tr}g_{xx}$, but the traces are equal:

$$\text{trace } (b^{tr}g_{xx}b) = \text{ trace } (bb^{tr}g_{xx}) = \sum_{i,j} \frac{\partial^2 g}{\partial x_i \partial x_j} \underbrace{\sum_{k=1}^{m} b_{ik}b_{jk}}_{=:c_{ij}} \ .$$

See also [Øk98].

Exercise: Let X be vector and Y scalar, where $dX = a_1dt + b_1dW$, $dY = a_2dt + b_2dW$, and consider $g(X,Y) := XY$. Show

$$\begin{aligned} d(XY) &= YdX + XdY + dXdY \\ &= (Xa_2 + Ya_1 + b_1b_2)dt + (Xb_2 + Yb_1)dW \ . \end{aligned} \tag{B2.2}$$

Application:

$$dS = rSdt + \sigma Sd\hat{W} \quad \Rightarrow \quad d(e^{-rt}S) = e^{-rt}\sigma Sd\hat{W} \tag{B2.3}$$

for any Wiener process $\hat{W}$.

Filtration of a Brownian motion

$$\mathcal{F}_t^W := \sigma\{W_s \mid 0 \le s \le t\} \tag{B2.4}$$

Here $\sigma\{.\}$ denotes the smallest σ-algebra containing the sets put in braces. $\mathcal{F}_t^W$ is a model of the information available at time t, since it includes every event based on the history of W_s, $0 \le s \le t$. The null sets $\mathcal{N}$ are included in the sense $\mathcal{F}_t := \sigma(\mathcal{F}_t^W \cup \mathcal{N})$ ("augmented").

Conditional Expectation

We recall conditional expectation because it is required for martingales. Let $\mathcal{G}$ be a sub σ-algebra of $\mathcal{F}$.

$E(X|\mathcal{G})$ is defined to be the (unique) $\mathcal{G}$-measurable random variable Y with the property

$$E(XZ) = E(YZ)$$

for all $\mathcal{G}$-measurable Z (such that $E(XZ) < \infty$). This is the conditional expectation of X given $\mathcal{G}$. Or, following [Doob53], an equivalent definition is via

$$\int_A E(Y|\mathcal{G})d\mathsf{P} = \int_A Y\,d\mathsf{P} \quad \text{for all} \quad A \in \mathcal{G} .$$

In case $E(X|Y)$, set $\mathcal{G} = \sigma(Y)$.

For properties of conditional expectation see, for example, [Mik98], [Shr04].

Martingales

Assume the standard scenario $(\Omega, \mathcal{F}, \mathcal{F}_t, \mathsf{P})$ with a filtration $\mathcal{F}_t \subset \mathcal{F}$.

Definition: $\mathcal{F}_t$-Martingale M_t is a process, which is "adapted" (that is, $\mathcal{F}_t$-measurable), $E(|M_t|) < \infty$, and

$$E(M_t|\mathcal{F}_s) = M_s \quad \text{(P-a.s.) for } s \leq t . \tag{B2.5}$$

The martingale property means that at time instant s with given information set $\mathcal{F}_s$ all variations of M_t for $t > s$ are unpredictable; M_s is the best forecast. The SDE of a martingale has no drift term.

Examples:

any Wiener process W_t ,

$W_t^2 - t$ for any Wiener process W_t ,

$\exp(\lambda W_t - \frac{1}{2}\lambda^2 t)$ for any $\lambda \in \mathbb{R}$ and any Wiener process W_t ,

$J_t - \lambda t$ for any Poisson process J_t with intensity λ.

For martingales, see for instance [Doob53], [Ne96], [Øk98], [Shi99], [Pro04], [Shr04].

For an adapted process γ define a process Z_t^γ by

$$Z_t^\gamma := \exp\left(-\frac{1}{2}\int_0^t \gamma_s^2 ds - \int_0^t \gamma_s dW_s\right) . \tag{B2.6}$$

Since $Z_0 = 1$, the integral equation

$$\log Z_t = \log Z_0 - \frac{1}{2}\int_0^t \gamma_s^2 ds - \int_0^t \gamma_s dW_s$$

follows, which is the SDE

$$d(\log Z_t) = (0 - \frac{1}{2}\gamma_t^2)dt - \gamma_t dW_t .$$

This is the Itô SDE for $\log Z_t$ when Z solves the drift-free $dZ_t = -Z_t\gamma_t dW_t$, $Z_0 = 1$. In summary, Z_t is the unique Itô process such that $dZ_t = -Z_t\gamma dW_t$, $Z_0 = 1$. Let Z^γ be a martingale. From the martingale properties, $\mathsf{E}(Z_T^\gamma) = 1$. Hence the Radon-Nikodym framework assures that an equivalent probability measure $\mathsf{Q}(\gamma)$ can be defined by

$$\frac{d\mathsf{Q}(\gamma)}{d\mathsf{P}} = Z_T^\gamma \quad \text{or} \quad \mathsf{Q}(A) := \int_A Z_T^\gamma d\mathsf{P} \tag{B2.7}$$

Girsanov's Theorem

Suppose a process γ is such that Z^γ is a martingale. Then

$$W_t^\gamma := W_t + \int_0^t \gamma_s ds \tag{B2.8}$$

is a Brownian motion and martingale under $\mathsf{Q}(\gamma)$.

B3 State-Price Process

Normalizing

A fundamental result of Harrison and Pliska [HP81] states that the existence of a martingale implies an arbitrage-free market. This motivates searching for a martingale. Since martingales have no drift term, we attempt to construct SDEs without drift.

Definition: A scalar positive Itô process Y_t with the property that the product $Y_t X_t$ has zero drift is called **state-price process** or *pricing kernel* or *deflator* for X_t.

The importance of state-price processes is highlighted by the following theorem.

Theorem: Assume that for X_t a state-price process Y_t exists, b is self-financing, and $Yb^{trp}X$ is bounded below. Then
(a) $Yb^{trp}X$ is a martingale, and
(b) the market does not admit self-financing arbitrage strategies.

([Nie99], p.148)

Sketch of Proof:
(a) Y is a state-price process, hence there exists σ such that $d(Y_tX_t) = \sigma dW_t$ (zero drift). By Itô's lemma,

$$d(Yb^{tr}X) = Yd(b^{trp}X) + dYb^{tr}X + dYd(b^{tr}X) .$$

(B2.2) and self-financing imply

$$d(Yb^{tr}X) = Yb^{tr}dX + dYb^{tr}X + dYb^{tr}dX$$
$$= b^{tr}[YdX + dYX + dYdX]$$
$$= b^{tr}d(XY) = b^{tr}\sigma dW =: \hat{\sigma}dW ,$$

hence zero drift of $Yb^{tr}X$.

It remains to show that $Yb^{tr}X$ is a martingale.
Because of the boundedness, $\tilde{Z} := Yb^{tr}X - c$ is a positive scalar Itô process for some c, with zero drift. For every such process there is a $\tilde{\gamma}$ such that $\tilde{Z}$ has the form

$$\tilde{Z}_t = \tilde{Z}_0 Z_t^{\tilde{\gamma}} .$$

Hence $Yb^{tr}X = \tilde{Z} + c$ has the same properties as $Z^{\tilde{\gamma}}$, namely it is a supermartingale. The final step is to show $\mathsf{E}(Z_t) =$constant. Now Q is defined via (B2.7). (The last arguments are from martingale theory.)
(b) Assume arbitrage in the sense

$$b_0^{tr}X_0 = 0 , \quad \mathsf{P}(b_t^{tr}X_t \geq 0) = 1$$
$$\mathsf{P}(b_t^{tr}X_t > 0) > 0 \quad \text{for some fixed } t .$$

For that t:

$$b^{tr}X > 0 \quad \Rightarrow \quad Yb^{tr}X > 0$$

Now $\mathsf{E}_\mathsf{Q}(Yb^{tr}X) > 0$ is intuitive. This amounts to

$$\mathsf{E}_\mathsf{Q}(Yb^{tr}X \mid \mathcal{F}_0) > 0$$

Because it is a martingale, $Y_0 b_0^{tr} X_0 > 0$. This contradicts $b_0^{tr}X_0 = 0$, so the market is free of arbitrage.

Existence of a State-Price Process

In order to discuss the existence of a state-price process we investigate the drift term of the product $Y_t X_t$. To this end take X as satisfying the vector SDE

$$dX = \mu^X dt + \sigma^X dW .$$

The coefficient functions μ^X and σ^X may vary with X. If no confusion arises, we drop the superscript X. Recall ($\longrightarrow$ Exercise 1.18) that each scalar positive Itô process must satisfy

$$dY = Y\alpha dt + Y\beta dW$$

for some α and β, where β and W can be vectors (β a one-row matrix). Without loss of generality, we take the SDE for Y in the form

$$dY = -rY dt - Y\gamma dW \ . \tag{B3.1}$$

(We leave the choice of the one-row matrix γ still open.) Itô's lemma (B2.1) allows to calculate the drift of YX. By (B2.2) the result is the vector

$$Y(\mu - rX - \sigma\gamma^{tr}) \ .$$

Hence Y is a state-price process for X if and only if

$$\mu^X - rX = \sigma^X \gamma^{tr} \tag{B3.2}$$

holds. This is a system of n equations for the m components of γ.

Special case geometric Brownian motion: For scalar $X = S$ and W, $\mu^X = \mu S$, $\sigma^X = \sigma S$, (B3.2) reduces to

$$\mu - r = \sigma\gamma \ .$$

Given μ, σ, r, the equation (B3.2) determines γ. (As explained in Section 1.7.3, γ is called the market price of risk.)

Discussion whether (B3.2) admits a (unique) solution:

Case I: unique solution: The market is complete. Further results below.

Case II: no solution: The market admits arbitrage.

Case III: multiple solutions: no arbitrage, but there are contingent claims that cannot be hedged; the market is said to be incomplete.

A solution of (B3.2) for full rank of the matrix σ is given by

$$\gamma^* := (\mu - rX)^{tr}(\sigma\sigma^{tr})^{-1}\sigma \ ,$$

which satisfies minimal length $\gamma^*\gamma^{*tr} \leq \gamma\gamma^{tr}$ for any other solution γ of (B3.2), see [Nie99].

Note that (B3.2) provides zero drift of YX but is not sufficient for YX to be a martingale. But it is "almost" a martingale; a small additional conditon suffices. Those trading strategies b are said to be *admissible* if $Yb^{tr}X$ is a martingale. (Sufficient is that $Yb^{tr}X$ be bounded below, such that it can not become arbitrarily negative. This rules out the "doubling strategy." For our purpose, we may consider the criterion as technical. [Gla04] on p.551: "It is common in applied work to assume that" a solution to an SDE with no drift term is a martingale.) There is ample literature on these topics; we just name [RY91], [BaR96], [Du96], [MR97], [Nie99].

Application: Derivative Pricing Formula for European Options

Let X_t be a vector price process, and b a self-financing trading strategy such that a European claim C is replicated. That is, for $V_t = b_t^{tr} X_t$ the payoff is reached: $V_T = b_T^{tr} X_T = C$. (Compare Appendix A4 for this argument.) We conclude from the above Theorem and from (B2.5)

$$Y_t b_t^{tr} X_t = \mathsf{E}_\mathsf{Q}(Y_T b_T^{tr} X_T \mid \mathcal{F}_t) \,,$$

or

$$V_t = \frac{1}{Y_t} \mathsf{E}_\mathsf{Q}(Y_T C \mid \mathcal{F}_t) \,.$$

Specifically for $t = 0$ the relation $\mathsf{E}_\mathsf{Q}(Y_T C \mid \mathcal{F}_0) = \mathsf{E}_\mathsf{Q}(Y_T C)$ holds, see [HuK00] p.136. This gives the value of European options as

$$V_0 = \frac{1}{Y_0} \mathsf{E}_\mathsf{Q}(Y_T C) \,.$$

This result is basic for Monte Carlo simulation, compare Subsection 3.5.1. Y_t represents a discounting process, for example, e^{-rt}. (Other discounting processes are possible, as long as they are tradable. They are called *numeraires*.) For a variable interest rate r_s,

$$V_t = \mathsf{E}_\mathsf{Q}(\exp(-\int_t^T r_s ds) C \mid \mathcal{F}_t)$$

In the special case r and γ constant, $Z_t = \exp(-\frac{1}{2}\gamma^2 t - \gamma W_t)$ and

$$\frac{V(t)}{e^{rt}} = \mathsf{E}_\mathsf{Q}\left(\frac{C}{e^{rT}} \mid \mathcal{F}_t\right)$$

$$\Rightarrow V(t) = e^{-r(T-t)} \mathsf{E}_\mathsf{Q}(C \mid \mathcal{F}_t) \,.$$

Appendix C Numerical Methods

C1 Basic Numerical Tools

This appendix briefly describes numerical methods used in this text. For additional information and detailed discussion we refer to the literature, for example to [Sc89], [HH91], [PTVF92], [SB96], [GV96], [QSS00].

Interpolation

Suppose $n+1$ pairs of numbers (x_i, y_i), $i = 0, 1, ..., n$ are given. These points in the (x, y)-plane are to be connected by a curve. An interpolating function $\Phi(x)$ satisfies

$$\Phi(x_i) = y_i \quad \text{for} \quad i = 0, 1, ..., n.$$

Depending on the choice of the class of Φ we distinguish different types of interpolation. A prominent example is furnished by polynomials,

$$\Phi(x) = P_n(x) = a_0 + a_1 x + ... + a_n x^n;$$

the degree n matches the number $n+1$ of points. The evaluation of a polynomial is done by the *nested multiplication* given by

$$P_n(x) = (...((a_n x + a_{n-1})x + a_{n-2})x + ... + a_1)x + a_0,$$

which is also called *Horner's method*. In case many points are given, the interpolation with one polynomial is generally not advisable since the high degree goes along with strong oscillations. A piecewise approach is preferred where low-degree polynomials are defined locally on one or more subintervals $x_i \leq x \leq x_{i+1}$ such that globally certain smoothness requirements are met. The simplest example is obtained when the points (x_i, y_i) are joined by straight-line segments in the order $x_0 < x_1 < ... < x_n$. The resulting *polygon* is globally continuous and linear over each subinterval. For the error of polygon approximation of a function we refer to Lemma 5.9. A C^2-smooth interpolation is given by the cubic *spline* using locally defined third-degree polynomials

$$S_i(x) := a_i + b_i(x - x_i) + c_i(x - x_i)^2 + d_i(x - x_i)^3 \quad \text{for} \quad x_i \leq x < x_{i+1}$$

that interpolate the points and are C^2-smooth at the nodes x_i.

Interpolation is applied for graphical illustration, numerical integration, and for solving differential equations. Generally interpolation is used to approximate functions.

Rational Approximation

Rational approximation is based on

$$\Phi(x) = \frac{a_0 + a_1 x + ... + a_n x^n}{b_0 + b_1 x + ... + b_m x^m}. \tag{C1.1}$$

Rational functions are advantageous in that they can approximate functions with poles. If the function that is to be approximated has a pole at $x = \xi$, then ξ must be zero of the denominator of Φ.

Integration

Approximating the definite integral

$$\int_a^b f(x)dx$$

is a classic problem of numerical analysis. Simple approaches replace the integral by

$$\int_a^b P_m(x)dx,$$

where the polynomial $P_m(x)$ approximates the function $f(x)$. The resulting formulas are called *quadrature* formulas. For example, an equidistant partition of the interval $[a, b]$ into m subintervals defines nodes x_i and support points $(x_i, f(x_i))$, $i = 0, ..., m$ for interpolation. After integrating the resulting polynomial $P_m(x)$ the *Newton-Cotes formulas* result. The simplest case $m = 1$ defines the *trapezoidal rule*.

A partition of the interval can be used more favorably. Applying the trapezoidal rule in each of n subintervals of length

$$h = \frac{b - a}{n}$$

leads to the composite formula of the *trapezoidal sum*

$$T(h) = h \left[\frac{f(a)}{2} + f(a + h) + ... + f(b - h) + \frac{f(b)}{2} \right]. \tag{C1.2}$$

The error of $T(h)$ satisfies a quadratic expansion

$$T(h) = \int_a^b f(x)dx + c_1 h^2 + c_2 h^4 + ...,$$

with a number of terms depending on the differentiability of f, and with constants c_i independent of h. This asymptotic expansion is fundamental for the high accuracy that can be achieved by *extrapolation*. Extrapolation evaluates $T(h)$ for a few h, for example, obtained by h_0, $h_1 = \frac{h_0}{2}$, $h_i = \frac{h_{i-1}}{2}$. Based on the values $T_i := T(h_i)$, an interpolating polynomial $\widetilde{T}(h^2)$ is calculated with $\widetilde{T}(0)$ serving as approximation to the exact value $T(0)$ of the integral.

The error behavior reflected by the above expansion can be simplified to

$$|T(h) - \int_a^b f(x)dx| \leq ch^2 ,$$

or written even shorter with the Landau symbol:

The error is of the order $O(h^2)$.

Zeros of Functions

The aim is to calculate a zero x^* of a function $f(x)$. An approximation is constructed in an iterative manner. Starting from some suitable initial guess x_0 a sequence $x_1, x_2, \ldots$ is calculated such that the sequence converges to x^*. Newton's method calculates the iterates by

$$x_{k+1} = x_k - \frac{f(x_k)}{f'(x_k)}.$$

In the vector case a system of linear equations needs to be solved in each step,

$$Df(x_k)(x_{k+1} - x_k) = -f(x_k), \tag{C1.3}$$

where Df denotes the Jacobian matrix of all first-order partial derivatives.

Example from Finance

Suppose a three-year bond with a principal of \$100 that pays a 6% coupon annually. Further assume zero rates of 5.8% for the first year, 6.3% for a two-year investment, and 6.4% for the three-year maturity. Then the *present value* (sum of all discounted future cashflows) is

$$6e^{-0.058} + 6e^{-0.063*2} + 106e^{-0.064*3} = 98.434$$

The *yield to maturity* (YTM) is the percentage rate of return y of the bond, when it is bought for the present value and is held to maturity. The YTM for the above example is the zero y of the cubic equation

$$0 = 98.434 - 6e^{-y} - 6e^{-2y} - 106e^{-3y}$$

which is 0.06384, or 6.384%, obtained with one iteration of Newton's method (C1.3), when started with 0.06 .

Convergence

Convergence is not guaranteed for any arbitrary choice of x_0. There are modifications and alternatives to Newton's method. Different methods are distinguished by their convergence speed. In the scalar case, *bisection* is a safe but slowly converging method. Newton's method for sufficiently regular problems shows fast convergence *locally*. That is, the error decays quadratically in a neighborhood of x^*,

$$\|x_{k+1} - x^*\| \le C\|x_k - x^*\|^p \quad \text{for } p = 2$$

for some constant C. This holds for an arbitrary vector norm $\|x\|$ such as

$$
\begin{aligned}
\|x\|_2 &:= \left(\sum_i x_i^2\right)^{1/2} & \textit{(Euclidian norm)} \\
\|x\|_\infty &:= \max_i |x_i| & \textit{(maximum norm)},
\end{aligned}
\tag{C1.4}
$$

$i = 1, \ldots, n$ for $x \in \mathbb{R}^n$.

The derivative $f'(x_k)$ can be approximated by difference quotients. If the difference quotient is based on $f(x_k)$ and $f(x_{k-1})$, in the scalar case, the *secant method* results. The secant method is generally faster than Newton's method if the speed is measured with respect to costs in evaluating $f(x)$ or $f'(x)$.

Gerschgorin's Theorem

A criterion for localizing the eigenvalues of a matrix $A = (a_{ij})$ is given by Gerschgorin's theorem: Each eigenvalue lies in the union of the discs

$$\mathcal{D}_j := \{z \text{ complex and } |z - a_{jj}| \le \sum_{\substack{k=1 \\ k \ne j}}^{n} |a_{jk}|\}$$

$(j = 1, \ldots, n)$. The centers of the discs $\mathcal{D}_j$ are the diagonal elements of A and the radii are given by the off-diagonal row sums (absolute values).

Triangular Decomposition

Let L denote a lower-triangular matrix (where the elements l_{ij} satisfy $l_{ij} = 0$ for $i < j$) and R an upper-triangular matrix ($r_{ij} = 0$ for $i > j$); the diagonal elements of L satisfy $l_{11} = \ldots = l_{nn} = 1$. Matrices A, L, R are supposed to be of size $n \times n$ and vectors x, b, ... have n components. Frequently, numerical methods must solve one or more systems of linear equations

$$Ax = b.$$

A well-known direct method to solve this system is Gaussian elimination. First, in a "forward"-phase, an equivalent system

$$Rx = \hat{b}$$

is calculated. Then, in a "backward"-phase starting with the last component x_n, all components of x are calculated one by one in the order $x_n, x_{n-1}, \ldots, x_1$. Gaussian elimination requires $\frac{2}{3}n^3 + O(n^2)$ arithmetic operations for full matrices A. With this count of $O(n^3)$, Gaussian elimination must be considered as an expensive endeavor, and is prohibitive for large values of n. (For alternatives, see iterative methods below in Appendix A5.) The forward phase of Gaussian elimination is equivalent to an *LR-decomposition*. This means the factorization into the product of two triangular matrices L, R in the form

$$PA = LR.$$

Here P is a permutation matrix arranging for the exchange of rows that corresponds to the pivoting of the Gaussian algorithm. The LR-decomposition exists for all nonsingular A. After the LR-decomposition is calculated, only two equations with triangular matrices need to be solved,

$$Ly = Pb \quad \text{and} \quad Rx = y.$$

Tridiagonal Matrices

For tridiagonal matrices the LR-decomposition specializes to an algorithm that requires only $O(n)$ operations, which is inexpensive. Since several of the matrices in this book are tridiagonal, we include the algorithm. Let the tridiagonal system $Ax = b$ be in the form

$$
\begin{pmatrix}
\alpha_1 & \beta_1 & & & 0 \\
\gamma_2 & \alpha_2 & \beta_2 & & \\
& \ddots & \ddots & \ddots & \\
& & \gamma_{n-1} & \alpha_{n-1} & \beta_{n-1} \\
0 & & & \gamma_n & \alpha_n
\end{pmatrix}
\begin{pmatrix}
x_1 \\ x_2 \\ \vdots \\ x_{n-1} \\ x_n
\end{pmatrix}
=
\begin{pmatrix}
b_1 \\ b_2 \\ \vdots \\ b_{n-1} \\ b_n
\end{pmatrix}
\tag{C1.5}
$$

Starting the Gaussian elimination with the first row to produce zeros in the subdiagonal during a forward loop, the algorithm is as follows:

$$
\begin{aligned}
& \hat{\alpha}_1 := \alpha_1, \ \hat{b}_1 := b_1 \\
& \text{(forward loop) for } i = 2, \ldots, n: \\
& \qquad \hat{\alpha}_i = \alpha_i - \beta_{i-1}\frac{\gamma_i}{\hat{\alpha}_{i-1}}, \quad \hat{b}_i = b_i - \hat{b}_{i-1}\frac{\gamma_i}{\hat{\alpha}_{i-1}} \\
& x_n := \frac{\hat{b}_n}{\hat{\alpha}_n} \\
& \text{(backward loop) for } i = n-1, \ldots, 1: \\
& \qquad x_i = \frac{1}{\hat{\alpha}_i}(\hat{b}_i - \beta_i x_{i+1})
\end{aligned}
\tag{C1.6}
$$

Here the "new" elements of the equivalent triangular system are indicated with a "hat;" the necessary checks for nonsingularity ($\hat{\alpha}_{i-1} \neq 0$) are omitted.

The algorithm (C1.6) needs about $8n$ operations. If one would start Gaussian elimination from the last row and produces zeros in the superdiagonal, an RL-decomposition results. The reader may wish to formulate the related backward/forward algorithm as an exercise.

Cholesky Decomposition

For *positive-definite* matrices A (means symmetric or Hermitian and $x^H A x > 0$ for all $x \neq 0$) there is exactly one lower-triangular matrix L with positive diagonal elements such that

$$ A = LL^H. $$

Here a normalization of the diagonal elements of L is not required. For real matrices A also L is real, hence $A = LL^{tr}$. (Hint: The Hermitian matrix A^H of A is defined as $\bar{A}^{tr}$, where $\bar{A}$ means elementwise complex conjugate.) For a computer program of Cholesky decomposition see [PTVF92].

C2 Iterative Methods for $Ax = b$

The system of linear equations $Ax = b$ in $\mathbb{R}^n$ can be written

$$ Mx = (M - A)x + b, $$

where M is a suitable matrix. For nonsingular M the system $Ax = b$ is equivalent to the fixed-point equation

$$ x = (I - M^{-1}A)x + M^{-1}b, $$

which leads to the iteration

$$ x^{(k+1)} = \underbrace{(I - M^{-1}A)}_{=:B}x^{(k)} + M^{-1}b. \tag{C2.1} $$

The computation of $x^{(k+1)}$ is done by solving the system of equations $Mx^{(k+1)} = (M - A)x^{(k)} + b$. Subtracting the fixed-point equation and applying Lemma 4.2 shows

$$ \text{convergence} \quad \Longleftrightarrow \quad \rho(B) < 1; $$

$\rho(B)$ is the spectral radius of matrix B. For this convergence criterion there is a sufficient criterion that is easy to check. Natural matrix norms satisfy $\|B\| \geq \rho(B)$. Hence $\|B\| < 1$ implies convergence. Application to the matrix norms

$$ \|B\|_\infty = \max_i \sum_{j=1}^n |b_{ij}|, $$

$$ \|B\|_1 = \max_j \sum_{i=1}^n |b_{ij}|, $$

produces sufficient convergence criteria: The iteration converges if

$$\sum_{j=1}^{n} |b_{ij}| < 1 \quad \text{for } 1 \leq i \leq n$$

or if

$$\sum_{i=1}^{n} |b_{ij}| < 1 \quad \text{for } 1 \leq j \leq n.$$

By obvious reasons these criteria are called row sum criterion and column sum criterion. The *preconditioner* matrix M is constructed such that rapid convergence of (C2.1) is achieved. Further, the structure of M must be simple so that the linear system is easily solved for $x^{(k+1)}$.

Simple examples are obtained by additive splitting of A into the form $A = D - L - U$, with
  D diagonal matrix
  L strict lower-triangular matrix
  U strict upper-triangular matrix

Jacobi's Method

Choosing $M := D$ implies $M - A = L + U$ and establishes the iteration

$$Dx^{(k+1)} = (L + U)x^{(k)} + b.$$

By the above convergence criteria a strict diagonal dominance of A is sufficient for the convergence of Jacobi's method.

Gauß–Seidel Method

Here the choice is $M := D - L$. This leads via $M - A = U$ to the iteration

$$(D - L)x^{(k+1)} = Ux^{(k)} + b.$$

SOR (Successive Overrelaxation)

The SOR method can be seen as a modification of the Gauß-Seidel method, where a *relaxation parameter* ω_R is introduced and chosen in a way that speeds up the convergence:

$$M := \frac{1}{\omega_R}D - L \implies M - A = \left(\frac{1}{\omega_R} - 1\right)D + U$$

$$\left(\frac{1}{\omega_R}D - L\right)x^{(k+1)} = \left(\left(\frac{1}{\omega_R} - 1\right)D + U\right)x^{(k)} + b$$

The SOR-method can be written as follows:

$$\begin{cases} B_{\mathrm{R}} := \left(\frac{1}{\omega_{\mathrm{R}}} D - L \right)^{-1} \left(\left(\frac{1}{\omega_{\mathrm{R}}} - 1 \right) D + U \right) \\ x^{(k+1)} = B_{\mathrm{R}} x^{(k)} + \left(\frac{1}{\omega_{\mathrm{R}}} D - L \right)^{-1} b \end{cases}$$

The Gauß–Seidel method is obtained as special case for $\omega_{\mathrm{R}} = 1$.

Choosing ω_{R}

The difference vectors $d^{(k+1)} := x^{(k+1)} - x^{(k)}$ satisfiy

$$d^{(k+1)} = B_{\mathrm{R}} d^{(k)}. \tag{$*$}$$

This is the power method for eigenvalue problems. Hence the $d^{(k)}$ converge to the eigenvector of the dominant eigenvalue $\rho(B_{\mathrm{R}})$. Consequently, if $(*)$ converges then

$$d^{(k+1)} = B_{\mathrm{R}} d^{(k)} \approx \rho(B_{\mathrm{R}}) d^{(k)}.$$

Then $|\rho(B_{\mathrm{R}})| \approx \frac{\|d^{(k+1)}\|}{\|d^{(k)}\|}$ for arbitrary vector norms. There is a class of matrices A with

$$\rho(B_{\mathrm{GS}}) = (\rho(B_{\mathrm{J}}))^2, \quad B_{\mathrm{J}} := D^{-1}(L + U)$$

$$\omega_{\mathrm{opt}} = \frac{2}{1 + \sqrt{1 - \rho(B_{\mathrm{J}})^2}},$$

see [Va62], [SB96]. Here B_{J} denotes the iteration matrix of the Jacobi method and B_{GS} that of the Gauß-Seidel method. For matrices A of that kind a few iterations with $\omega_{\mathrm{R}} = 1$ suffice to estimate the value $\rho(B_{\mathrm{GS}})$, which in turn gives an approximation to ω_{opt}. With our experience with Cryer's projected SOR applied to the valuation of options (Section 4.6) the simple strategy $\omega_{\mathrm{R}} = 1$ is frequently recommendable.

This appendix has merely introduced classic iterative solvers, which are stationary in the sense that the preconditioner matrix M does not vary with k. For an overview on advanced nonstationary iterative methods see [Ba94].

C3 Function Spaces

Let real-valued functions u, v, w be defined on $\mathcal{D} \subseteq \mathbb{R}^n$. We assume that $\mathcal{D}$ is a *domain*. That is, $\mathcal{D}$ is open, bounded and connected. The space of continuous functions is denoted $\mathcal{C}^0(\mathcal{D})$ or $\mathcal{C}(\mathcal{D})$. The functions in $\mathcal{C}^k(\mathcal{D})$ are k times continuously differentiable: All partial derivatives up to order k exist and are continuous on $\mathcal{D}$. The sets $\mathcal{C}^k(\mathcal{D})$ are examples of function spaces. Functions in $\mathcal{C}^k(\bar{\mathcal{D}})$ have in addition bounded and uniformly continuous derivatives and consequently can be extended to $\bar{\mathcal{D}}$.

Apart from being distinguished by differentiability, functions are also characterized by their integrability. The proper type of integral is the Lebesgue integral. The space of square-integrable functions is

$$\mathcal{L}^2(\mathcal{D}) := \left\{ v : \int_\mathcal{D} v^2 \, dx < \infty \right\}. \tag{C3.1}$$

For example, $v(x) = x^{-1/4} \in \mathcal{L}^2(0,1)$ but $v(x) = x^{-1/2} \notin \mathcal{L}^2(0,1)$. More general, for $p > 0$ the $\mathcal{L}^p$-spaces are defined by

$$\mathcal{L}^p(\mathcal{D}) := \left\{ v : \int_\mathcal{D} |v(x)|^p \, dx < \infty \right\}.$$

For $p \geq 1$ these spaces have several important properties [Ad75]. For example,

$$\|v\|_p := \left(\int_\mathcal{D} |v(x)|^p dx \right)^{1/p} \tag{C3.2}$$

is a norm.

In order to establish the existence of integrals such as

$$\int_a^b uv \, dx, \quad \int_a^b u'v' \, dx$$

we might be tempted to use a simple approach, defining a function space

$$\mathcal{H}^1(a,b) := \left\{ u \in \mathcal{L}^2(a,b) : \ u' \in \mathcal{L}^2(a,b) \right\}, \tag{C3.3}$$

with $\mathcal{D} = (a,b)$. But a classical derivative u' may not exist for $u \in \mathcal{L}^2$ or needs not be square integrable. What is needed is a weaker notion of derivative.

Weak Derivatives

In $\mathcal{C}^k$-spaces classical derivatives are defined in the usual way. For $\mathcal{L}^2$-spaces *weak derivatives* are defined. For motivation let us review standard integration by parts

$$\int_a^b uv' \, dx = - \int_a^b u'v \, dx, \tag{C3.4}$$

which is correct for all $u, v \in \mathcal{C}^1(a,b)$ with $v(a) = v(b) = 0$. For $u \notin \mathcal{C}^1$ the equation (C3.4) can be used to define a weak derivative u' provided smoothness is transferred to v. For this purpose define

$$\mathcal{C}_0^\infty(\mathcal{D}) := \{ v \in \mathcal{C}^\infty(\mathcal{D}) : \ \mathrm{supp}(v) \text{ is a compact subset of } \mathcal{D} \}.$$

$v \in \mathcal{C}_0^\infty(\mathcal{D})$ implies $v = 0$ at the boundary of $\mathcal{D}$. For $\mathcal{D} \subseteq \mathbb{R}^n$ one uses the multi-index notation

$$\alpha := (\alpha_1, ..., \alpha_n), \quad \alpha_i \in \mathbb{N} \cup \{0\}$$

with

$$|\alpha| := \sum_{i=1}^{n} \alpha_i.$$

Then the partial derivative of order $|\alpha|$ is defined as

$$D^\alpha v := \frac{\partial^{|\alpha|}}{\partial x_1^{\alpha_1} ... \partial x_n^{\alpha_n}} v(x_1, ..., x_n).$$

If a $w \in \mathcal{L}^2$ exists with

$$\int_{\mathcal{D}} u D^\alpha v \, dx = (-1)^{|\alpha|} \int_{\mathcal{D}} wv \, dx \quad \text{for all} \;\; v \in \mathcal{C}_0^\infty(\mathcal{D}),$$

the weak derivative of u with multiindex α is defined by $D^\alpha u := w$.

Sobolev Spaces

The definition (C3.3) is meaningful if u' is considered as weak derivative in the above sense. More general, one defines the *Sobolev spaces*

$$\mathcal{H}^k(\mathcal{D}) := \{ v \in \mathcal{L}^2(\mathcal{D}) : \; D^\alpha v \in \mathcal{L}^2(\mathcal{D}) \quad \text{für} \;\; |\alpha| \leq k \}. \tag{C3.5}$$

The index $_0$ specifies the subspace of $\mathcal{H}^1$ that consists of those functions that vanish at the boundary of $\mathcal{D}$. For example,

$$\mathcal{H}_0^1(a, b) := \{ v \in \mathcal{H}^1(a, b) : \; v(a) = v(b) = 0 \}.$$

The Sobolev spaces $\mathcal{H}^k$ are equipped with the norm

$$\|v\|_k := \left(\sum_{|\alpha| \leq k} \int_{\mathcal{D}} |D^\alpha v|^2 \, dx \right)^{1/2}, \tag{C3.6}$$

which is the sum of $\mathcal{L}^2$-norms of (C3.2). For the special case discussed in Chapter 5 with $k = 1$, $n = 1$, $\mathcal{D} = (a, b)$, the norm is

$$\|v\|_1 := \left(\int_a^b (v^2 + (v')^2) \, dx \right)^{1/2}.$$

Embedding theorems state which function spaces are subsets of other function spaces. In this way, elements of Sobolev spaces can be characterized and distinguished with respect to smoothness and integrability. For instance, the space $\mathcal{H}^1$ includes those functions that are globally continuous on all of $\mathcal{D}$ and its boundary and are *piecewise* $\mathcal{C}^1$-functions.

Hilbert Spaces

The function spaces $\mathcal{L}^2$ and $\mathcal{H}^k$ have numerous properties. Here we just mention that both spaces are *Hilbert spaces*. Hilbert spaces have an inner product

(,) such that the space is *complete* with respect to the norm $\|v\| := \sqrt{(v,v)}$. In complete spaces every Cauchy sequence converges. In Hilbert spaces the *Schwarzian inequality*

$$|(u,v)| \leq \|u\| \, \|v\| \qquad\qquad (C3.7)$$

holds. Examples of Hilbert spaces and their inner products are

$$\mathcal{L}^2(\mathcal{D}) \text{ with } (u,v)_0 := \int_{\mathcal{D}} u(x)v(x) \, dx$$

$$\mathcal{H}^k(\mathcal{D}) \text{ with } (u,v)_k := \sum_{|\alpha|\leq k} (D^\alpha u, D^\alpha v)_0$$

For further discussion of function spaces we refer, for instance, to [Ad75], [KA64], [Ha92], [Wl87].

Appendix D Complementary Material

This appendix lists useful formula without further explanation.

D1 Bounds for Options

The following bounds can be derived based on arbitrage arguments, see [Mer73], [CR85], [In87], [Kwok98], [Hull00]. If neither the subscript $_C$ nor $_P$ is listed, the inequality holds for both put and call. If neither the ^{eur} nor the ^{am} is listed, the inequality holds for both American and European options. We always assume $r > 0$.

a) Bounds valid for both American and European options, no matter whether dividends are paid or not:

$$
\begin{aligned}
0 &\leq V_C(S_t, t) \leq S_t \\
0 &\leq V_P(S_t, t) \leq K
\end{aligned}
$$

$$V^{eur}(S_t, t) \leq V^{am}(S_t, t)$$

$$S_t - K \leq V_C^{am}(S_t, t)$$

$$K - S_t \leq V_P^{am}(S_t, t)$$

$$V_P^{eur}(S_t, t) \leq Ke^{-r(T-t)}$$

b) Bounds valid provided that no dividend is paid for $0 \leq t \leq T$:

$$S_t - Ke^{-r(T-t)} \leq V_C^{eur}(S_t, t)$$

$$Ke^{-r(T-t)} - S_t \leq V_P^{eur}(S_t, t)$$

$V^{am} \geq V^{eur}$ allows to improve the lower the bound on V_C^{am} to

$$S_t - Ke^{-r(T-t)} \leq V_C^{am}(S_t, t) \,.$$

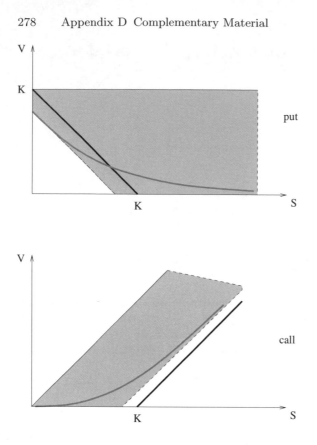

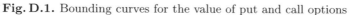

Fig. D.1. Bounding curves for the value of put and call options

c) Monotonicity:

Monotonicity with respect to S:

$$V_C(S_1, t) < V_C(S_2, t) \quad \text{for } S_1 < S_2$$
$$V_P(S_1, t) > V_P(S_2, t) \quad \text{for } S_1 < S_2,$$

which implies

$$\frac{\partial V_C}{\partial S} > 0 \ , \ \frac{\partial V_P}{\partial S} < 0 \ .$$

Monotonicity of American options with respect to time:

$$V_C^{am}(S, t_1) \geq V_C^{am}(S, t_2) \quad \text{for } t_1 < t_2$$
$$V_P^{am}(S, t_1) \geq V_P^{am}(S, t_2) \quad \text{for } t_1 < t_2,$$

which implies

$$\frac{\partial V^{am}}{\partial t} \leq 0 \ .$$

Options are convex with respect to K and with respect to S.

To express monotonicity with respect to the strike K or to the time to expiration T, we indicate dependencies by writing $V(S, t; T, K)$, where we only quote the parameter that is changed.

$$
\begin{array}{llll}
V^{\mathrm{am}}(\,.\,;T_1) & \leq & V^{\mathrm{am}}(\,.\,;T_2) & \text{for } T_1 < T_2 \\
V_{\mathrm{C}}(\,.\,;K_1) & \geq & V_{\mathrm{C}}(\,.\,;K_2) & \text{for } K_1 < K_2 \\
V_{\mathrm{P}}(\,.\,;K_1) & \leq & V_{\mathrm{P}}(\,.\,;K_2) & \text{for } K_1 < K_2
\end{array}
$$

d) Put-call parity relation for American options:

$$
Ke^{-r(T-t)} + V_{\mathrm{C}}^{\mathrm{am}}(S, t) \leq S + V_{\mathrm{P}}^{\mathrm{am}}(S, t).
$$

This holds no matter whether dividends are paid or not. If the asset pays no dividends, then also the upper bound

$$
S + V_{\mathrm{P}}^{\mathrm{am}}(S, t) - V_{\mathrm{C}}^{\mathrm{am}}(S, t) \leq K
$$

holds.

D2 Approximation Formula

Distribution Function of the Standard Normal Distribution

$$
f(x) := \frac{1}{\sqrt{2\pi}} \exp\left(-\frac{x^2}{2}\right)
$$

$$
F(x) := \int_{-\infty}^{x} f(t)\, dt
$$

Let us define

$$
z := \frac{1}{1 + 0.2316419x}
$$

and the coefficients

$$
\begin{array}{ll}
a_1 = 0.319381530 & a_4 = -1.821255978 \\
a_2 = -0.356563782 & a_5 = 1.330274429 \\
a_3 = 1.781477937.
\end{array}
$$

Then

$$
F(x) = 1 - f(x) \left(a_1 z + a_2 z^2 + a_3 z^3 + a_4 z^4 + a_5 z^5\right) + \varepsilon(x) ,
$$

for $0 \leq x < \infty$ with an absolute error ε bounded by

$$
|\varepsilon(x)| < 7.5 * 10^{-8}
$$

(see [AS68]). Hence we have the approximating formula

$$F(x) \approx 1 - f(x)z((((a_5 z + a_4)z + a_3)z + a_2)z + a_1) \ ,$$

which requires 17 arithmetic operations and the evaluation of the exponential function to obtain an accuracy of about 7 decimals. For $x < 0$ apply $F(x) = 1 - F(-x)$. Higher accuracy can be achieved with quadrature methods ($\longrightarrow$ Exercise 1.3).

Inversion Formula

A FORTRAN code for the inversion of the normal distribution can be found in
$$\texttt{http://lib.stat.cmu.edu/apstat/111} .$$
(Many other codes relevant for statistical computation can be obtained via the   .../apstat page.) Here we report the formula of [Moro95] to approximate the inverse function of the standard normal distribution

$$F(x) := \frac{1}{\sqrt{2\pi}} \int\limits_{-\infty}^{x} \exp\left(-\frac{t^2}{2}\right) dt \ .$$

That is, we calculate $x = G(u)$ such that $G(u) \approx F^{-1}(u)$. The interval $0 < u < 1$ is truncated to $10^{-12} \le u \le 1 - 10^{-12}$. Symmetry with respect to $(x, u) = (0, 0.5)$ is exploited. The interval is subdivided into two relevant parts, namely

$$0.08 < u < 0.92 \quad \text{and} \quad 0.92 \le u \le 1 - 10^{-12} \ .$$

The part $10^{-12} \le u \le 0.08$ is obtained by symmetry. For each of the two subintervals an appropriate approximation is given. In the middle part of the interval a rational approximation in the form

$$(u - 0.5)\frac{\sum\limits_{j=0}^{3} a_j (u - 0.5)^{2j}}{1 + \sum\limits_{j=0}^{3} b_j (u - 0.5)^{2j}}$$

is used, whereas the tails are approximated by a polynomial in $\log(-\log r)$, where $10^{-12} \le r \le 0.08$.

Algorithm   (inversion of the standard normal distribution)

input: u, drawn from $\mathcal{U}(0,1)$

$y := u - 0.5$

in case $|y| < 0.42$:

$r := y^2$

$x := y \dfrac{((a_3 r + a_2)r + a_1)r + a_0}{(((b_3 r + b_2)r + b_1)r + b_0)r + 1}$

in case $|y| \geq 0.42$:

$r := u$, in case $y > 0$ set $r := 1 - u$

$r := \log(-\log r)$

$x := c_0 + r(c_1 + r(c_2 + r(c_3 + r(c_4 + r(c_5 + r(c_6 + r(c_7 + rc_8)))))))$

in case $y < 0$ set $x := -x$

output: x

The coefficients of the above algorithm are given by[3]

$a_0 = 2.50662823884,$
$a_1 = -18.61500062529,$
$a_2 = 41.39119773534,$
$a_3 = -25.44106049637$

$b_0 = -8.47351093090,$
$b_1 = 23.08336743743,$
$b_2 = -21.06224101826,$
$b_3 = 3.13082909833$

$c_0 = 0.3374754822726147,$
$c_1 = 0.9761690190917186,$
$c_2 = 0.1607979714918209,$
$c_3 = 0.0276438810333863,$
$c_4 = 0.0038405729373609,$
$c_5 = 0.0003951896511919,$
$c_6 = 0.0000321767881768,$
$c_7 = 0.0000002888167364,$
$c_8 = 0.0000003960315187$

The rational approximation formula for $|y| < 0.42$ (that is, $0.08 < u < 0.92$) is reported to have a largest absolute error of $3 \cdot 10^{-9}$.

D3 Software

A dedicated computer person will program the mathematics such that the resulting codes run with utmost possible speed. Such a person will probably use compilers like C, C++, or FORTRAN to create production codes, where the speed counts.

---

[3] These digits are listed in [Moro95].

But there are packages available that make programming, implementing, testing, and graphics more comfortable. For example, MATLAB offers a platform for scientific computation and numerical experiments. This software package is widely used to develop algorithms, and is assisted by good graphics.

Programs related to finance have been published in the literature. For MATLAB codes see [Hig04], for MATHEMATICA codes see [Sto03]. For elementary computations, spreadsheets are also used.

For partial differential equations, the finite-element program PDE2D is recommended. This package is available via the University of Texas, El Paso. The PREMIA project offers codes via `www-rocq.inria.fr/mathfi` . For further hints and test algorithms see the platform `www.compfin.de` .

References

[AS68]   M. Abramowitz, I. Stegun: Handbook of Mathematical Functions. With Formulas, Graphs, and Mathematical Tables. Dover Publications, New York (1968).

[Ad75]   R.A. Adams: Sobolev Spaces. Academic Press, New York (1975).

[AiC97]   F. AitSahlia, P. Carr: American options: A comparison of numerical methods. In [RT97] (1997) p. 67-87.

[AnA00]   L. Andersen, J. Andreasen: Jump diffusion process: Volatility smile fitting and numerical methods for option pricing. Review Derivatives Research 4 (2000) 231-262.

[AB97]   L.B.G. Andersen, R. Brotherton-Ratcliffe: The equity option volatility smile: an implicit finite-difference approach. J. Computational Finance 1,2 (1997/1998) 5–38.

[AnéG00]   T. Ané, H. Geman: Order flow, transaction clock, and normality of asset returns. J. of Finance 55 (2000) 2259-2284.

[Ar74]   L. Arnold: Stochastic Differential Equations (Theory and Applications). Wiley, New York (1974).

[ArDEH99] P. Artzner, F. Delbaen, J.-M. Eber, D. Heath: Coherent measures of risk. Math. Finance 9 (1999) 203-228.

[BaS01]   I. Babuška, T. Strouboulis: The Finite Element Method and its Reliability. Oxford Science Publications, Oxford (2001).

[BaR94]   C.A. Ball, A. Roma: Stochastic volatility option pricing. J. Financial Quantitative Analysis 29 (1994) 589-607.

[Bar97]   G. Barles: Convergence of numerical schemes for degenerate parabolic equations arising in finance theory. in [RT97] (1997) 1-21.

[BaN97]   O.E. Barndorff-Nielsen: Processes of normal inverse Gaussian type. Finance & Stochastics 2 (1997) 41–68.

[BaW87]   G. Barone-Adesi, R.E. Whaley: Efficient analytic approximation of American option values. J. Finance 42 (1987) 301-320.

[BaP96]   J. Barraquand, T. Pudet: Pricing of American path-dependent contingent claims. Mathematical Finance 6 (1996) 17–51.

[Ba94]   R. Barrett et al.: Templates for the Solution of Linear Systems: Building Blocks for Iterative Methods. SIAM, Philadelphia (1994).

[BaR96]   M. Baxter, A. Rennie: Financial Calculus. An Introduction to Derivative Pricing. Cambridge University Press, Cambridge (1996).

[Beh00]   E. Behrends: Introduction to Markov Chains. Vieweg, Braunschweig (2000).

[Ben84]   A. Bensoussan: On the theory of option pricing. Acta Applicandae Math. 2 (1984) 139-158.

[Bi79]   P. Billingsley: Probability and Measure. John Wiley, New York (1979).

[BV00]   G.I. Bischi, V. Valori: Nonlinear effects in a discrete-time dynamic
           model of a stock market. Chaos, Solitons and Fractals **11** (2000) 2103-
           2121.

[BS73]   F. Black, M. Scholes: The pricing of options and corporate liabilities.
           J. Political Economy **81** (1973) 637–659.

[Blo86]   E.C. Blomeyer: An analytic approximation for the American put price
           for options with dividends. J. Financial Quantitative Analysis **21**
           (1986) 229-233.

[BP00]   J.-P. Bouchaud, M. Potters: Theory of Financial Risks. From Statisti-
           cal Physics to Risk Management. Cambridge Univ. Press, Cambridge
           (2000).

[Bo98]   N. Bouleau: Martingales et Marchés Financiers. Edition Odile Jacob
           (1998).

[BoM58]   G.E.P. Box, M.E. Muller: A note on the generation of random normal
           deviates. Annals Math.Statistics **29** (1958) 610-611.

[BBG97]   P. Boyle, M. Broadie, P. Glasserman: Monte Carlo methods for security
           pricing. J. Economic Dynamics and Control **21** (1997) 1267-1321.

[BTT00]   M.-E. Brachet, E. Taflin, J.M. Tcheou: Scaling transformation and
           probability distributions for time series. Chaos, Solitons and Fractals
           11 (2000) 2343-2348.

[Br91]   R. Breen: The accelerated binomial option pricing model. J. Financial
           and Quantitative Analysis **26** (1991) 153–164.

[BrS77]   M. Brennan, E. Schwartz: The valuation of American put options. J.
           of Finance **32** (1977) 449-462.

[BrS02]   S.C. Brenner, L.R. Scott: The Mathematical Theory of Finite Element
           Methods. Second Edition. Springer, New York (2002).

[Br94]   R.P. Brent: On the periods of generalized Fibonacci recurrences. Math.
           Comput. **63** (1994) 389–401.

[BrD97]   M. Broadie, J. Detemple: Recent advances in numerical methods for
           pricing derivative securities. in [RT97] (1997) 43-66.

[BrG97]   M. Broadie, P. Glasserman: Pricing American-style securities using
           simulation. J. Economic Dynamics and Control **21** (1997) 1323-1352.

[BrG04]   M. Broadie, P. Glasserman: A stochastic mesh method for pricing high-
           dimensional American options. J. Computational Finance **7**,4 (2004)
           35-72.

[BH98]   W.A. Brock, C.H. Hommes: Heterogeneous beliefs and routes to chaos
           in a simple asset pricing model. J. Economic Dynamics and Control
           22 (1998) 1235–1274.

[BuJ92]   D.S. Bunch, H. Johnson: A simple and numerically efficient valuation
           method for American puts using a modified Geske-Johnson approach.
           J. Finance **47** (1992) 809-816.

[CaMO97]   R.E. Caflisch, W. Morokoff, A. Owen: Valuation of mortgaged-backed
           securities using Brownian bridges to reduce effective dimension. J.
           Computational Finance **1**,1 (1997) 27-46.

[CaF95]   P. Carr, D. Faguet: Fast accurate valuation of American options. Wor-
           king paper, Cornell University (1995).

[CaM99]   P. Carr, D.B. Madan: Option valuation using the fast Fourier trans-
           form. J. Computational Finance **2**,4 (1999) 61-73.

[Cash84]   J.R. Cash: Two new finite difference schemes for parabolic equations.
           SIAM J.Numer.Anal. **21** (1984) 433-446.

[CDG00]   C. Chiarella, R. Dieci, L. Gardini: Speculative behaviour and complex
           asset price dynamics. Proceedings Urbino 2000, Ed.: G.I. Bischi (2000).

[CW83]   K.L. Chung, R.J. Williams: Introduction to Stochastic Integration. Birkhäuser, Boston (1983).
[Ci91]   P.G. Ciarlet: Basic Error Estimates for Elliptic Problems. in: Handbook of Numerical Analysis, Vol. II (Eds. P.G. Ciarlet, J.L. Lions) Elsevier/North-Holland, Amsterdam (1991) 19–351.
[CL90]   P. Ciarlet, J.L. Lions: Finite Difference Methods (Part 1) Solution of equations in $\mathbb{R}^n$. North-Holland Elsevier, Amsterdam (1990).
[CoLV02]   T.F. Coleman, Y. Li, Y. Verma: A Newton method for American option pricing. J. Computational Finance 5,3 (2002) 51-78.
[ConT04]   R. Cont, P. Tankov: Financial Modelling with Jump Processes. Chapman & Hall, Boca Raton (2004).
[CRR79]   J.C. Cox, S. Ross, M. Rubinstein: Option pricing: A simplified approach. Journal of Financial Economics 7 (1979) 229–264.
[CR85]   J.C. Cox, M. Rubinstein: Options Markets. Prentice Hall, Englewood Cliffs (1985).
[Cra84]   J. Crank: Free and Moving Boundary Problems. Clarendon Press, Oxford (1984).
[CN47]   J.C. Crank, P. Nicolson: A practical method for numerical evaluation of solutions of partial differential equations of the heat-conductive type. Proc. Cambr. Phil. Soc. 43 (1947) 50–67.
[Cr71]   C. Cryer: The solution of a quadratic programming problem using systematic overrelaxation. SIAM J. Control 9 (1971) 385–392.
[CKO01]   S. Cyganowski, P. Kloeden, J. Ombach: From Elementary Probability to Stochastic Differential Equations with MAPLE. Springer (2001).
[Dai00]   M. Dai: A closed-form solution for perpetual American floating strike lookback options. J. Computational Finance 4,2 (2000) 63-68.
[DaJ03]   R.-A. Dana, M. Jeanblanc: Financial Markets in Continuous Time. Springer, Berlin (2003).
[DH99]   M.A.H. Dempster, J.P. Hutton: Pricing American stock options by linear programming. Mathematical Finance 9 (1999) 229–254.
[DeHR98]   M.A.H. Dempster, J.P. Hutton, D.G. Richards: LP valuation of exotic American options exploiting structure. J. Computational Finance 2,1 (1998) 61-84.
[Deu02]   H.-P. Deutsch: Derivatives and Internal Models. Palgrave, Houndmills (2002).
[Dev86]   L. Devroye: Non–Uniform Random Variate Generation. Springer, New York (1986).
[DBG01]   R. Dieci, G.-I. Bischi, L. Gardini: From bi-stability to chaotic oscillations in a macroeconomic model. Chaos, Solitons and Fractals 12 (2001) 805-822.
[Doob53]   J.L. Doob: Stochastic Processes. John Wiley, New York (1953).
[Dowd98]   K. Dowd: Beyond Value at Risk: The New Science of Risk Management. Wiley & Sons, Chichester (1998).
[Du96]   D. Duffie: Dynamic Asset Pricing Theory. Second Edition. Princeton University Press, Princeton (1996).
[EK95]   E. Eberlein, U. Keller: Hyperbolic distributions in finance. Bernoulli 1 (1995) 281–299.
[EO82]   C.M. Elliott, J.R. Ockendon: Weak and Variational Methods for Moving Boundary Problems. Pitman, Boston (1982).
[ElK99]   R.J. Elliott, P.E. Kopp: Mathematics of Financial Markets. Springer, New York (1999).
[EKM97]   P. Embrechts, C. Klüppelberg, T. Mikosch: Modelling Extremal Events. Springer, Berlin (1997).

286   References

[Epps00]   T.W. Epps: Pricing Derivative Securities. World Scientific, Singapore (2000).
[Fe50]   W. Feller: An Introduction to Probability Theory and its Applications. Wiley, New York (1950).
[Fi96]   G.S. Fishman: Monte Carlo. Concepts, Algorithms, and Applications. Springer, New York (1996).
[Fisz63]   M. Fisz: Probability Theory and Mathematical Statistics. John Wiley, New York (1963).
[FöS02]   H. Föllmer, A. Schied: Stochastic Finance: An Introduction to Discrete Time. de Gruyter, Berlin (2002).
[FV02]   P.A. Forsyth, K.R. Vetzal: Quadratic convergence of a penalty method for valuing American options. SIAM J. Sci. Comp. **23** (2002) 2095-2122.
[FVZ99]   P.A. Forsyth, K.R. Vetzal, R. Zvan: A finite element approach to the pricing of discrete lookbacks with stochastic volatility. Applied Math. Finance **6** (1999) 87–106.
[FoLLLT99] E. Fournié, J.-M. Lasry, J. Lebuchoux, P.-L. Lions, N. Touzi: An application of Malliavin calculus to Monte Carlo methods in finance. Finance & Stochastics **3** (1999) 391-412.
[FHH04]   J. Franke, W. Härdle, C.M. Hafner: Statistics of Financial Markets. Springer, Berlin (2004).
[Fr71]   D. Freedman: Brownian Motion and Diffusion. Holden Day, San Francisco (1971).
[Fu01]   M.C. Fu (et al): Pricing American options: a comparison of Monte Carlo simulation approaches. J. Computational Finance **4**,3 (2001) 39-88.
[FuST02]   G. Fusai, S. Sanfelici, A. Tagliani: Practical problems in the numerical solution of PDEs in finance. Rend. Studi Econ. Quant. 2001 (2002) 105-132.
[Gem00]   H. Geman et al. (Eds.): Mathematical Finance. Bachelier Congress 2000. Springer, Berlin (2002).
[Ge98]   J.E. Gentle: Random Number Generation and Monte Carlo Methods. Springer, New York (1998)
[GeJ84]   R. Geske, H.E. Johnson: The American put option valued analytically. J.Finance **39** (1984) 1511-1524.
[GeG98]   T. Gerstner, M. Griebel: Numerical integration using sparse grids. Numer. Algorithms **18** (1998) 209-232.
[GeG03]   T. Gerstner, M. Griebel: Dimension-adaptive tensor-product quadrature. Computing **70**,4 (2003).
[GiRS96]   W.R. Gilks, S. Richardson, D.J. Spiegelhalter (Eds.): Markov Chain Monte Carlo in Practice. Chapman & Hall, Boca Raton (1996).
[Gla04]   P. Glasserman: Monte Carlo Methods in Financial Engineering. Springer, New York (2004).
[GV96]   G. H. Golub, C. F. Van Loan: Matrix Computations. Third Edition. The John Hopkins University Press, Baltimore (1996).
[GK01]   L. Grüne, P.E. Kloeden: Pathwise approximation of random ODEs. BIT **41** (2001) 710-721.
[Ha85]   W. Hackbusch: Multi-Grid Methods and Applications. Springer, Berlin (1985).
[Ha92]   W. Hackbusch: Elliptic Differential Equations: Theory and Numerical Treatment. Springer Series in Computational Mathematics **18**, Berlin, Springer (1992).

[Häg02]   O. Häggström: Finite Markov Chains and Algorithmic Applications. Cambridge University Press, Cambridge (2002).

[HNW93]   E. Hairer, S.P. Nørsett, G. Wanner: Solving Ordinary Differential Equations I. Nonstiff Problems. Springer, Berlin (1993).

[Ha60]   J.H. Halton: On the efficiency of certain quasi-random sequences of points in evaluating multi-dimensional integrals. Numer. Math. 2 (1960) 84–90.

[HH64]   J.M. Hammersley, D.C. Handscomb: Monte Carlo Methods. Methuen, London (1964).

[HH91]   G. Hämmerlin, K.-H. Hoffmann: Numerical Mathematics. Springer, Berlin (1991).

[HP81]   J.M. Harrison, S.R. Pliska: Martingales and stochastic integrals in the theory of continuous trading. Stoch. Processes and their Applications 11 (1981) 215-260.

[Hei05]   P. Heider: A condition number for the integral representation of American options. Report Mathem. Inst., Univ. Köln (2005).

[Hig01]   D.J. Higham: An algorithmic introduction to numerical solution of stochastic differential equations. SIAM Review 43 (2001) 525-546.

[Hig04]   D.J. Higham: An Introduction to Financial Option Valuation. Cambridge, Univ. Press, Cambridge (2004).

[HiMS04]   N. Hilber, A.-M. Matache, C. Schwab: Sparse Wavelet Methods for Option Pricing under Stochastic Volatility. Report, ETH-Zürch (2004).

[HPS92]   N. Hofmann, E. Platen, M. Schweizer: Option pricing under incompleteness and stochastic volatility. Mathem. Finance 2 (1992) 153–187.

[HoP02]   P. Honoré, R. Poulsen: Option pricing with EXCEL. in [Nie02].

[Hull00]   J.C. Hull: Options, Futures, and Other Derivatives. Fourth Edition. Prentice Hall International Editions, Upper Saddle River (2000).

[HuK00]   P.J. Hunt, J.E. Kennedy: Financial Derivatives in Theory and Practice. Wiley, Chichester (2000).

[IkT04]   S. Ikonen, J. Toivanen: Pricing American options using LU decomposition. Report Univ. Jyvaskyla (2004).

[In87]   J.E. Ingersoll: Theory of Financial Decision Making. Rowmann and Littlefield, Savage (1987).

[Int05]   R. Int-Veen: Avoiding numerical dispersion in option valuation. Computing and Visualization in Science. to appear (2005).

[Irle98]   A. Irle: Finanzmathematik — Die Bewertung von Derivaten. Teubner, Stuttgart 1998.

[IK66]   E. Isaacson, H.B. Keller: Analysis of Numerical Methods. John Wiley, New York (1966).

[JaP03]   J. Jacod, P. Protter: Probability Essentials. Second Edition. Springer, Berlin (2003).

[Jam92]   F. Jamshidian: An analysis of American options. Review of Futures Markets 11 (1992) 72-80.

[JiD04]   L. Jiang, M. Dai: Convergence of binomial tree method for European/American path-dependent options. SIAM J. Numerical Analysis 42 (2004) 1094-1109.

[Joh83]   H.E. Johnson: An analytic approximation for the American put price. J. Financial Quantitative Analysis 18 (1983) 141-148.

[KMN89]   D. Kahaner, C. Moler, S. Nash: Numerical Methods and Software. Prentice Hall Series in Computational Mathematics, Englewood Cliffs (1989).

[KaN00]   R. Kangro, R. Nicolaides: Far field boundary conditions for Black-Scholes equations. SIAM J.Numer.Anal. 38 (2000) 1357-1368.

[KA64]   L.W. Kantorovich, G.P. Akilov: Functional Analysis in Normed
           Spaces. Pergamon Press, Elmsford (1964).
[KS91]   I. Karatzas, S.E. Shreve: Brownian Motion and Stochastic Calculus.
           Second Edition. Springer Graduate Texts, New York (1991).
[KS98]   I. Karatzas, S.E. Shreve: Methods of Mathematical Finance. Springer,
           New York (1998).
[Kat95]   H.M. Kat: Pricing Lookback options using binomial trees: An evalua-
           tion. J. Financial Engineering 4 (1995) 375–397.
[Kl01]   T.R. Klassen: Simple, fast and flexible pricing of Asian options. J.
           Computational Finance 4,3 (2001) 89-124.
[KP92]   P.E. Kloeden, D. Platen: Numerical Solution of Stochastic Differential
           Equations. Springer, Berlin (1992).
[KPS94]   P.E. Kloeden, E. Platen, H. Schurz: Numerical Solution of SDE
           Through Computer Experiments. Springer, Berlin (1994).
[Kn95]   D. Knuth: The Art of Computer Programming, Vol 2. Addison–Wiley,
           Reading (1995).
[Korn01]   R. Korn, E. Korn: Option Pricing and Portfolio Optimization. Ame-
           rican Mathem.Soc., Providence 2001.
[Kr97]   D. Kröner: Numerical Schemes for Conservation Laws. Wiley Teubner,
           Chichester (1997).
[Kwok98]   Y.K. Kwok: Mathematical Models of Financial Derivatives. Springer,
           Singapore (1998).
[La91]   J.D. Lambert: Numerical Methods for Ordinary Differential Systems.
           The Initial Value Problem. John Wiley, Chichester (1991).
[LL96]   D. Lamberton, B. Lapeyre: Introduction to Stochastic Calculus App-
           lied to Finance. Chapman & Hall, London (1996).
[La99]   K. Lange: Numerical Analysis for Statisticians. Springer, New York
           (1999).
[LEc99]   P. L'Ecuyer: Tables of linear congruential generators of different sizes
           and good lattice structure. Mathematics of Computation 68 (1999)
           249-260.
[Lehn02]   J. Lehn: Random Number Generators. GAMM-Mitteilungen 25 (2002)
           35-45.
[LonS01]   F.A. Longstaff, E.S. Schwartz: Valuing American options by simula-
           tion: a simple least-squares approach. Review Financial Studies 14
           (2001) 113-147.
[Los01]   C.A. Los: Computational Finance: A Scientific Perspective. World
           Scientific, Singapore (2001).
[Lux98]   T. Lux: The socio-economic dynamics of speculative markets: interac-
           ting agents, chaos, and the fat tails of return distributions. J. Economic
           Behavior & Organization 33 (1998) 143-165.
[Lyuu02]   Y.-D. Lyuu: Financial Engineering and Computation. Principles, Ma-
           thematics, Algorithms. Cambridge University Press, Cambridge (2002).
[MaM86]   L.W. MacMillan: Analytic approximation for the American put option.
           Advances in Futures and Options Research 1 (1986) 119-139.
[MRGS00]   R. Mainardi, M. Roberto, R. Gorenflo, E. Scalas: Fractional calculus
           and continuous-time finance II: the waiting-time distribution. Physica
           A 287 (2000) 468-481.
[Man99]   B.B. Mandelbrot: A multifractal walk down Wall Street. Scientific
           American, Febr. 1999, 50–53.
[MCFR00]   M. Marchesi, S. Cinotti, S. Focardi, M. Raberto: Development and
           testing of an artificial stock market. Proceedings Urbino 2000, Ed.
           I.-G. Bischi (2000).

[Mar78]   W. Margrabe: The value of an option to exchange one asset for another. J. Finance **33** (1978) 177-186.
[Ma68]   G. Marsaglia: Random numbers fall mainly in the planes. Proc. Nat. Acad. Sci. USA **61** (1968) 23–28.
[MaB64]   G. Marsaglia, T.A. Bray: A convenient method for generating normal variables. SIAM Rev. **6** (1964) 260-264.
[Mas99]   M. Mascagni: Parallel pseudorandom number generation. SIAM News **32**, 5 (1999).
[MaPS02] A.-M. Matache, T. von Petersdorff, C. Schwab: Fast deterministic pricing of options on Lévy driven assets. Report 2002-11, Seminar for Applied Mathematics, ETH Zürich (2002).
[MaN98]   M. Matsumoto, T. Nishimura: Mersenne Twister: A 623-dimensionally equidistributed uniform pseudorandom number generator. ACM Transactions on Modeling and Computer Simulations **8** (1998) 3-30.
[Mayo00] A. Mayo: Fourth order accurate implicit finite difference method for evaluating American options. Proceedings of Computational Finance, London (2000).
[McW01]   L.A. McCarthy, N.J. Webber: Pricing in three-factor models using icosahedral lattices. J. Computational Finance **5**,2 (2001/02) 1-33.
[MeVN02] A.V. Mel'nikov, S.N. Volkov, M.L. Nechaev: Mathematics of Financial Obligations. Amer. Math. Soc., Providence (2002).
[Mer73]   R.C. Merton: Theory of rational option pricing. Bell J. Economics and Management Science **4** (1973) 141-183.
[Mer76]   R. Merton: Option pricing when underlying stock returns are discontinous. J. Financial Economics **3** (1976) 125-144.
[Me90]   R.C. Merton: Continuous-Time Finance. Blackwell, Cambridge (1990).
[Mik98]   T. Mikosch: Elementary Stochastic Calculus, with Finance in View. World Scientific, Singapore (1998).
[Mi74]   G.N. Mil'shtein: Approximate integration of stochastic differential equations. Theory Prob. Appl. **19** (1974) 557–562.
[Moe76]   P. van Moerbeke: On optimal stopping and free boundary problems. Archive Rat.Mech.Anal. **60** (1976) 101-148.
[Moro95] B. Moro: The full Monte. Risk **8** (1995) 57–58.
[MC94]   W.J. Morokoff, R.E. Caflisch: Quasi-random sequences and their discrepancies. SIAM J. Sci. Comput. **15** (1994) 1251–1279.
[Mo98]   W.J. Morokoff: Generating quasi-random paths for stochastic processes. SIAM Review **40** (1998) 765–788.
[Mo96]   K.W. Morton: Numerical Solution of Convection-Diffusion Problems. Chapman & Hall, London (1996).
[MR97]   M. Musiela, M. Rutkowski: Martingale Methods in Financial Modelling. (Second Edition 2005) Springer, Berlin (1997).
[Ne96]   S.N. Neftci: An Introduction to the Mathematics of Financial Dervatives. Academic Press, San Diego (1996).
[New97]   N.J. Newton: Continuous-time Monte Carlo methods and variance reduction. in [RT97] (1997) 22-42.
[Ni78]   H. Niederreiter: Quasi-Monte Carlo methods and pseudo-random numbers. Bull. Am. Math. Soc. **84** (1978) 957–1041.
[Ni92]   H. Niederreiter: Random Number Generation and Quasi–Monte Carlo Methods. Society for Industrial and Applied Mathematics, Philadelphia (1992).

[Ni95]   H. Niederreiter, P. Jau–Shyong Shiue (Eds.): Monte Carlo and Quasi–
            Monte Carlo Methods in Scientific Computing. Proceedings of a Con-
            ference at the University of Nevada, Las Vegas, Nevada, USA, June
            23–25, 1994. Springer, New York (1995).

[NiST02]   B.F. Nielsen, O. Skavhaug, A. Tveito: Penalty and front-fixing me-
            thods for the numrical solution of American option problems. J. Com-
            putational Finance **5**,4 (2002) 69-97.

[Nie99]   L.T. Nielsen: Pricing and Hedging of Derivative Securities. Oxford
            University Press, Oxford (1999).

[Nie02]   S. Nielsen (Ed.): Programming Languages and Systems in Computa-
            tional Economics and Finance. Kluwer, Amsterdam (2002).

[Øk98]   B. Øksendal: Stochastic Differential Equations. Springer, Berlin (1998).

[Oo03]   C.W. Oosterlee: On multigrid for linear complementarity problems
            with application to American-style options. Electronic Transactions
            on Numerical Analysis **15** (2003) 165-185.

[PT96]   A. Papageorgiou, J.F. Traub: New results on deterministic pricing of fi-
            nancial derivatives. Columbia University Report CUCS-028-96 (1996).

[PT95]   S. Paskov, J. Traub: Faster valuation of financial derivatives. J. Port-
            folio Management **22** (1995) 113–120.

[PT83]   R. Peyret, T.D. Taylor: Computational Methods for Fluid Flow. Sprin-
            ger, New York (1983).

[Pl99]   E. Platen: An introduction to numerical methods for stochastic diffe-
            rential equations. Acta Numerica (1999) 197-246.

[Pli97]   S.R. Pliska: Introduction to Mathematical Finance. Discrete Time Mo-
            dels. Blackwell, Malden (1997).

[PFVS00]   D. Pooley, P.A. Forsyth, K. Vetzal, R.B. Simpson: Unstructured mes-
            hing for two asset barrier options. Appl. Mathematical Finance **7**
            (2000) 33-60.

[PTVF92]   W.H. Press, S.A. Teukolsky, W.T. Vetterling, B.P. Flannery: Numeri-
            cal Recipes in FORTRAN. The Art of Scientific Computing. Second
            Edition. Cambridge University Press, Cambridge (1992).

[Pro04]   P.E. Protter: Stochastic Integration and Differential Equations. Sprin-
            ger, Berlin (2004).

[QSS00]   A. Quarteroni, R. Sacco, F. Saleri: Numerical Mathematics. Springer,
            New York (2000).

[Ran84]   R. Rannacher: Finite element solution of diffusion problems with irre-
            gular data. Numer. Math. **43** (1984) 309-327.

[Re96]   R. Rebonato: Interest-Rate Option Models: Understanding, Analysing
            and Using Models for Exotic Interest-Rate Options. John Wiley &
            Sons, Chichester (1996).

[RY91]   D. Revuz, M. Yor: Continuous Martingales and Brownian Motion.
            Springer, Berlin 1991.

[RiW02]   C. Ribeiro, N. Webber: Valuing path dependent options in the Variance-
            Gamma model by Monte Carlo with a gamma bridge. Working paper,
            City University, London (2002).

[RiW03]   C. Ribeiro, N. Webber: A Monte Carlo method for the normal inverse
            Gaussian option valuation model using an inverse Gaussian bridge. J.
            Computational Finance **7**,2 (2003/04) 81-100.

[Ri87]   B.D. Ripley: Stochastic Simulation. Wiley Series in Probability and
            Mathematical Statistics, New York (1987).

[Ro00]   L.C.G. Rogers: Monte Carlo valuation of American options. Manus-
            cript, University of Bath (2000).

[RT97]   L.C.G. Rogers, D. Talay (Eds.): Numerical Methods in Finance. Cambridge University Press, Cambridge (1997).
[Ru94]   M. Rubinstein: Implied binomial trees. J. Finance **69** (1994) 771-818.
[Ru81]   R.Y. Rubinstein: Simulation and the Monte Carlo Method. Wiley, New York (1981).
[Rup04]   D. Ruppert: Statistics and Finance. An Introduction. Springer, New York (2004).
[SM96]   Y. Saito, T. Mitsui: Stability analysis of numerical schemes for stochastic differential equations. SIAM J. Numer. Anal. **33** (1996) 2254–2267.
[Sa01]   K. Sandmann: Einführung in die Stochastik der Finanzmärkte. Second edition. Springer, Berlin (2001).
[Sato99]   K.-I. Sato: Lévy Processes and Infinitely Divisible Distributions. Cambridge University Press, Cambridge (1999).
[SH97]   J.G.M. Schoenmakers, A.W. Heemink: Fast Valuation of Financial Derivatives. J. Comp. Finance **1** (1997) 47–62.
[Sc80]   Z. Schuss: Theory and Applications of Stochastic Differential Equations. Wiley Series in Probability and Mathematical Statistics, New York (1980).
[Sc89]   H.R. Schwarz: Numerical Analysis. John Wiley & Sons, Chichester (1989).
[Sc91]   H.R. Schwarz: Methode der finiten Elemente. Teubner, Stuttgart (1991).
[Se94]   R. Seydel: Practical Bifurcation and Stability Analysis. From Equilibrium to Chaos. Second Edition. Springer Interdisciplinary Applied Mathematics Vol. 5, New York (1994).
[Shi99]   A.N. Shiryaev: Essentials of Stochastic Finance. Facts, Models, Theory. World Scientific, Singapore (1999).
[Shr04]   S.E. Shreve: Stochastic Calculus for Finance. Springer, New York (2004).
[Sm78]   G.D. Smith: Numerical Solution of Partial Differential Equations: Finite Difference Methods. Second Edition. Clarendon Press, Oxford (1978).
[Smi97]   C. Smithson: Multifactor options. Risk **10**,5 (1997) 43-45.
[SM94]   J. Spanier, E.H. Maize: Quasi-random methods for estimating integrals using relatively small samples. SIAM Review **36** (1994) 18–44.
[Sta01]   D. Stauffer: Percolation models of financial market dynamics. Advances in Complex Systems **4** (2001) 19–27.
[Ste01]   J.M. Steele: Stochastic Calculus and Financial Applications. Springer, New York (2001).
[SWH99]   M. Steiner, M. Wallmeier, R. Hafner: Baumverfahren zur Bewertung diskreter Knock-Out-Optionen. OR Spektrum **21** (1999) 147–181.
[SB96]   J. Stoer, R. Bulirsch: Introduction to Numerical Analysis. Springer, Berlin (1996).
[SW70]   J. Stoer, C. Witzgall: Convexity and Optimization in Finite Dimensions I. Springer, Berlin (1970).
[Sto03]   S. Stojanovic: Computational Financial Mathematics using MATHEMATICA. Birkhäuser, Boston (2003).
[St86]   G. Strang: Introduction to Applied Mathematics. Wellesley, Cambridge (1986).
[SF73]   G. Strang, G. Fix: An Analysis of the Finite Element Method. Prentice-Hall, Englewood Cliffs (1973).
[Sw84]   P.K. Sweby: High resolution schemes using flux limiters for hyperbolic conservation laws. SIAM J. Numer. Anal. **21** (1984) 995–1011.

[TR00]   D. Tavella, C. Randall: Pricing Financial Instruments. The Finite Difference Method. John Wiley, New York (2000).

[Te95]   S. Tezuka: Uniform Random Numbers: Theory and Practice. Kluwer Academic Publishers, Dordrecht (1995).

[Th95]   J.W. Thomas: Numerical Partial Differential Equations: Finite Difference Methods. Springer, New York (1995).

[Th99]   J.W. Thomas: Numerical Partial Differential Equations. Conservation Laws and Elliptic Equations. Springer, New York (1999).

[Top00]   J. Topper: Finite element modeling of exotic options. OR Proceedings (2000) 336–341.

[Top05]   J. Topper: Financial Engineering with Finite Elements. Wiley (2005).

[TW92]   J.F. Traub, H. Wozniakowski: The Monte Carlo algorithm with a pseudo-random generator. Math. Computation **58** (1992) 323–339.

[TOS01]   U. Trottenberg, C. Oosterlee, A. Schüller: Multigrid. Academic Press, San Diego (2001).

[Tsay02]   R.S. Tsay: Analysis of Financial Time Series. Wiley, New York (2002).

[Va62]   R.S. Varga: Matrix Iterative Analysis. Prentice Hall, Englewood Cliffs (1962).

[Vi81]   R. Vichnevetsky: Computer Methods for Partial Differential Equations. Volume I. Prentice-Hall, Englewood Cliffs (1981).

[We01]   P. Wesseling: Principles of Computational Fluid Dynamics. Springer, Berlin (2001).

[Wi98]   P. Wilmott: Derivatives. John Wiley, Chichester (1998).

[WDH96]   P. Wilmott, J. Dewynne, S. Howison: Option Pricing. Mathematical Models and Computation. Oxford Financial Press, Oxford (1996).

[Wl87]   J. Wloka: Partial Differential Equations. Cambridge University Press, Cambridge (1987).

[Zag02]   R. Zagst: Interest-Rate Management. Springer, Berlin (2002).

[Zi77]   O.C. Zienkiewicz: The Finite Element Method in Engineering Science. McGraw-Hill, London (1977).

[ZhWC04]   Y.-I. Zhu, X. Wu, I.-L. Chern: Derivative Securities and Difference Methods. Springer (2004).

[ZFV98]   R. Zvan, P.A. Forsyth, K.R. Vetzal: Robust numerical methods for PDE models of Asian options. J. Computational Finance **1**,2 (1997/98) 39–78.

[ZvFV98a]   R. Zvan, P.A. Forsyth, K.R. Vetzal: Penalty methods for American options with stochastic volatility. J. Comp. Appl. Math. **91** (1998) 199-218.

[ZFV99]   R. Zvan, P.A. Forsyth, K.R. Vetzal: Discrete Asian barrier options. J. Computational Finance **3**,1 (1999) 41–67.

[ZVF00]   R. Zvan, K.R. Vetzal, P.A. Forsyth: PDE methods for pricing barrier options. J. Econ. Dynamics & Control **24** (2000) 1563-1590.

Index

Universitext

Aguilar, M.; Gitler, S.; Prieto, C.: Algebraic Topology from a Homotopical Viewpoint

Aksoy, A.; Khamsi, M. A.: Methods in Fixed Point Theory

Alevras, D.; Padberg M. W.: Linear Optimization and Extensions

Andersson, M.: Topics in Complex Analysis

Aoki, M.: State Space Modeling of Time Series

Arnold, V. I.: Lectures on Partial Differential Equations

Audin, M.: Geometry

Aupetit, B.: A Primer on Spectral Theory

Bachem, A.; Kern, W.: Linear Programming Duality

Bachmann, G.; Narici, L.; Beckenstein, E.: Fourier and Wavelet Analysis

Badescu, L.: Algebraic Surfaces

Balakrishnan, R.; Ranganathan, K.: A Textbook of Graph Theory

Balser, W.: Formal Power Series and Linear Systems of Meromorphic Ordinary Differential Equations

Bapat, R.B.: Linear Algebra and Linear Models

Benedetti, R.; Petronio, C.: Lectures on Hyperbolic Geometry

Benth, F. E.: Option Theory with Stochastic Analysis

Berberian, S. K.: Fundamentals of Real Analysis

Berger, M.: Geometry I, and II

Bliedtner, J.; Hansen, W.: Potential Theory

Blowey, J. F.; Coleman, J. P.; Craig, A. W. (Eds.): Theory and Numerics of Differential Equations

Blowey, J.; Craig, A.: Frontiers in Numerical Analysis. Durham 2004

Blyth, T. S.: Lattices and Ordered Algebraic Structures

Börger, E.; Grädel, E.; Gurevich, Y.: The Classical Decision Problem

Böttcher, A; Silbermann, B.: Introduction to Large Truncated Toeplitz Matrices

Boltyanski, V.; Martini, H.; Soltan, P. S.: Excursions into Combinatorial Geometry

Boltyanskii, V. G.; Efremovich, V. A.: Intuitive Combinatorial Topology

Bonnans, J. F.; Gilbert, J. C.; Lemaréchal, C.; Sagastizábal, C. A.: Numerical Optimization

Booss, B.; Bleecker, D. D.: Topology and Analysis

Borkar, V. S.: Probability Theory

Brunt B. van: The Calculus of Variations

Bühlmann, H.; Gisler, A.: A Course in Credibility Theory and its Applications

Carleson, L.; Gamelin, T. W.: Complex Dynamics

Cecil, T. E.: Lie Sphere Geometry: With Applications of Submanifolds

Chae, S. B.: Lebesgue Integration

Chandrasekharan, K.: Classical Fourier Transform

Charlap, L. S.: Bieberbach Groups and Flat Manifolds

Chern, S.: Complex Manifolds without Potential Theory

Chorin, A. J.; Marsden, J. E.: Mathematical Introduction to Fluid Mechanics

Cohn, H.: A Classical Invitation to Algebraic Numbers and Class Fields

Curtis, M. L.: Abstract Linear Algebra

Curtis, M. L.: Matrix Groups

Cyganowski, S.; Kloeden, P.; Ombach, J.: From Elementary Probability to Stochastic Differential Equations with MAPLE

Dalen, D. van: Logic and Structure

Das, A.: The Special Theory of Relativity: A Mathematical Exposition

Debarre, O.: Higher-Dimensional Algebraic Geometry

Deitmar, A.: A First Course in Harmonic Analysis

Demazure, M.: Bifurcations and Catastrophes

Devlin, K. J.: Fundamentals of Contemporary Set Theory

DiBenedetto, E.: Degenerate Parabolic Equations

Diener, F.; Diener, M.(Eds.): Nonstandard Analysis in Practice

Dimca, A.: Sheaves in Topology

Dimca, A.: Singularities and Topology of Hypersurfaces

DoCarmo, M. P.: Differential Forms and Applications